Telecommunications Engineering Group
Research School of Information Sciences and Engineering
The Australian National University
Canberra ACT 0200
AUSTRALIA

Land-Mobile Radio System Engineering

The Artech House Mobile Communications Library

John Walker, *Series Editor*

Mobile Information Systems, John Walker, editor

Narrowband Land-Mobile Radio Networks, Jean-Paul Linnartz

Land-Mobile Radio System Engineering, Garry C. Hess

The Evolution of Mobile Communications in the U. S. and Europe: Regulation, Technology, and Markets, Michael Paetsch

For a complete listing of *The Artech House Telecommunications Library*, turn to the back of this book

Land-Mobile Radio System Engineering

Garry C. Hess

Artech House
Boston • London

Library of Congress Cataloging-in-Publication Data

Hess, Garry C.
Land-mobile radio system engineering/Garry C. Hess
Includes bibliographical references and index.
ISBN 0-89006-680-9
1. Mobile communications systems. 2. Mobile radio stations.
I. Title
TK6570.M6H45   1993   93-12356
621.3845—dc20   CIP

British Library Cataloguing in Publication Data

Hess, Garry C.
Land-mobile radio system engineering
I. Title
621.3845

ISBN 0-89006-680-9

© 1993 ARTECH HOUSE, INC.
685 Canton Street
Norwood, MA 02062

All rights reserved. Printed and bound in the United States of America. No part of this book may be reproduced or utilized in any form or by any means, electronic or mechanical, including photocopying, recording, or by any information storage and retrieval system, without permission in writing from the publisher.

International Standard Book Number: 0-89006-680-9
Library of Congress Catalog Card Number: 93-12356

10 9 8 7 6 5 4 3 2

*Though not a believer in heaven,
I know two most deserving of such a place,
and to my parents, Glenn and Virginia,
I dedicate this book*

Contents

Preface	xv
Acknowledgments	xvii
Chapter 1 Introduction	1
1.1 A Brief History of Land-Mobile Radio Communication	1
1.2 Outline of Chapter Contents	3
Chapter 2 Free-Space Propagation	7
2.1 Free-Space Transmission Loss	7
2.2 Far-Field Qualification	8
2.3 Additional Line-of-Sight Attenuation Factors	8
2.4 Antenna Gain Units	10
2.5 Additional Antenna Terminology	10
2.6 Field Strength	12
2.7 Example Problem: Spurious Power Emission	13
2.8 Homework Problems	14
2.8.1 Problem 1	14
2.8.2 Problem 2	14
2.8.3 Problem 3	15
Chapter 3 Additional Transmission Loss Relations	19
3.1 Plane Earth Propagation Model	19
3.2 Meteor Scatter Propagation Model	21
3.3 Power-Law Propagation	22
3.4 Homework Problem	23
Chapter 4 Noise Considerations	25
4.1 Additive White Gaussian Noise	25
4.2 Noise Equivalent Bandwidth	26
4.3 Noise Figure	26
4.4 Y-Factor Measurement	28
4.5 System Noise Figure and G/T	29

4.6	An Example of Nonflat (Colored) Noise	30
4.7	Homework Problem	33

Chapter 5 Sample Radio Link Analyses   **35**
- 5.1 Satellite TVRO Link Analysis   35
- 5.2 Land-Mobile Radio Link Analysis   38
  - 5.2.1 Example 1   44
  - 5.2.2 Example 2   45
- 5.3 Homework Problem   45

Chapter 6 Character of Land-Mobile Radio Propagation   **49**
- 6.1 Power-Law Propagation Regime   49
- 6.2 Lognormal-Shadowing Propagation Regime   49
- 6.3 Rayleigh-Fading Propagation Regime   51
- 6.4 Summary   53
- 6.5 Homework Problem   54

Chapter 7 Probability Theory Refresher   **57**
- 7.1 What Is Probability?   57
  - 7.1.1 Axiomatic   57
  - 7.1.2 Relative Frequency   58
  - 7.1.3 Classical   58
  - 7.1.4 Measure of Belief   58
- 7.2 Operations with Events   58
- 7.3 Probability Laws   59
- 7.4 Concept of the Random Variable   59
  - 7.4.1 Definitions   59
  - 7.4.2 Distribution Function   60
  - 7.4.3 Density Function   60
- 7.5 Mathematical Expectation   61
  - 7.5.1 Expected Value   61
  - 7.5.2 Conditional Expected Value   61
  - 7.5.3 Variance   62
  - 7.5.4 General Moments   62
  - 7.5.5 Characteristic Functions   63
- 7.6 Sample Applications   64
  - 7.6.1 Gaussian (Normal) Density Function   64
  - 7.6.2 Gaussian Distribution Function   65
  - 7.6.3 Mean of Gaussian Random Variable   66
  - 7.6.4 Variance of Gaussian Random Variable   66
  - 7.6.5 Linear Transformation of a Gaussian Random Variable   67
  - 7.6.6 Moments of a Rayleigh Random Variable   67

	7.6.7	Moments of an Exponential Random Variable	69
	7.6.8	Conditional Expectation Example	69
	7.6.9	Analysis of Peak-to-Average Power Ratio for Speech	70
	7.6.10	Moments of a Rician-Distributed Random Variable	72
	7.6.11	Analysis of Battery Savings Potential Using Power Control	75
7.7	Homework Problems		78
	7.7.1	Problem 1	78
	7.7.2	Problem 2	79
	7.7.3	Problem 3	80

Chapter 8 Central Limit Theorem   85
8.1 Sum of Many Independent, Identically Distributed Random Variables   85
8.2 Rapidity of Convergence   87
8.3 Central Limit Theorem   89
8.4 Homework Problems   89
    8.4.1 Problem 1   89
    8.4.2 Problem 2   90
    8.4.3 Problem 3   92

Chapter 9 Probability Theory Refresher, Continued   93
9.1 Functions of a Single Continuous Random Variable   93
9.2 Extension of Random Variable Concept to Two Variables   94
9.3 Sample Applications   95
    9.3.1 Sum of Two Random Variables   95
    9.3.2 Square Root of Sum of Squares of Two Random Variables   97
    9.3.3 Maximum of Two Random Variables   97
9.4 Estimation of Distribution Parameters   99
    9.4.1 Rayleigh-Distribution Parameter   99
    9.4.2 Lognormal Distribution Parameters   101
    9.4.3 Maximum Likelihood Estimation in General   102
9.5 Other Aspects of Rayleigh Fading   103
    9.5.1 Level Crossing Rate   103
    9.5.2 Average Fade Duration   104
    9.5.3 Test of Distribution   106
9.6 Homework Problems   110
    9.6.1 Problem 1   110
    9.6.2 Problem 2   111
    9.6.3 Problem 3   112
    9.6.4 Problem 4   114

Chapter 10 Coverage Analysis and Simulation   121

10.1	Area Coverage with Power-Law Propagation and Lognormal Shadowing	121
10.2	Impact of Rayleigh Fading on Coverage	122
10.3	Analysis of Composite Rayleigh-Lognormal Distribution	124
	10.3.1 Mean and Standard Deviation of Power in Decibels for Rayleigh-Faded Signal	124
	10.3.2 Mean and Standard Deviation of Power in Decibels for Rayleigh plus Lognormal-Faded Signal	125
	10.3.3 Composite Distribution Function	125
	10.3.4 Monte Carlo Simulation Results	128
	10.3.5 Regression Analysis	130
10.4	Measurement of Trunked System Subscriber Unit Signal Characteristics	133
	10.4.1 Acquiring Information about the ISW	133
	10.4.2 Estimating the Subscriber Unit Transmit Frequency	133
	10.4.3 Estimating the Subscriber Unit Range	134
	10.4.4 Measured Results	136
10.5	Adjacent Channel Interference Considerations in the Land-Mobile Radio Service	138
10.6	Spectrum Efficiency Potential of 25-kHz Offset Channel Assignments in the 821-to 824- and 866- to 869-MHz Public Safety Bands	142
	10.6.1 Geographic Separation Requirements	142
	10.6.2 Channel Assignment Procedure	145
10.7	Interference Potential of Cordless Telephone Sharing with Public Safety Band Users	147
10.8	Further Sharing of UHF Television by Private Land-Mobile Radio Services	149
	10.8.1 Cochannel Sharing	150
	10.8.2 Noncochannel Sharing	158
10.9	800-MHz SMR Cochannel Spacing	159
	10.9.1 Summary	159
	10.9.2 FCC Spacing Rules and Engineering Justification	162
	10.9.3 Probability of Interference Analysis	163
	10.9.4 Discussion of the Shortcomings	165
	10.9.5 Alternative Interference Analysis	167
	10.9.6 Equivalent Short-Spacing Methodologies	168
10.10	Cellular System Operation	170
	10.10.1 The Cellular Concept	170
	10.10.2 Call Control and Handoff	171
	10.10.3 Reuse Patterns and Performance	175
	10.10.4 Cochannel Isolation Characteristics	176
	10.10.5 Spectrum Efficiency Measures	179

	10.10.6 Generation of Correlated Signal Strength Draws for Multiple Site Reuse System Simulations	180
10.11	Analysis of Intermodulation Interference	183
	10.11.1 Wide-Area Coverage Site Receiver Intermodulation Probability	184
	10.11.2 Small-Area Coverage Site (Cellular) Receiver Intermodulation Probability	185
	10.11.3 Transmitter IM	188
10.12	Homework Problems	189
	10.12.1 Problem 1	189
	10.12.2 Problem 2	190
	10.12.3 Problem 3	192
	10.12.4 Problem 4	192
	10.12.5 Problem 5	193
	10.12.6 Problem 6	195
	10.12.7 Problem 7	196
	10.12.8 Problem 8	197

Chapter 11 Diversity 205

11.1	Classification of Techniques	205
	11.1.1 Space Diversity	205
	11.1.2 Polarization Diversity	207
	11.1.3 Angle Diversity	207
	11.1.4 Frequency Diversity	207
	11.1.5 Time Diversity	208
	11.1.6 Other Techniques	208
11.2	Selection Diversity Combining	208
11.3	Maximal-Ratio Combining	209
11.4	Analysis of a Special Form of Selection Diversity	212
11.5	Homework Problems	216
	11.5.1 Problem 1	216
	11.5.2 Problem 2	216

Chapter 12 Simulcast 221

12.1	Description	221
12.2	Two-Signal Frequency Modulation Simulcast Analysis	222
12.3	Capture Analyses	224
	12.3.1 Independent Rayleigh Fading, Fixed Average Power, Envelope Analysis	224
	12.3.2 Independent Rayleigh Fading, Fixed Average Power, Power Analysis	225

12.3.3 Independent Rayleigh Fading, Fixed Average Power, Envelope Analysis Including Threshold 226
12.3.4 Independent Lognormal Fading, Power Analysis Including Threshold 227
12.3.5 Independent Rayleigh Fading with Independent Lognormal Fading of Average Power, Power Analysis Including Threshold 228
12.4 Computer Simulation of Analog FM Simulcast Performance 229
12.5 Rapid, Robust Method for Establishing Signal Quality 234
12.5.1 Algorithm for Estimating Amplitude and Phase of a Sinusoid 234
12.5.2 Simulated Performance 235
12.5.3 Application: Automated Measurements of Delivered Audio Quality 237
12.6 Computer Simulation of Digital Simulcast Coverage 240
12.6.1 Approximate Methods for Reducing Multisite Problems to Two-Site Problems 240
12.6.2 Multipath Spread Model 241
12.6.3 Comparisons of Model Performance 245
12.6.4 Use of the Multipath Spread Model in Coverage Predictions 246
12.7 Homework Problem 246

Chapter 13 Traffic Engineering 249
13.1 Message Length Characteristics 249
13.2 Traffic Load Estimation 253
13.2.1 Nominal System Peak Load 253
13.2.2 System-to-System Variation of Peak Load 255
13.2.3 Day-to-Day Variation of Peak Load 257
13.2.4 Interconnect and Mixed Traffic Loads 257
13.3 Grade-of-Service Evaluation 260
13.3.1 Blocked Calls Lost, Erlang-B Viewpoint 260
13.3.2 Blocked Calls Delayed, Erlang-C Viewpoint 263
13.3.3 Poisson Viewpoint 264
13.3.4 Application to Trunked Dispatch Grade of Service 265
13.4 Trunked System Control Channel Performance 267
13.4.1 Introduction 267
13.4.2 Example Control Channel Operation 267
13.4.3 Simulation Description 268
13.4.4 Simulation Results 270
13.4.5 Experiment Description 270
13.4.6 Experiment Results 272
13.5 Analysis of Trunked System Performance with Load Shedding 273
13.6 Analysis of Group Call Access Delay 275
13.6.1 Introduction 275

	13.6.2 Average Group Call Delay Relative to Individual Call Delay	276
	13.6.3 Group Call Excess Traffic Load	277
	13.6.4 Application	278
13.7	Homework Problems	279
	13.7.1 Problem 1	279
	13.7.2 Problem 2	280
	13.7.3 Problem 3	280

Chapter 14 Land-Mobile Satellite Systems 287

14.1	History	287
14.2	Multiple Beam, Geostationary Satellite System Cost and Performance Optimization	288
	14.2.1 Cost-Performance Model	289
	14.2.2 LaGrangian Multiplier Optimization	290
14.3	LEO Satellite-to-Subscriber Unit Fade-Margin Analysis	291
	14.3.1 Preliminaries	291
	14.3.2 Ideal Conditions	292
	14.3.3 Unshadowed Conditions	292
	14.3.4 Unshadowed Conditions, Enhanced Scatter	292
	14.3.5 Mixed Shadowed and Unshadowed Conditions	292
	14.3.6 In-Building Conditions	293
	14.3.7 Summary	293
	14.3.8 Impact of Elevation Angle on Fade Margin	293
14.4	Sample LEO Mobile Satellite Link Budget	294
	14.4.1 Assumptions	294
	14.4.2 Spacecraft-to-Land Mobile Terminal	294
	14.4.3 Land Mobile Terminal-to-Spacecraft	298
14.5	Homework Problem	301

Chapter 15 Frequency Modulation Performance 309

15.1	Fundamentals of Frequency Modulation	309
15.2	Spectrum	311
15.3	Transmission Bandwidth	313
15.4	Demodulation in the Presence of Noise and Nonfaded Conditions	315
	15.4.1 Signal Suppression Noise	315
	15.4.2 Postdetection Baseband Noise Spectrum	318
	15.4.3 Pre-emphasis and De-emphasis Filtering	320
	15.4.4 Click Analysis	322
15.5	Performance in the Presence of Noise and Rayleigh Fading	326
	15.5.1 Faded Signal Power	326
	15.5.2 Signal-Suppression Noise Power	326
	15.5.3 Above-Threshold Noise Power	327

15.5.4 Threshold-and-Below Noise	327
15.5.5 Overall Fading Performance	328
15.5.6 Random FM Noise	328
15.5.7 Alternative View of Overall Fading Performance	330
15.6 Performance in the Presence of Noise, Interference, and Rayleigh Fading	331
15.7 Homework Problems	332
15.7.1 Problem 1	332
15.7.2 Problem 2	333

Chapter 16 Building Shadowing Adjustment Model Investigation   **341**

16.1 Introduction	341
16.1.1 The Problem	341
16.1.2 Prior Art	342
16.1.3 Chicago Building Database	343
16.2 Artificial Neural Network Models	345
16.2.1 Brief Tutorial on Artificial Neural Networks	345
16.2.2 Artificial Neural Network Software	349
16.2.3 Artificial Neural Network Model Performance	351
16.3 Group Classification Models	354
16.4 Linear Regression Models	357
16.5 Conclusions	358
16.6 Homework Problem	358

Index   **365**

Preface

The Audience

This book is intended primarily for engineers practicing in the land-mobile industry; particularly those engineers who have recently joined this rapidly expanding field. Hopefully it will be of use to more than just those directly engaged in system design. Both hardware and software engineers can benefit from the big picture viewpoint and the practical applications orientation. Readers are assumed to have had the equivalent of an undergraduate course in basic communication theory. Familiarity with probability and statistics is also helpful and several chapters in the book are devoted to reviewing these topics.

The Subject

Federal Communication Commission (FCC) rules define a land-mobile radio system as a regularly interacting group of base, mobile, and associated control and fixed relay stations intended to provide land-mobile radio communications service over a single area of operation. The term mobile refers to movement of the radio, rather than association with a vehicle; hence mobile radio also encompasses handheld portable radios. Examples include public safety dispatch communications conducted by police and fire departments, personal pagers, cellular radiotelephones, and packetized data radio networks. In frequency allocation matters, distinctions may be made among land mobile, maritime mobile, and aeronautical mobile. While the practical system engineering aspects of this book focus on land-based systems, the theory and analytical procedures apply to radio links in general. By system engineering we mean the arrangement and specification of various radio equipment to accomplish some task; for example, to allow communication throughout some area with a given reliability in the face of both noise and interference and with a given average access delay.

The Approach

This is not a step-by-step cookbook of how to lay out a mobile radio communications system. Such a rigid approach would leave the reader ill prepared to do much other than blindly reproduce what had already been done before, yet the engineer's task is generally to do better in some sense than what has been done before. Thus, we have attempted to present the science behind the art of key system design activities (such as coverage and capacity prediction) and key techniques (such as diversity and simulcast) and prepare the reader to address not only today's systems, but the emerging digital and microcellular systems of tomorrow.

This is not a stand-alone book. Certain topics of major importance in the design of radio systems (such as probability and statistics, queueing theory, and reliability theory) are the subjects themselves of many textbooks. Other more specific topics are the subjects of numerous articles published in technical journals. Where deemed appropriate such sources are referenced and important results are simply cited, rather than redeveloped in detail. The reader is thus at least pointed in a direction to pursue if further detail is of interest.

This leaves us free to concentrate on "original" analysis and applications. Quotations are used because some of the items have perhaps been carried out earlier by others, though either the results or the details behind them have not to our knowledge been published. In the problems we address, we are very interested in the details. Too often we find theoretical development compressed with comments like "after some algebraic manipulation ..." This may be fine for advanced readers, but it is not helpful for teaching beginners what is really going on or for preparing them to extend the theory to their particular problems and interests.

Several textbooks have already been written on land-mobile radio, some of which are still in print. They generally cover more territory than this text. However, in the opinion of the author, they tend to be lacking in matters of practical application. While substantial text herein is theoretical in nature, the less mathematically inclined readers should not be intimidated. The purpose behind the theory is real-world engineering and the reader is shown numerous examples of how the theory is applied in practice. Assumptions associated with the theory are also discussed in terms of real-world observations.

The homework problems included at the end of most topics highlight subtleties of the material covered, and in many cases serve as an excuse to introduce new material. Thus solutions are included. Hopefully, readers will find these problems challenging and gain insight into how the theory and various analytical procedures can be used in practice.

Acknowledgments

Thanks are expressed to William Turney and Dr. J. Earl Foster for general inspiration and guidance in completing this manuscript. The support of management, in particular Drs. Jona Cohn and Robert Janc, and a wide spectrum of challenging assignments during our tenure at Motorola, Inc. are gratefully acknowledged. We are indebted to Ross Lillie and Loren Rittle for assistance on word processing issues and Barry Leung and Mark Marsan for help with artwork generation. Many helpful review comments have been received, but in particular we want to acknowledge those of Dr. Faramaz Davarian, Dr. Wolfhard Vogel, and Dr. Kenneth Zdunek. Finally, we acknowledge the following specific contributions: Chapter 1, Karen Brailean and Kenneth Crisler; Chapter 2, Allen Davidson; Chapter 4, Peter Walter; Chapter 10, James Butler, Bradley Hiben, and Mark Marsan; Chapter 12, Mark Birchler, Christopher Kurby, Michael Needham, and Carl B. Olson; Chapter 13, Kenneth Crisler; Chapter 14, John Matz; and Chapter 16, Karen Brailean and Mark Marsan.

Chapter 1

Introduction

1.1 A Brief History of Land-Mobile Radio Communication

The possibility of radio communications was established in 1864 by James Clerk Maxwell, then a professor of physics at Cambridge University. Maxwell showed theoretically that an electrical disturbance, propagating at the speed of light, could produce an effect at a distance. Theory was first put into practice by Hertz, who demonstrated spark-gap communications over distances of several feet in the 1880s. The distance was rapidly extended by Marconi, who by 1901 succeeded in transmitting Morse code across the Atlantic ocean. The vacuum tube made speech transmissions practical and by 1915 the American Telephone & Telegraph company had sent speech transmissions from Washington, D.C., to Paris and Honolulu.

The first practical land-mobile communications occurred in 1928 when the Detroit Police Department finally succeeded in solving the instability and low sensitivity problems that had plagued their mobile receiver designs for 7 years [1]. By 1933, a mobile transmitter had been developed, allowing the first two-way police system to operate in Bayonne, New Jersey. The 1939 success of a statewide Connecticut highway patrol system using *frequency modulation* (FM) led to a nationwide phaseout of amplitude modulated equipment.

By 1933 the need for radio regulation was also apparent and the first operating rules were mandated by the Federal Radio Commission. The Federal Communications Commission was established 1 year later. Twenty-nine *very high frequency* (VHF) channels between 30 and 40 MHz (known today as low band) were allocated for police use. In 1946 the initial rules for the Domestic Public Land Mobile Radio Service were established and high band frequencies between 152 and 162 MHz were allocated. Small businesses could now purchase airtime from common carriers, thus avoiding the large startup costs of a private system. Previously only certain industries had access to mobile radio frequencies; for example, public safety, public utilities, transportation, and the media.

The rapid growth of the land-mobile radio industry, as evidenced by Figure 1.1 [2], has been accompanied by substantial additional spectrum allocations in the *ultra high frequency* (UHF) band (406 to 512 MHz), and the 800- and 900-MHz bands. Mobile satellite spectrum has been allocated at L-band (1.5 GHz), along with supporting point-to-point microwave frequencies. As of this writing 2 MHz of spectrum at 220 MHz is being readied for narrowband (5-kHz channelization) use. Nonetheless, it appears that practical mobile spectrum (i.e., spectrum at L-band and lower frequencies[1]) is unlikely to significantly increase because there is nothing left to allocate; hence, today's strong push toward more efficient use of the spectrum already allocated.

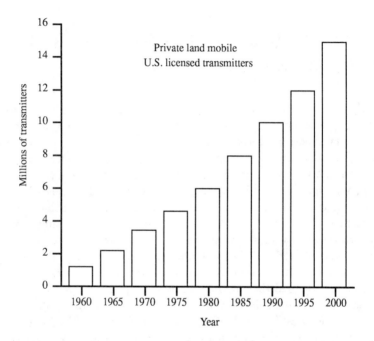

Figure 1.1 Growth in licensed land-mobile transmitters.
After: Brailean, K. A., "Land-Mobile Communications: Predicting the Future Based on Trends of the Past," *SWE National Convention*, San Diego, CA, June 1991, p. I-31.

---

[1]One should not forget, however, that as recently as the 800-MHz allocation deliberations, a substantial portion of the technical community argued that frequencies above UHF were impractical. The issue of practicality is probably more a matter of the economics of technology than the ability of technology to surmount problems that arise with increasing frequency.

Introduction

Increased efficiency can be achieved in a variety of ways. Technological advancements in frequency control have allowed the 900-MHz band to be channelized at 12.5 kHz, rather than 25 kHz (Figure 1.2 [3]). Figures 1.3 and 1.4 [3] illustrate the major improvements that have been occurring in speech coding and modulation efficiency. Trunking of groups of radio channels allows operation at much higher loading levels than single channels can handle with acceptable access delay. Finally, the cellular radiotelephone service introduced in 1981 has boosted spectrum efficiency through geographic reuse of channels in the same metropolitan coverage area.

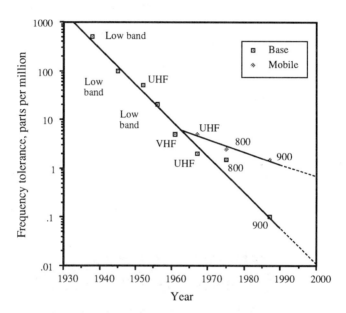

Figure 1.2 Land-mobile frequency stability trend.
After: Crisler, K. and A. Davidson, "Impact of 90's Technology on Spectrum Management," *International Symposium on Electromagnetic Compatibility*, Washington, D.C., August 21-23, 1990, p. 403.

1.2 Outline of Chapter Contents

The book proper begins in Chapter 2 by considering the simplest propagation scenario, that of free space. Development of the transmission loss formula for this situation requires some discussion of basic antenna concepts. Chapter 3 considers two other propagation models, then notes the unifying concept of power-law behavior. Absolute signal level is often less important than signal level relative to

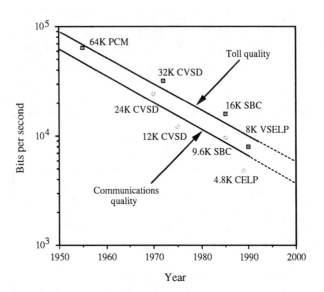

Figure 1.3 Speech coding trends.
After: Crisler, K. and A. Davidson, "Impact of 90's Technology on Spectrum Management," *International Symposium on Electromagnetic Compatibility*, Washington, D.C., August 21-23, 1990, p. 403.

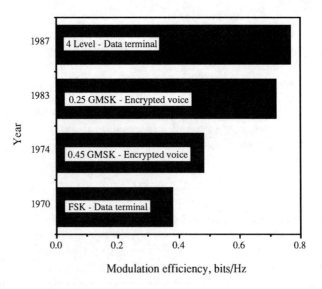

Figure 1.4 Modulation efficiency of 25-kHz systems.
After: Crisler, K. and A. Davidson, *ibid.*, p. 403.

unwanted signals such as noise and interference, a matter treated by Chapters 4 and 5.

The first chapter specific to land-mobile propagation is Chapter 6. Here the character of the propagation is examined in terms of three propagation regimes, differing by the distance scale over which the signal is averaged. The finest scale regime involves Rayleigh fading and the application of probability theory. Hence, Chapters 7 through 9 review the theory and illustrate its application to a variety of land-mobile radio problems.

Chapter 10 tackles the important topic of coverage, a concept that is inherently probabilistic. Analyses are included that address not only noise outage, but outage caused by interference as well. Examples involving adjacent channel interference, offset channel interference, cochannel interference, and intermodulation interference are offered.

Chapters 11 and 12 address two important techniques for improving system performance, diversity and simulcast. The former improves system coverage by reducing the impact of Rayleigh fading; the latter improves it by filling in coverage holes through the simultaneous use of distributed transmitters.

Chapter 13 covers the topic of traffic engineering, particularly as it applies to trunked radio systems. The application of classical queueing theory is discussed, along with traffic load estimation procedures. Control channel performance and the impact of group calls on cellular systems are also discussed.

Chapter 14 deals with the "far out" topic of direct mobile-to-satellite communications. This has been appreciated as a real possibility since the late 1970s and it now appears that operational systems are only a few years away.

Because FM is currently such a dominant factor in land-mobile communications, Chapter 15 is included to review the fundamentals of FM and examine how it performs under Rayleigh-fading conditions.

The book ends with Chapter 16, devoted to a current research topic; namely, the impact of buildings along the propagation path on received signal levels. Certainly some means for improving signal strength predictions is needed to accompany the imminent introduction of microcellular personal communications systems into major metropolitan areas. Traditional regression and classification techniques are discussed, along with nontraditional modeling using artificial neural networks.

References

[1] Noble, D. E., "The History of Land-Mobile Radio Communications," *Proc. of IRE*, Vol. 50, No. 5, May 1962.

[2] Brailean, K. A., "Land-Mobile Communications: Predicting the Future Based on Trends of the Past," *SWE National Convention*, San Diego, CA, June 1991.

[3] Crisler, K. and A., Davidson, "Impact of 90's Technology on Spectrum Management," *International Symposium on Electromagnetic Compatibility*, Washington, D.C., August 21-23, 1990, pp. 401-405.

Chapter 2

Free-Space Propagation

The attenuation of radio signals between transmitters and receivers is a key concern of system engineering. The logical starting point for addressing such attenuation is radio propagation in free space. Free-space transmission loss generally represents the least possible loss versus distance and thus in a sense can be considered the ideal loss. It is also the practical loss for many situations of interest; for example, satellite links and certain interference scenarios. Even when significant attenuation above and beyond that of free space occurs, it is common to separate out the free-space loss and deal only with the remainder, which is termed excess path loss. Free-space transmission loss is intimately tied to antenna concepts. The amount of loss depends on the units used to express antenna gain and is valid only at distances sufficiently far from the antennas.

2.1 Free-Space Transmission Loss

Consider an isotropic point source fed by a transmitter of P_t watts. At an arbitrary, large distance r from the source, the radiated power is uniformly distributed over the surface area of a sphere of that radius. Thus, assuming the radiator is lossless, the power density at distance r is given by:

$$S_r = \frac{P_t}{4\pi r^2} \qquad (2.1)$$

If a receiving antenna is located at distance r from the point source, it will intercept an amount of power proportional to its effective aperture A_e. The power delivered to a receiver matched in impedance to the antenna can be written as:

$$P_r = \frac{P_t A_e}{4\pi r^2} \qquad (2.2)$$

The relationship between effective aperture and antenna gain is derived in Reference [1] (see Equation (10) of Section 2-22) and is:

$$G = \frac{4\pi}{\lambda^2} A_e \qquad (2.3)$$

where λ equals the wavelength of the electromagnetic field. By substitution one obtains:

$$P_r = \frac{P_t G_r}{[4\pi(r/\lambda)]^2} \qquad (2.4)$$

where the antenna gain subscript r is used to associate the gain with the receive end of the link. Finally, if the isotropic point source is replaced by a transmit antenna of gain G_t toward the receiving antenna, the received power expression becomes:

$$\begin{aligned} P_r &= \frac{P_t G_t G_r}{[4\pi(r/\lambda)]^2} \\ &= \frac{(\text{ERP}_t) G_r}{[4\pi(r/\lambda)]^2} \\ &= \frac{(\text{ERP}_t) G_r}{L_{fs}} \end{aligned} \qquad (2.5)$$

where ERP_t equals the *effective radiated power* (ERP) of the transmitter and L_{fs} is called the free-space path loss. Since wavelength equals speed of propagation (essentially light speed, $3.0E8$ m/s) divided by frequency, free-space path loss in decibels can be written as:

$$\begin{aligned} L_{fs}(\text{dBi}) &= 20\log(F_{\text{MHz}}) + 20\log(r_{\text{km}}) + 32.44 \\ &= 20\log(F_{\text{MHz}}) + 20\log(r_{\text{mi}}) + 36.6 \end{aligned} \qquad (2.6)$$

2.2 Far-Field Qualification

The preceding relations do not apply to arbitrarily small path lengths. For applicability, the transmitting antenna must be located in the far field of the receiving antenna. Section 34-14 of Reference [2] elaborates on the criteria necessary to ensure far field operation. The commonly specified criterion, which applies to antennas whose physical sizes exceed a few wavelengths, is $r \geq (2d^2/\lambda)$, where d is the major antenna dimension (diameter for parabolic dishes, length for land-mobile base station "sticks"). This criterion is based on limiting the phase difference over the aperture to one-sixteenth of the wavelength (22.5 deg).

2.3 Additional Line-of-Sight Attenuation Factors

Excess path loss over that given by Equation (2.6) can occur even with line-of-sight conditions due to tropospheric effects. Table 2.1 [3] shows representative tropo-

Free-Space Propagation

spheric attenuations for earth-to-satellite links as a function of frequency for four effects: (1) clear air absorption, (2) cloud attenuation, (3) fog attenuation, and (4) rain attenuation. None of those effects is significant for present U.S. land-mobile spectrum; i.e., below 1.6 GHz.

Table 2.1 Estimated Tropospheric Attenuation for 30-Deg Elevation Angle and One-Way Traversal

After: Interim Working Party 5/2, CCIR Study Groups Period 1982-1986, "New Report: Propagation Data for Land Mobile Satellite Systems for Frequencies Above 100 MHz," Study Program 7C-1/5, Doc. 5/12-E, March 1983, p. 3.

Effect	Magnitudes (dB) for selected operating frequencies				
	0.85 GHz	1.6 GHz	20 GHz	45 GHz	94 GHz
Clear air absorption (dB)[a]					
3 g/m^3 (dry)	0.06	0.07	0.3	0.1	1.5
7.5 g/m^3 (average)	0.06	0.07	0.6	1.3	2.7
17 g/m^3 (moist)	0.06	0.07	1.2	1.9	5.2
Cloud attenuation (dB)[b]					
0.5 g/m^3, 1 km thick	<0.01	<0.01	0.4	1.7	5
1 g/m^3, 2 km thick	<0.01	<0.01	1.6	6.8	20
Fog attenuation (dB)[c]					
0.05 g/m^3 (aver.), 0-75 m ht	-	-	-	-	0.03
0.5 g/m^3 (hvy), 0-150 m ht	-	-	-	0.1	0.07
Rain attenuation (dB)[d]					
5 mm/hr	<0.01	<0.01	3.3	12.2	25.1
25 mm/hr	<0.1	<0.1	19.01	55	81.5

a. Derived from Report 719, Annex II using t = 15 deg. C. Values for other elevation angles above theta=10 degrees may be obtained by dividing the magnitudes (dB) by 2 times the sine of theta. Worldwide values of water vapor concentration are available from Report 563, section 3.

b. Derived from specific attenuation values given in Report 721, Table I, using models from Slobin. For other elevation angles the attenuation may be estimated for theta greater than or equal to ten degrees by dividing the magnitudes (dB) by 2 times the sine of theta.

c. Derived from section 9 of Report 721 for average fog and heavy fog.

d. Derived from Report 564, section 2.2.1.

2.4 Antenna Gain Units

Notice that free-space path loss has been quoted in units of dBi. This means the antenna gains must be referenced to an isotropic antenna. Common FCC[1] practice instead references antenna gains to a half-wave dipole. Such an antenna has a gain of 2.15 dB over isotropic, assuming equal antenna efficiencies. Thus, path loss must be decreased by 4.30 dB when using dBd gain units because two antennas are involved, one for transmitting and one for receiving.

Yet another type of gain unit is often used in satellite links, dBic. Linear polarization of the electromagnetic wave places a constraint on transmit and receive antenna alignment. This is not a problem for terrestrial land-mobile applications, but it is untenable for many satellite links where the receiver, the transmitter, or both may be moving. In such cases circular polarization is used on at least one end of the link. A linearly polarized antenna can still be used on the other end of the link, but a 3-dB loss in received power is incurred.

2.5 Additional Antenna Terminology

The gain of an antenna, referenced to a lossless isotropic source, equals the product of its directivity and its efficiency:

$$G = kD \tag{2.7}$$

Here of course the term gain and the symbol G are used to represent the maximum nose-on main-beam value. Gain at an arbitrary angle is found by multiplying G by the normalized power pattern. Antenna efficiency runs from 0 to 1 and has to do only with ohmic losses in the antenna. In transmitting, these losses involve power fed to the antenna that is not radiated but instead heats the antenna structure. k equals the ratio of actual effective antenna aperture to maximum effective antenna aperture.

Directivity is formally defined as the ratio of maximum radiation intensity (power per unit solid angle) to average radiation intensity; or, at some distance from the antenna, it equals the ratio of the maximum to average Poynting vector magnitudes (each in units of watts per area). Using the latter definition and a spherical coordinate system with angles θ and ϕ (Figure 2.1 [1]), one obtains:

$$\begin{aligned} D &= S(\theta,\phi)_{\max}/S_{\mathrm{av}} \\ &= S(\theta,\phi)_{\max} / \frac{1}{4\pi} \int_0^{2\pi} \int_0^{\pi} S(\theta,\phi) d\Omega \end{aligned}$$

---

[1]The Federal Communications Commission licenses and regulates non-federal government radio activity in the United States (the *National Telecommunication and Information Administration* (NTIA) handles federal government users). FCC rules are contained in Title 47 of the Code of Federal Regulations.

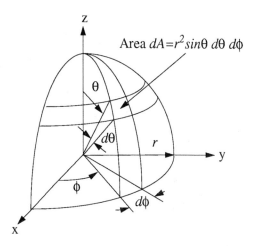

Figure 2.1 Spherical coordinates in relation to the differential area of solid angle. After: Kraus, J. D., *Antennas*, McGraw-Hill Book Co., New York, NY, 1988, p. 24.

$$
\begin{aligned}
&= \frac{1}{(1/4\pi) \int\int [S(\theta,\phi)/S(\theta,\phi)_{\max}]d\Omega} \\
&= \frac{1}{(1/4\pi) \int\int P_n(\theta,\phi)d\Omega} \\
&= \frac{4\pi}{\Omega_A}
\end{aligned}
\qquad (2.8)
$$

where $P_n(\theta,\phi)$ is the normalized power pattern, Ω_A represents the beam solid angle in sr (3282.8064 deg²), and $d\Omega = \sin(\theta)\,d\theta\,d\phi$.

Figure 2.2 is helpful for conceptualizing how an actual antenna power pattern and the equivalent solid angle are related.

A useful approximation is that beam solid angle is the product of the -3-dB (half-power) main-lobe beamwidths in the two principal planes. The directivity of an antenna can thus be approximated by:

$$
\begin{aligned}
D &\approx \frac{4\pi}{\theta_{-3\text{dB}}\phi_{-3\text{dB}}} \\
&\approx \frac{41000}{\theta^\circ_{-3\text{dB}}\phi^\circ_{-3\text{dB}}}
\end{aligned}
\qquad (2.9)
$$

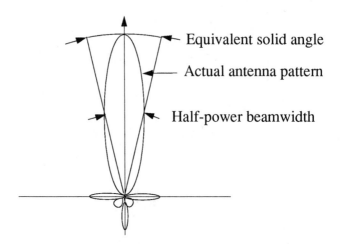

Figure 2.2 Cross-section of symmetrical power pattern of antenna showing equivalent solid angle for a cone-shaped (triangular) pattern.

2.6 Field Strength

Another FCC practice that should be mentioned is its preference for quoting field strength in units of microvolts per meter rather than power. Consider a traveling plane wave of *root mean square* (rms) field strength E volts per meter. The average power density $S = E^2/Z_0$, where Z_0 equals the intrinsic impedance of free space, about 377 Ω [4]. The maximum effective aperture of a linear half-wave dipole with sinusoidal current distribution is 0.13 square wavelengths (see Table 2-24 in Reference [1]) and the wavelength in meters equals $300/F$, where F is frequency in megahertz. Thus, the power delivered to a matched load can be expressed as:

$$P = \left(\frac{E^2}{377}\right) 0.13 \left(\frac{300}{F_{\text{MHz}}}\right)^2 \qquad (2.10)$$

Power in dBm is then given by:

$$\begin{aligned} P_{\text{dBm}} &= 10\log(P) + 30 \\ &= 10\log\left[\frac{E^2_{\mu\text{V/m}} 10^{-12}\, 0.13\,(300)^2}{377\, F^2_{\text{MHz}}}\right] + 30 \\ &= 20\log(E_{\mu\text{V/m}}) - 75 - 20\log(F_{\text{MHz}}) \\ &= P_{\text{dB}\mu} - 75 - 20\log(F_{\text{MHz}}) \qquad (2.11) \end{aligned}$$

Note that dBμ stands for dB relative to a field strength of 1 microvolt per meter; it does *not* mean dB below 1 microvolt of rms voltage (which equals -107 dBm in a

2.7 Example Problem: Spurious Power Emission

FCC regulations generally pertain to transmitters, not receivers. However, because receivers typically contain oscillators (for frequency translation between *radio frequency* (RF) and *intermediate frequency* (IF); for clocking control microprocessors, digital signal processors, etc.) they can emit signals as well. Part 15.63 of the FCC Rules requires such spurious emissions to be less than 500 μV/m at a distance of 30 m (100 ft) for the frequency range of 470 to 1000 MHz. What apparent source power does this permit?

To answer this question for an 800-MHz mobile radio, one solves Equation (2.6) using $F_{\text{MHz}} = 800$ and $r_{\text{mi}} = (100/5280)$ to obtain an isotropic path loss of 60.2 dBi. Next, Equation (2.11) is solved for the allowable received power in dBm: 20log(500) − 75 − 20log(800) = -79.1. The equivalent source power for a half-wave dipole received by a half-wave dipole is thus $-79.1 + 60.2 - 4.3 = -23.3$ dBm, or about -53 dBW. This is in agreement with section 15.3 of Reference [5], a document that details precisely how such tests are to be conducted[2]. Interestingly, the *Electronic Industries Association* (EIA)/*Telecommunications Industry Association* (TIA) standard for conducted spurious emissions (i.e., those emitted through the receiver antenna) is much more stringent. Section 16.3 of Reference [5] states that the limit is 1000 μV across 50 Ω (or an equivalent output power). This is only -47 dBm. *European Conference of Posts and Telecommunications* (CEPT) spurious specifications are even more stringent and do not differentiate among the leakage mechanisms; they simply disallow equivalent power sources in excess of -57 dBm.

Should one expect system problems from radios that just barely meet the radiated spurious specification of -23 dBm? Measurements on many land-mobile systems show that 80 dB is representative of the minimum path loss encountered. This number includes antenna gains for both ends of the link and implies a spurious received power as high as -103 dBm might occasionally occur. This is substantially above the typical base station receiver static sensitivity of -113 dBm (0.5 μV) and comparable to the faded sensitivity. In cases where low-noise preamplifiers are used right at the base receive antenna outputs to increase sensitivity, interference is even more likely.

Worse yet, most trunked systems [6] use control channels that subscriber radios monitor when idle. Thus, most of the time, most of the subscriber radios are listening to a few (or only one) channels outbound from the trunked site. Base transmit/mobile receive channels are in the range of 851 to 866 MHz. A typical receiver *local oscillator* (LO) operates in the range of (851 to 866) − 53.9 = 797.1 to

---

[2]Because the test methodology includes both reflected and direct paths, the equivalent source power is overstated by 6 dB.

812.1 MHz, where 53.9 MHz represents the IF and the minus sign is used because LO injection is "low side." Unfortunately, the base receive/mobile transmit band ranges from 806 to 821 MHz so it is possible for receiver local oscillator leakage to appear on the inbound channel of another system, perhaps even an inbound control channel. Because so many units have purposely congregated on each control channel their total leakage power can be substantial.

2.8 Homework Problems

2.8.1 Problem 1

Can free-space path loss ever exceed actual path loss?

Solution: Yes. For example, a coherent ground reflection can boost the signal by 6 dB. This phenomenon was used by Webb et al. to receive the first radar echoes from the Moon. Also, serendipitous focusing is possible from buildings, mountains, and atmospheric bending, thus making the apparent aperture larger than the physical antenna aperture. Finally, one must be careful in measuring the line-of-sight reference level. For example, the land mobile satellite service field tests in San Francisco described in Reference [7] involved an elevation angle such that the receive antenna depressed line-of-sight signal strength noticeably, yet certain building reflections came in nose-on and thus several decibels stronger.

2.8.2 Problem 2

Comment on the maximum signal strength likely to be input to an 860-MHz base station receiver located beside a major highway. Assume a worst-case subscriber unit (the "mobile") power of 35 W, with 2-dB feedline loss and a 3-dBd antenna mounted on the vehicle roof at 5 ft. Take the base parameters to be representative of cellular sites (see Section 10.10); that is, a 12-dBd 120-deg horizontal beamwidth sectored antenna at 100 ft with 2-dB feedline loss to the receiver distribution amplifier.

Solution: The subscriber unit ERP equals 44 W relative to a dipole (46.4 dBmd). At a horizontal distance of 600 ft the lookdown angle for the base antenna equals the arctangent of $(100 - 5)/600 = 9.1$ deg. At this angle a 15-deg vertical beamwidth base antenna will provide about 5 dB of protection, yielding a net base antenna gain of $12 - 5 = 7$ dBd (note that several decibels of additional path isolation can be obtained at the cost of 1 dB system gain by uptilting the sectored antennas). Free-space loss for a path equal to the square root of 95 squared and 600 squared (608 ft) is 72.8 dB on a dipole-to-dipole basis (65.8 dB on a port-to-port basis). Hence, the free-space received power at the receiver distribution amplifier input would be $46.4 - 72.8 + 7 - 2 = -21.4$ dBm.

Real-world examples of port-to-port path loss versus path length are shown in Figure 2.3. The minimum values observed are 67 dB at 16 ft, 74 dB at 40 ft, and 74

Free-Space Propagation 15

dB at 140 ft. These data pertain to a Phelps Dodge PD1109 7.5-dB omnidirectional antenna (8-deg vertical beamwidth) as the base antenna; the mobile used a 3-dBd vertical antenna, mounted on the center of the test car roof. The impact of vertical directivity is clearly evident as close-in loss diverges from free-space predictions. The larger scatter in the 140-ft case is the result of driving among other vehicles and buildings. The 40- and 16-ft collection runs were pretty much in the clear.

This means that even a mobile with 100 W ERP (50 dBmd) should produce no more than 50 - 74 = -24 dBm average power at the output of omnidirectional base receive antennas. Thus, -20 dBm seems a reasonable design specification for receiver "blocking" tolerance with omnidirectional antennas. Unfortunately, physically small sectored antennas (desired to minimize tower loading and for aesthetics) have wide vertical pattern beamwidths and afford less discrimination against close-in signal sources. Vertical directivity is crucial in setting the minimum path loss. Notice that at the ridiculous height of 16 ft, the subscriber unit can begin to enter the main beam of the PD1109 at very modest path length; hence, only 67 dB of isolation is assured. A common antenna used in 120-deg sectored cells provides a port-to-port path loss under 70 dB for horizontal distances between about 430 ft and 1650 ft (the minimum of 65 dB occurs at 600 ft). This implies that 100 W ERP subscriber units between these distances might overload a receiver designed to tolerate only -20 dBm.

A crude, conservative estimate of the likelihood of such signal levels is as follows. Consider Los Angeles to encompass a coverage area of about 100 mi^2 in which the entire 800-MHz land mobile allocation (15 MHz x 40 channels/MHz = 600 channels) is occupied with a reuse factor of 1.4, implying 140 subscriber units/channel (this excludes control stations, which can be delt with on an individual basis, if necessary). The result is an average user density of 600 x 140 / 100 x 100 = 8 units per mi^2. Assume that user density can peak 20 times greater, yielding as many as 160 units per mi^2, five of which could fall in the overload region (remember the cell involves three sectors). Further assume that half the units are of sufficient power to cause overload and free-space propagation conditions exist half the time. Dispatch busy-period traffic loads are on the order of 1.5 calls per unit per hour x 20 seconds per call = 0.008 Erl[3], indicating a probability of about 1% for receiver overload.

2.8.3 Problem 3

What size dish antennas are required, back-to-back, to provide valley-to-valley communication when located atop an intervening mountain? Assume the link is at 960 MHz and involves a free-space path length of 10 mi from valley to mountaintop. Further assume a transmitter *effective isotropic radiated power* (EIRP) of 10 W and

---

[3]One erlang, the standard unit of traffic (we will abbreviate it as Erl.), implies 3600 seconds worth of calls to be handled per hour, or 60 seconds worth of calls to be handled per minute, etc.; traffic matters are discussed in Chapter 13.

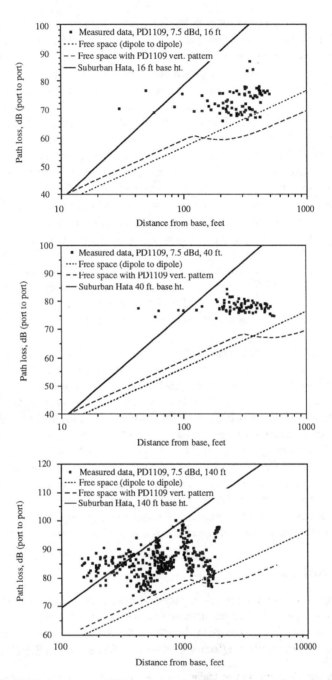

Figure 2.3 Path loss versus distance scatter plots.

receive sensitivity of 0.5 µV with 9 dBd of antenna gain. Allow for a 20-dB fade margin.

Solution: The numeric link equation is:

$$P_R = \frac{[(\text{EIRP})_T G_x] G_x G_R}{PL^2} \tag{2.12}$$

where PL represents the one-way path loss. The dBm equivalent, including 20-dB fade margin is $-113 + 20 = 40 + 2G_x + 11.15 - 2[20\log(960) + 20\log(10) + 36.6]$. Thus, $G_x = 44.2$ dBi. Assuming an aperture efficiency of 55% this implies a dish of about 71 feet in diameter, hardly practical. However, if a 40-dB gain block is inserted between the two antennas, then a maximum gain of just 24.2 dBi will suffice. This can be achieved with 7.1 foot dishes, which are practical. Hence, the importance of active boosters in such applications [8]. However, passive repeaters are quite useful in other applications [9].

References

[1] Kraus, J. D., *Antennas*, New York, NY, McGraw-Hill Book Co., 1988.

[2] Jasik, H. (editor), *Antenna Engineering Handbook*, New York, NY, McGraw-Hill Book Co., 1961.

[3] Interim Working Party 5/2, CCIR Study Groups Period 1982-1986, "New Report: Propagation Data for Land Mobile Satellite Systems for Frequencies Above 100 MHz," Study Program 7C-1/5, Doc. 5/12-E, March 30, 1983.

[4] Kraus, J. D., *Electromagnetics*, New York, NY, McGraw-Hill Book Co., 1953.

[5] EIA/TIA Standard, "Minimum Standards for Land Mobile Communications FM or PM Receivers, 25-866 MHz," EIA/TIA-204-D, April 1989.

[6] Thro, S., "Trunking: A New Dimension in Fleet Dispatch Communications," *INTELCOM '79*, Dallas, TX, 1979, pp. 277-281.

[7] Hess, G., "Land-Mobile Satellite Excess Path Loss Measurements," *IEEE Tr. on Veh. Tech.*, Vol. VT-29, No. 2, May 1980, pp. 290-297.

[8] Jakubowski, R. J., "Propagation Considerations of Low Power Cellular Boosters and Case Histories," *39th IEEE Vehicular Technology Conference*, San Francisco, CA, May 1989, pp. 523-527.

[9] Isberg, R. A. and R. L. Chufo, "Passive Reflectors as a Means for Extending UHF Signals Down Intersecting Cross Cuts in Mines or Large Corridors," *28th IEEE Vehicular Technology Conference*, Denver, CO, March 22-24, 1978, pp. 267-272.

Chapter 3

Additional Transmission Loss Relations

This chapter discusses transmission loss formulas for propagation over a plane earth and for signals scattered by meteor trails. In both cases the received power is inversely proportional to the path length raised to a power, just as was found for free-space propagation. The exponents of 4 and 3, respectively, are greater than the exponent of 2, which holds for free-space propagation, thus implying more rapid attenuation with distance. It is not surprising then that a simple power-law propagation model is useful for describing land-mobile radio propagation too, albeit over limited distances. Lest one mistakenly conclude that power-law propagation is universal, an exception is noted: tunnel propagation loss in decibels can vary with linear distance rather than the logarithm of distance.

3.1 Plane Earth Propagation Model

The free-space transmission formula developed in Chapter 2 is inverse square law in nature. That is, for each doubling of path length the received power decreases by a factor of two squared. In decibels this means 6 dB additional loss per doubling of path length. A useful generalization of previous work is to consider propagation over a plane earth as in Figure 3.1 (see Section 2.1 of Reference [1]).

Based on work by Norton and Bullington, the formula relating transmitted power to received power is given by:

$$P_r = \frac{P_t G_t G_r}{[4\pi(\frac{d}{\lambda})]^2} |1 + R\exp(j\delta) + (1-R)A\exp(j\delta) + \cdots|^2 \qquad (3.1)$$

where the first term in the summation (unity) represents the direct wave, the second term represents the reflected wave, the third term represents the surface wave, and additional terms account for the inductive field and secondary effects of the ground.

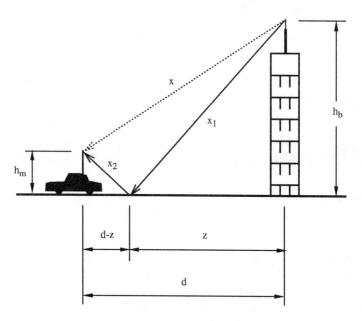

Figure 3.1 Propagation paths over a plane earth.

The reflection coefficient of the ground, R, depends on the angle of incidence, θ, the polarization of the wave, and the ground characteristics. It is given by the formula:

$$R = \frac{\sin(\theta) - z}{\sin(\theta) + z} \tag{3.2}$$

where $z = \sqrt{(\epsilon_0 - \cos^2(\theta))}/\epsilon_0$ for vertical polarization, $\epsilon_0 = \epsilon - j\,60\,\sigma\,\lambda$, ϵ equals the dielectric constant of the ground relative to free space (unity), σ equals the conductivity of the earth in mhos per meter, and λ equals the wavelength. In the limit of grazing angle incidence, the value of R approaches -1 independent of the polarization. For frequencies above 100 MHz and for average soil conditions, the reflection magnitude of vertically polarized signals exceeds 0.9 when θ is less than 10 degrees.

The phase difference between the direct and reflected paths is given by δ. For the geometry of Figure 3.1, δ can be found as follows. First, note that through the Pythagorean relation the direct path distance is given by $x_d = \sqrt{d^2 + (h_b - h_m)^2}$. In terms of phase units (i.e., 2π radians per distance λ), one obtains:

$$x_{dp} = \frac{2\pi d}{\lambda}\sqrt{1 + (\frac{h_b - h_m}{d})^2} \tag{3.3}$$

Additional Transmission Loss Relations

The reflected path is comprised of two segments, x_1 and x_2. Again using the geometry of Figure 3.1 one finds:

$$\begin{aligned} x_r &= x_1 + x_2 \\ &= \sqrt{h_b^2 + z^2} + \sqrt{h_m^2 + (d-z)^2} \end{aligned} \quad (3.4)$$

Also we have $\tan(\theta) = h_b/z = h_m/(d-z)$, which allows us to solve for z in terms of d: $z = dh_b/(h_m + h_b)$. Substituting this result into the former result and simplifying leads to:

$$x_{rp} = \frac{2\pi d}{\lambda}\sqrt{1 + \left(\frac{h_b + h_m}{d}\right)^2} \quad (3.5)$$

and thus

$$\begin{aligned} \delta &= x_{rp} - x_{dp} \\ &= \frac{2\pi d}{\lambda}\sqrt{1 + \left(\frac{h_b + h_m}{d}\right)^2} - \frac{2\pi d}{\lambda}\sqrt{1 + \left(\frac{h_b - h_m}{d}\right)^2} \end{aligned} \quad (3.6)$$

If d is large relative to the sum of the antenna heights, then a Taylor series expansion [2] can be used to show [1]:

$$\delta \approx \frac{4\pi h_b h_m}{\lambda d} \quad (3.7)$$

The effect of surface waves is significant only in the region a few wavelengths above the ground; hence, it can often be ignored, even in mobile communication problems. Under that assumption, and with $R = -1$, Equation (3.1) becomes (with the help of Euler's identity, $\exp(i\alpha) = \cos(\alpha) + i\sin(\alpha)$, and series expansions of the sine function, $\sin(x) \approx x$, and cosine function, $\cos(x) \approx 1$, for small arguments):

$$P_r \approx P_t G_t G_r \left(\frac{h_b h_m}{d^2}\right)^2 \quad (3.8)$$

Notice that inverse fourth-law propagation is implied by the plane earth propagation model. Thus, doubling the path length decreases the received signal by 12 dB, twice as much as for free-space conditions.

3.2 Meteor Scatter Propagation Model

When meteors burn up in the atmosphere they leave behind a trail of free electrons that can serve as a substantial means of reflecting electromagnetic radiation. Such trails have been used for a number of years to accomplish over-the-horizon communication at very high frequencies [3]. More recently such trails have been used to relay location information of vehicles spread throughout the nation [4].

---

[1] $\sqrt{1 + x^2} \approx 1 + \frac{1}{2}x^2$, for small x.

If the electron density is sufficiently low, incident radio waves are scattered by the individual free electrons as if no others were present; i.e., secondary radiative and absorptive effects can be neglected. This condition defines the underdense meteor trail. The scattering cross section of a free electron is given by $\sigma_e = 4\pi r_e^2 \sin\gamma$, where r_e is the classical radius of an electron and γ is the angle between the electric field strength vector of the incident wave and the line-of-sight direction to the receiver [5]. For backscatter conditions $\sigma_e \approx 1E-28$ m^2.

At a point on the trail a distance R from the transmitter, the power flux density of the incident wave is $P_t G_t / 4\pi R^2$ W/m^2. Assuming the receiver is collocated with the transmitter (monostatic radar situation), the received power is given by:

$$\begin{aligned} P_r &= \frac{P_t G_t}{4\pi R^2} \frac{\sigma_e}{4\pi R^2} \frac{G_r \lambda^2}{4\pi} \\ &= \frac{P_t G_t G_r \sigma_e \lambda^2}{(4\pi)^3 R^4} \end{aligned} \quad (3.9)$$

This is in the form of the classic radar equation. Notice that again fourth-law variation with range occurs.

However, the preceding analysis is for just a single electron. When the combined effects of all electrons along the trail are considered, a third-law variation with range is found. Additional factors affect the received power. The trail laid down as the meteor ablates is of finite radius. Furthermore, the trail radius increases with time due to diffusion after being formed. Accounting for these aspects leads to the following meteor scatter link equation [6]:

$$P_r = \frac{P_t G_t G_r \lambda^3 q^2 r_e^2}{32\pi^2 R^3} \exp\left(\frac{-8\pi^2 r_0^2}{\lambda^2}\right) \exp\left(\frac{-32\pi^2 Dt}{\lambda^2}\right) \quad (3.10)$$

where q equals the electron density of the trail in electrons per unit length, r_0 equals the initial radius of the meteor trail, D equals the ambipolar diffusion coeffcient, and t equals the time measured from the formation of the trail, which is assumed instantaneous. This of course is not really true. To account for real trail formation one must introduce an additional time variation whose form involves Fresnel integrals just like in optical diffraction theory [5].

3.3 Power-Law Propagation

A number of propagation scenarios have been analyzed thus far, all of which yielded a path loss varying according to some power of the range. It should not be surprising then that field tests of actual land-mobile propagation paths yield results reasonably well described by a simple power-law model: the typical (median) received signal strength is given by a constant divided by the range raised to some power (for example, see Figure 3.2).

Additional Transmission Loss Relations

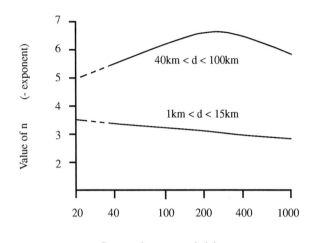

Figure 3.2 Distance dependence of median field strength in an urban area. After: Okumura, Y., et al., "Field Strength and Its Variability in VHF and UHF Land-Mobile Radio Service," *Rev. of Elec. Comm. Lab.*, Vol. 16, Nos. 9-10, September-October, 1968, p. 841.

Of course at short distances free-space conditions must prevail so, to avoid the predicament of less than free-space loss, one can calibrate the power-law model by equating its loss to free-space loss at some short range and only using it when the range is larger. The crossover range to use is a matter of judgement, but it is a function of base antenna height; greater heights implying greater crossover ranges. Also, beyond some maximum range (related to the radio horizon) the power-law model will become optimistic.

One circumstance where power-law propagation does not hold is in tunnels. Here propagation at land-mobile frequencies is generally like that for a waveguide below cutoff [7] and losses increase in decibels per linear distance. In addition, there is an appreciable insertion loss in coupling transmit power to the dominant waveguide mode.

3.4 Homework Problem

Consider two portable users attempting to communicate through a 1/4-mi grove of trees at 850 MHz. Is success very likely? Assume −4-dBd antenna gain at the transmitting end (a practical value for portable radios in the talking position), −9-dBd antenna gain at the receiving end (a practical value for the portable radio when stowed on the hip), 10 mW transmitter power (in line with cordless telephone power levels), and 0.5 μV radio sensitivity (-113 dBm, in line with typical commercial mobile and portable radio performance).

Solution: Free-space loss at 850 MHz over a 1/4-mi path is about 83.1 dB between isotropic antennas. Due to the lossy portable antennas, the actual loss would be 13 dB greater, minus 4.3 dB to convert between dBd and dBi, or 91.8-dB loss. With 10 mW of transmit power, a free-space receive power of -81.8 dBm would result, well above the required -113 dBm. However, we have yet to account for foliage loss. An empirical relation for this loss is [8] $L = 1.33 f^{0.284} d^{0.588}$, where L is the foliage loss in decibels along a path blocked by dense, dry, in-leaf temperate-climate trees, f is the link frequency in gigahertz, and d is the path length in meters. In our problem the extra loss equals 43.2 dB, dropping the received signal power to -125.0 dBm, well below receive sensitivity. Interestingly, this does not mean that communication is impossible. It simply means that it is unlikely directly through the foliage. If both antennas are sufficiently in the clear, the principal mode of propagation will be diffraction over the tops of the nearest trees and surface waves along the intervening foliage.

References

[1] Jakes, Jr., W. C. (editor), *Microwave Mobile Communications*, New York, NY, John Wiley & Sons, 1974.

[2] Protter, M. H. and C. B. Morrey, Jr., *Modern Mathematical Analysis*, Reading, MA, Addison-Wesley Publishing Co., 1964.

[3] Ince, A. N., "Spatial Properties of Meteor-Burst Propagation," *IEEE Tr. on Comm.*, Vol. COM-28, No. 6, June 1980, pp. 841-849.

[4] Mickelson, K. D., "Tracking 64,000 Vehicles with Meteor-Scatter Radio," *Mobile Radio Technology*, January 1989, pp. 24-38.

[5] McKinley, D.W.R., *Meteor Science and Engineering*, New York, NY, McGraw-Hill Book Co., 1961.

[6] Sugar, G. R., "Radio Propagation by Reflection from Meteor Trails," *Proc. IEEE*, Vol. 52, 1964, pp. 116-136.

[7] Emslie, A. G., et al., "Theory of the Propagation of UHF Radio Waves in Coal Mine Tunnels," *IEEE Tr. on Ant. and Prop.*, Vol. AP-23, No. 2, March 1975, pp. 192-205.

[8] Weissberger, M. A., "An Initial Critical Summary of Models for Predicting the Attenuation of Radio Waves by Trees," ESD-TR-81-101, Electromagnetic Compatibility Analysis Center, Annapolis, MD, August 1981.

Chapter 4
Noise Considerations

Equally important to path loss considerations are considerations of noise. This is because system performance generally hinges on the ratio of desired signal power to noise power rather than on absolute levels. This chapter discusses the most common type of noise encountered, additive white Gaussian. The concept of noise figure is introduced and some of the practical difficulties in measuring it are noted in a concluding exercise. A practical example involving nonflat (colored) noise is also given.

4.1 Additive White Gaussian Noise

Absolute signal level is often of less interest than the *signal-to-noise ratio* (SNR). Of the three natural noise mechanisms (thermal, 1/f [1], and shot noise), thermal noise is of most interest in communications system design [1]. Manmade noise sources also abound; see Reference [2] for an extensive study of such sources.

Thermal noise is caused by the random thermally excited vibrations of charge carriers in a conductor. It is in essence "Brownian motion of electricity." Thermal noise was first observed in 1927 by J. B. Johnson of Bell Laboratories, and first theoretically described by H. Nyquist in 1928. Thus, thermal noise is sometimes referred to as Johnson noise or Nyquist noise.

The available thermal noise power in a conductor is given by $N_t = kTB$, where k equals Boltzmann's constant, $1.38E-23$ W-s/K, T equals the conductor temperature in Kelvin, and B equals the noise bandwidth (one-sided viewpoint) of the measuring system in hertz. Because available power is the power that can be supplied by a source when feeding a resistance load equal to its internal resistance, the rms noise voltage of a resistance R is given by $E_t = \sqrt{4kTRB}$.

The preceding two relations stem from a classical physics viewpoint. Sections 3.1

---

[1] The intensity of this type noise increases with decreasing frequency, at least to a point (otherwise the total power would seem to be unbounded).

and 5.2c of Reference [3] offer proof, and also note that when the quantity hB/kT becomes too large relative to unity a quantum correction must be made (here h equals Planck's constant, $6.62E-34$ J-s). This is fortunate since otherwise an infinite amount of power would be implied as the measuring bandwidth increased without bound! At room temperatures, correction is unnecessary below optical frequencies; even at cryogenic temperatures correction is unnecessary for frequencies below tens of gigahertz.

Thermal noise is called white based on an analogy with white light, which is made up of many colors (as thermal noise is composed of many frequency components). White is synonymous with flat, meaning that the noise power spectral density is a constant, independent of frequency[2]. Thus, every bandwidth B, regardless of the absolute frequencies spanned, contains the same amount of noise power.

Thermal noise is also called Gaussian because the distribution of instantaneous noise voltage follows the Gaussian *probability density function* (PDF). This is a consequence of the central limit theorem (discussed in Chapter 8). Finally, the term additive is often used with noise, meaning that its power adds linearly along with desired signal power in a communications channel.

4.2 Noise Equivalent Bandwidth

Noise bandwidth is not necessarily the same as the commonly used 3-dB signal bandwidth. It is convenient to define a noise equivalent bandwidth as the frequency span of an ideal filter whose area equals the area under the actual power transfer function curve and whose gain equals the peak gain of the actual power transfer function (Figure 4.1) [4].

$$B_e = \frac{1}{H_0^2} \int_0^\infty |H(f)|^2 df \tag{4.1}$$

The filtered noise power can thus be simply written as $H_0^2 kTB_e$. For practical purposes, B_e is somewhat greater than the 3-dB signal bandwidth. However, as the filter becomes more selective the two bandwidths approach each other, and taking them as equal is not unreasonable. Such is the case for typical land-mobile FM receiver selectivity prior to detection.

4.3 Noise Figure

Assume that the noise present at the input of some arbitrary two-port "black box" is caused by a resistor at room temperature T_0 (standardized as 290 K). If the two-port was perfectly noiseless, then the one-sided output noise power spectral density would

---

[2] The noise power spectral density of colored noise is some function of the frequency.

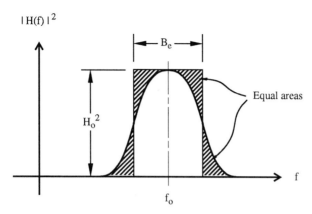

Figure 4.1 Noise equivalent bandwidth of a bandpass filter.

be $S_{ai,f} = g_{a,f} k T_0$, where $g_{a,f}$ equals the two-port available gain, possibly a function of frequency. For nonideal two-ports, however, additional noise will be added. To handle this fact it is convenient to treat the two-port as lossless and instead assign a higher temperature to the source; i.e., to represent the output noise-power spectral density as:

$$S_{a,f} = g_{a,f} k (T_0 + T_e) \qquad (4.2)$$

where T_e is called the effective input-noise temperature. The so-called spot noise figure of the two-port is defined by:

$$\begin{aligned} F(f) &= S_{a,f}/S_{ai,f} \\ &= 1 + (T_e/T_0) \end{aligned} \qquad (4.3)$$

If F is indeed a function of frequency, then the average noise figure is often quoted. For an arbitrary frequency span, it is given by:

$$F_{av} = \int_{f_1}^{f_2} g_{a,f} F(f) df \Big/ \int_{f_1}^{f_2} g_{a,f} df \qquad (4.4)$$

An alternative expression for noise figure is the ratio of input SNR to output SNR. Taking available gain as constant over the frequency range of interest means that $F = (S_i/N_i)/(S_o/N_o) = (1/g_a)(N_o/N_i)$, where S and N represent signal and noise powers, respectively. But $N_o = g_a N_i + N_{tp}$, where N_{tp} represents the output noise due to an arbitrary noisy two-port. Thus, one has $N_{tp} = g_a(F-1)N_i$. With the help of this expression, one can express the overall noise figure for a cascade of two-ports (Figure 4.2) [5] as follows.

For a two-stage cascade with input-noise power N_i, the associated output-noise power is given by $g_{a1} g_{a2} N_i$. The noise output of the first stage due to noise generated

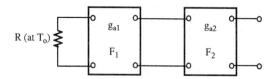

Figure 4.2 A noise source driving a cascade of two-ports.

within the first two-port equals $g_{a1}(F_1-1)N_i$. This is amplified by the second-stage gain, g_{a2}, before appearing at the output. Finally, the output noise due to noise generated in the second stage is $g_{a2}(F_2-1)N_i$. Thus, the total noise power at the output is given by:

$$N_o = g_{a1}g_{a2}N_i + g_{a1}g_{a2}(F_1-1)N_i + g_{a2}(F_2-1)N_i \qquad (4.5)$$

Hence,

$$\begin{aligned} F &= \frac{1}{g_a}\left(\frac{N_o}{N_i}\right) \\ &= \frac{1}{g_{a1}g_{a2}}\left(\frac{N_o}{N_i}\right) \\ &= F_1 + \frac{F_2-1}{g_{a1}} \end{aligned} \qquad (4.6)$$

and a first-stage amplifier with sufficient gain will make the overall noise figure effectively independent of second-stage performance.

The preceding analysis can be applied to an arbitrary number of stages, say k, to yield Friis' formula:

$$F = F_1 + \frac{F_2-1}{g_{a1}} + \frac{F_3-1}{g_{a1}g_{a2}} + \cdots + \frac{F_k-1}{g_{a1}g_{a2}\cdots g_{ak}} \qquad (4.7)$$

In terms of equivalent noise temperatures, this becomes (by substitution of Equation (4.3) into Equation (4.7)):

$$T_e = T_{e1} + \frac{T_{e2}}{g_{a1}} + \frac{T_{e3}}{g_{a1}g_{a2}} + \cdots + \frac{T_{ek}}{g_{a1}g_{a2}\cdots g_{ak}} \qquad (4.8)$$

4.4 Y-Factor Measurement

The Y-factor refers to the ratio of noise-power outputs corresponding to two different source temperatures [6]. Let T_h stand for the higher of the two temperatures (i.e., hot) and let T_c stand for the lower temperature (i.e., cold). Typically only T_c is a physical temperature; T_h is an electrical temperature related to the amount of noise generated by an avalanche diode noise source. Such sources are generally specified in terms of *excess noise ratio* (ENR), which is the base 10 logarithm of [(effective noise temperature in Kelvin/290) − 1]. Figure 4.3 shows a typical Y-factor test setup.

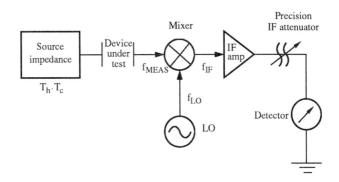

Figure 4.3 Y-factor noise figure measurement test setup.

The detected power with the noise diode off and the device under test in-line is first noted. Then the diode is turned on and the IF attenuator is varied until the original noise level is again obtained. The difference in attenuator settings equals the Y-factor. Noise figure can be deduced from it as follows:

The hot output-noise power equals $g_{ah}k(T_h + T_e)B$, where measurement system gain and bandwidth are g_{ah} and B, respectively. The cold output-noise power equals $g_{ac}k(T_c + T_e)B$. With an attenuator setting $Y = g_{ac}/g_{ah}$, the output hot and cold powers are equal. Thus, solving for T_e in terms of Y gives $T_e = (T_h - YT_c)/(Y - 1)$. But recall that noise figure $F = (T_e/T_0) + 1$, thus we have:

$$F = \frac{\left(\frac{T_h}{T_0} - 1\right) - Y\left(\frac{T_c}{T_0} - 1\right)}{(Y - 1)} \tag{4.9}$$

4.5 System Noise Figure and G/T

Reference [7] presents a detailed system noise analysis for the land mobile satellite service. In particular, the numerous contributors to antenna noise temperature are addressed. In satellite systems, receiving performance is commonly quantified by the figure of merit G/T, where G represents the antenna gain and T represents the overall noise temperature. This comes about as follows:

$$P_r/N = \frac{\text{ERP}_t}{L}\frac{G_r}{kTB}$$

$$P_r/N_0 = \frac{\text{ERP}_t}{Lk}\left(\frac{G_r}{T}\right) \tag{4.10}$$

where, in the final expression, bandwidth has been normalized to 1 Hz.

4.6 An Example of Nonflat (Colored) Noise

The FCC often controls interference by specifying a spectrum mask below which transmitted power must fall. For this purpose 0 dBr represents the unmodulated or carrier power level and the mask indicates the required relative power reduction for frequencies offset from the assigned channel center frequency (hence the unit dBr, where r stands for relative).

A spectrum mask proposed in the *Further Notice of Proposed Rulemaking*, General Docket 85-171 of the FCC, which is relevant to adjacent channel interference, is as follows:

$$\begin{aligned} f &\geq 25 \text{ kHz}, \ G_{\text{dB}}(f) = -70 \\ 11 \text{ kHz} \leq f &< 25 \text{ kHz}, \ G_{\text{dB}}(f) = -60 \\ 4 \text{ kHz} \leq f &< 11 \text{ kHz}, \ G_{\text{dB}}(f) = 23.286 - 7.571f \end{aligned} \tag{4.11}$$

where f represents the frequency difference between an arbitrary frequency of interest and the center frequency of the channel (channels are 25 kHz in this rulemaking) and $G_{\text{dB}}(f)$ represents the worst-case (maximum) power content in a 300-Hz bandwidth centered at that arbitrary frequency, relative to the unmodulated carrier power.

In general one can represent the power spectrum of the proposed mask with line segments. To calculate the unwanted power coupled into the adjacent channel (the "technical" term for this is splatter), one must first relate the spectra measured with 300-Hz bandwidth, denoted by G, to the true power spectral density, denoted by S. For this refer to the diagrams in Figure 4.4. The task is to relate the straight-line coefficients a' and b' describing the G function behavior to the S function coefficients a and b. The relevant mathematics is as follows:

$$\begin{aligned} G(f_c) &= \int_{f_c - \frac{B}{2}}^{f_c + \frac{B}{2}} S(f)\, df \\ &= \int 10^{[S_{\text{dB}}(f)/10]}\, df \\ &= \int \exp[k(af + b)]\, df \\ &= \exp[k(af_c + b)]\frac{\sinh(\alpha B)}{\alpha} \end{aligned} \tag{4.12}$$

where $k = [\ln(10)/10]$, $\alpha = (ka/2)$, and f_c equals the center frequency for bandwidth

Noise Considerations

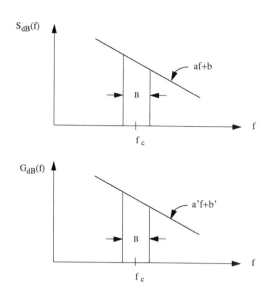

Figure 4.4 Sample power spectral density and spectrum analyzer power reading versus frequency.

B. Also,

$$G_{dB}(f_c) = 10\log[G(f_c)] = \frac{1}{k}\ln[G(f_c)] = a'f_c + b' \qquad (4.13)$$

Thus, equating frequency coefficients yields:

$$a = a' \qquad (4.14)$$

$$b = b' - \frac{1}{k}\ln[\frac{\sinh(\alpha B)}{\alpha}]$$

In the limit as a' approaches zero, the relation for b becomes (applying l'Hôpital's rule) [8]:

$$b = b' - \frac{1}{k}\ln(B) \qquad (4.15)$$

Now consider the power allowed to be coupled into the adjacent channel by the mask described above. For frequency differences between 11 and 25 kHz, the power per 300 Hz is independent of frequency and 60 dB below the unmodulated carrier reference level. Hence $a' = 0$ and $b' = -60$, which implies $a = 0$ and $b = -60 - 4.343\ln(0.3) = -54.77$. Similarly, for frequency differences between 25 and 37.5 kHz, the power per 300 Hz is independent of frequency and 70 dB below the reference. Here $a' = 0$ and $b' = -70$, implying $a = 0$ and $b = -70 - 4.343\ln(0.3) = -64.77$. The coupled power is thus:

$$P_c = \int_{18.75-6.25}^{18.75+6.25} 10^{-5.477}\, df + \int_{31.25-6.25}^{31.25+6.25} 10^{-6.477}\, df = 4.585E-5 \quad (4.16)$$

or $P_c(\text{dB}) = -43.4$. The amount of protection afforded an adjacent channel user equals the negative of cochannel coupling minus the minimum SNR necessary for communications. For 25-kHz FM land-mobile radios the latter is typically 5 dB (12-dB static SINAD[3] conditions). Hence, only 38.4 dB of protection is guaranteed by the mask, a far from sufficient value because the dynamic range of land-mobile radio signals is commonly on the order of 80 dB.

However, the preceding analysis is overly pessimistic because it fails to account for receiver selectivity. A model for selectivity that is both analytically tractable and representative of actual performance is:

$$H_{\text{dB}}(f) = -c|f_c - f|^2 \quad (4.17)$$

where typically $c = 0.25$ for frequency in kilohertz. Selectivity modifies the mathematics given in Equation (4.12) as follows:

$$\begin{aligned}
G(f) &= \int_{f_1}^{f_2} S(f)H(f)\, df \\
&= \int_{f_1}^{f_2} 10^{[S_{\text{dB}}(f)+H_{\text{dB}}(f)]/10}\, df \\
&= \int_{f_1}^{f_2} \exp k[S_{\text{dB}}(f)+H_{\text{dB}}(f)]\, df \\
&= \exp[k(b-cf_c^2)] \int_{f_1}^{f_2} \exp\{[k(a+2cf_c)]f - (kc)f^2\}\, df \\
&= \sqrt{\frac{\pi}{4kc}} \exp[k(b+a^2+af_c)][\operatorname{erf}(\sqrt{kc}(f_2-f_c) - \sqrt{\frac{k}{4c}}a) \\
&\quad -\operatorname{erf}(\sqrt{kc}(f_1-f_c) - \sqrt{\frac{k}{4c}}a)] \quad (4.18)
\end{aligned}$$

The solution of this integral in terms of the error function and the definition of that special function are found in Reference [9] (also, see Section 7.6.2).

The power coupled into an adjacent channel receiver permitted by the proposed mask can be estimated by the sum of two integrals. The first runs from $-\infty$ to 11 kHz and uses the third relation of Equation (4.11) for input (strictly speaking negative frequencies do not apply here, but this is unimportant because of filter attenuation); the parameters are thus $a = -7.571$ and $b = 28.51$. The second integral runs from 11 to 25 kHz and uses $G_{dB}(f) = -60$; hence $a = 0$ and $b = -54.77$. The contribution

---

[3]SINAD represents the ratio of signal power plus noise power and distortion power to the sum of noise and distortion powers.

Noise Considerations

beyond 25 kHz is insignificant compared to that of the second integral due to the 10-dB decrease in mask level. Thus for $c = 0.25$ one has:

$$\begin{aligned} P_c &= 3.694 \exp(-23.819)[\text{erf}(0.274) - \text{erf}(-\infty)] \\ &\quad + 3.694 \exp(-12.611)[\text{erf}(0) - \text{erf}(-3.359)] \\ &= 2.173E - 10 + 1.232E - 5 \end{aligned} \quad (4.19)$$

and $P_c(\text{dB}) = -49.1$. Clearly the mask floor of just -60 dB is limiting the benefit of adjacent channel receiver selectivity. The resultant protection of $49.1 - 5 = 44.1$ dB is still not sufficient for many land-mobile situations.

4.7 Homework Problem

A common noise figure measurement setup uses an automatic analyzer like the HP8970A covering up to 1500 MHz with a 15.2-dB ENR noise source. Could one use such a setup to measure the noise figure of a 3-GHz *low noise amplifier* (LNA) as shown in Figure 4.5? If not, what are some of the problems in attempting to do so?

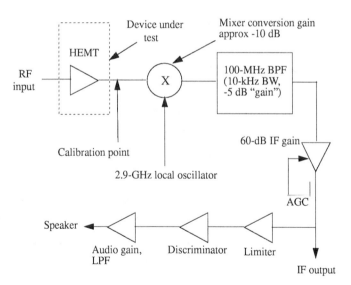

Figure 4.5 Candidate noise figure measurement arrangement.

Solution: Not without great difficulty. The *automatic gain control* (AGC) should be disabled. The noise power is spread over a much larger bandwidth (typically 4 MHz) than the specified IF bandwidth of just 10 kHz, implying about 26 dB less noise power than otherwise expected. The noise figure meter can be used above 1.5 GHz if the noise source performance is known there; the only requirement is that the IF be below 1.5 GHz. But the single-stage LNA gain is almost surely insufficient to "take over" the losses of the downconverter. Those losses would run 10 dB for the mixer, 5 dB for the IF filter, and whatever the IF gain block noise figure is. The lack of a preselector filter to reject the mixer image of 2.8 GHz implies double sideband noise operation rather than single sideband noise operation. The HP8970A corrects for this only if both high- and low-side noise contributions are equal. Finally, 15.2-dB ENR is too high for LNAs with very low noise figures. A better choice is a 5.2-dB ENR source, which can be obtained with the 15.2-dB source followed by a precision 10-dB attenuator. This greatly improves the return loss into the noise source.

References

[1] Motchenbacher, C. D. and F. C. Fitchen, *Low-Noise Electronic Design*, New York, NY, John Wiley & Sons, 1973.

[2] Skomal, E. N., *Man-Made Radio Noise*, New York, NY, Van Nostrand Reinhold Co., 1978.

[3] Van Der Ziel, A., *Noise in Measurements*, New York, NY, John Wiley & Sons, 1976.

[4] Carlson, A. B., *Communication Systems: An Introduction to Signals and Noise in Electrical Communication*, New York, NY, McGraw-Hill Book Co., 1968.

[5] Taub, H. and D. L. Schilling, *Principles of Communication Systems*, New York, NY, McGraw-Hill Book Co., 1971.

[6] Hewlett Packard Application Note 57-1. "Fundamentals of RF and Microwave Noise Figure Measurement," July 1983.

[7] Bell, D., et al., "MSAT-X Antennas: Noise Temperature and Mobile Receiver G/T," *MSAT-X Quarterly*, No. 16, Jet Propulsion Laboratory, August 1988.

[8] Protter, M. H. and C. B. Morrey, Jr., *Modern Mathematical Analysis*, Reading, MA, Addison-Wesley Publishing Co., 1964.

[9] Gradshteyn, I. S. and I. M. Ryzhik, *Table of Integrals, Series, and Products*, New York, NY, Academic Press, 1965.

Chapter 5

Sample Radio Link Analyses

This chapter illustrates how system engineers analyze radio links. Two cases are considered: the first involves reception of a satellite television broadcast; the second involves terrestrial land-mobile radio communications. For the latter, the empirical path loss methodology championed by Okumura is introduced and applied in two examples. The first example treats coverage from high repeater sites; the second example deals with interference by low-power units sharing a channel with normal power units. A concluding exercise highlights differences between point-to-point and broadcast systems.

5.1 Satellite TVRO Link Analysis

Consider the design of a satellite *television receive-only* (TVRO) station [1]. While such a task may seem far afield from mobile communication, it does illustrate the treatment of rebroadcast noise. This is an important consideration also generally required when mobile units do not communicate directly to their home station or into the public switched telephone network, but rather communicate via repeater stations.

The earth station figure of merit expressed in decibels is given by:

$$\begin{aligned}(G/T)_{\text{dB}} &= G_A - 10\log(T_S) \\ T_S &= T_A + T_{\text{LNA}} + \frac{(F_e - 1)T_0}{G_{\text{LNA}}}\end{aligned} \quad (5.1)$$

where G_A is the receive antenna gain in dB, T_S is the system noise temperature referenced to the antenna output port, T_A is the antenna noise temperature, T_{LNA} is the low noise amplifier noise temperature (this unit is located right at the antenna output port), G_{LNA} is the LNA gain (numeric), F_e is the effective noise figure (numeric) of the cable and receiver combination that follows the LNA, and T_0 is the reference noise temperature of 290 K.

The C-band satellite service operates over the downlink frequency range of 3.7 to 4.2 GHz. Assuming a 15-foot dish is available with 65% efficiency, the receive antenna gain can be about 43.6 dB. Typical antenna noise for such a dish elevated 30 deg is on the order of 22 K. LNA gain blocks commonly offer 55-dB gain and 0.75-dB noise figure (55 K temperature). Finally, if one allows for only a 16-dB effective noise figure following the LNA, an earth station figure of merit of 24.7 dB/K is obtained.

The next important calculation concerns the *carrier-to-temperature* (C/T) ratios for the broadcast transmitter to satellite receiver (the uplink) and the satellite transmitter to earth terminal (the downlink). This ratio in decibels is equivalent to the carrier-to-noise power ratio for a 1-Hz bandwidth expressed in decibels plus the decibel value of Boltzmann's constant (-228.6). The uplink relation in decibels is as follows:

$$(C/T)_u = \phi + A_i + (G/T)_{\text{sat}} \tag{5.2}$$

where ϕ is the uplink flux density (a representative value for geostationary satellites is -80 dBW/m^2), A_i is the area of an isotropic antenna (such an antenna has unity gain and directivity, and the aperture in square meters is -37 dB for an uplink frequency of 6 GHz), and $(G/T)_{\text{sat}}$ is the satellite receive figure of merit (see Equation (4.10); a representative value is -6 dB/K). Substituting these values into Equation (5.2) yields $(C/T)_u = $ -123 dBW/K.

The downlink relation in decibels can be written as:

$$(C/T)_d = \text{EIRP}_{\text{sat}} - PL + (G/T)_{\text{es}} \tag{5.3}$$

where EIRP$_{\text{sat}}$ is the satellite effective isotropic radiated power (a representative value for geostationary satellites at the antenna beam edge (-3 dB) is 33 dBW), PL is the path loss, including atmospheric absorption (195.7 dB is a reasonable value for geostationary satellites at 4 GHz; such satellites have a subsatellite distance from Earth of about 23,000 nautical miles, or 39,000 km), and $(G/T)_{\text{es}}$ has already been calculated to be 24.7 dB/K. Substituting these values into Equation (5.3) yields $(C/T)_d = $ -138 dBW/K.

Note how the uplink performance substantially exceeds that of the downlink. One reason for this is that the satellite service must share the 3.7- 4.2-GHz band with previously existing terrestrial users. To protect such users, CCIR[1] specifications limit satellite transmissions to power flux densities of between -142 and -152 dBW/m^2/ 4-kHz bandwidth, depending on the ground elevation angle to the satellite. If a satellite emits 33 dBW, then 39,000 km distant the power flux density will be about -129.8 dBW/m^2. However, satellite transponder bandwidths are typically on the order of 36 MHz. If the power is evenly spread across that bandwidth then only -169.3 dBW/m^2/4 kHz would occur, well below CCIR specifications.

---

[1] The International Radio Consultative Committee, a branch of the *International Telecommunication Union* (ITU) concerned with preparing documents dealing with the preparation, transmission, and reception of information via radio signals, in the broadest sense.

Sample Radio Link Analyses

The relation of the system C/T to the individual link C/T values can be derived with the help of Figure 5.1.

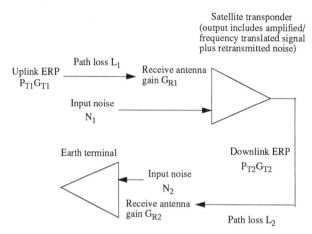

Figure 5.1 Satellite link signal flow diagram, including retransmitted noise.

At the input of the satellite receiver, the signal-to-noise power ratio can be written as:

$$(C_1/N_1) = \frac{P_{T1} G_{T1} G_{R1}}{L_1} \frac{1}{N_1} \tag{5.4}$$

For a linear transponder $P_{T2} = \alpha(N_1 + C_1)$, where α represents the gain of the satellite repeater system. Here the noise received at the transponder input is retransmitted along with the desired signal. Hence the signal received at a TVRO earth station is $\alpha C_1 G_{T2} G_{R2}/L_2$ and the total noise received there is the sum of N_2 and the retransmitted noise multiplied by the factor $G_{T2} G_{R2}/L_2$. After simplification one obtains the desired result of:

$$(C/N) = \frac{1}{\frac{1}{(C/N)_u} + \frac{1}{(C/N)_d}} \tag{5.5}$$

A similar equation results when noise temperature is used rather than noise power:

$$(C/T) = \frac{1}{\frac{1}{(C/T)_u} + \frac{1}{(C/T)_d}} \tag{5.6}$$

Thus, the system C/T is found to be -138.1 dBW/K.

For a frequency-modulated carrier, the video SNR is given by [2]:

$$(S/N)_w = 6 \frac{C}{kT} \frac{f_d^2}{f_m^3} q \tag{5.7}$$

where $(S/N)_w$ is the peak-to-peak luminance signal to rms-weighted noise power ratio, (C/kT) is the predetection carrier-to-noise power density, f_d is the peak frequency deviation, f_m is the highest modulation frequency, and q is an improvement factor related to emphasis and weighting. For the U.S. television system, NTSC[2], $f_m = 4.2$ MHz and $10\log(q) = 13$ dB. Substituting these values leads to the following equation in decibels:

$$(S/N)_w = 170.7 + \frac{C}{T}(\text{dBW/K}) + 20\log[f_d(\text{MHz})] \qquad (5.8)$$

A peak deviation of 12 MHz is feasible if the entire transponder is dedicated to a single television signal. This yields a weighted SNR of 54.2 dB. This is well above a common goal of 50 dB; also, the predetection carrier-to-noise ratio of 14.9 dB[3] is approximately 5 dB above conventional FM threshold, providing a modest fade margin.

5.2 Land-Mobile Radio Link Analysis

Numerous organizations have conducted field tests to measure the path loss between base station antenna sites, which are relatively high, and mobile radio units, whose antennas are relatively near ground level. Such data allow the generation of empirical models for predicting path loss. One of the most widely accepted models is that of Okumura [3]. It offers relatively good accuracy with relatively little complexity [4].

The basic procedure has been nicely summarized by four figures as in Jakes' textbook on microwave mobile communications [5]. A reference level of median attenuation beyond the free-space path loss (often referred to as excess path loss) is first obtained from Figure 5.2. The reference assumes an urban environment with base antenna height of 200 m and mobile antenna height of 3 m. Prediction curves are given for path lengths between 1 and 100 km and frequencies between 100 MHz and 3 GHz.

The reference loss is then adjusted according to the actual base antenna height (Figure 5.3), the actual mobile antenna height (Figure 5.4), and the actual environment type (Figure 5.5). The choices for the latter are urban, suburban, quasi-open, and open. Note that doubling the base antenna height typically decreases path loss by 6 dB, which is in agreement with the plane earth propagation model studied in Chapter 3, whereas doubling the mobile antenna height typically decreases path loss by just 3 dB.

The results obtained using Figures 5.2 through 5.5 have been fitted to simple mathematical formulas by Hata [6]. As published, the formulas are limited to the

---

[2] After the National Television System Committee, whose 1953 color television recommendations were later adopted by the FCC

[3] $[(\frac{C}{T})/kB]_{dB} = -138.1 - 10\log[(1.38E-23)(36E6)]$.

Sample Radio Link Analyses

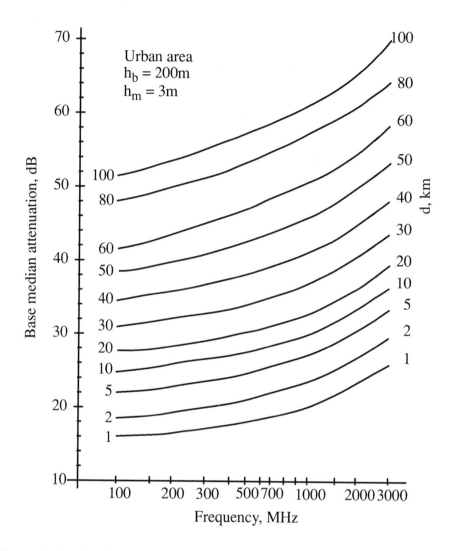

Figure 5.2 Prediction curves for basic median attenuation relative to free space in urban area over quasi-smooth terrain, referred to base antenna height of 200 m and mobile antenna height of 3 m.
After: Okumura, Y., et al., "Field Strength and Its Variability in VHF and UHF Land-Mobile Radio Service," *Rev. of Elec. Comm. Lab.*, Vol. 16, Nos. 9-10, September-October 1968, p. 842.

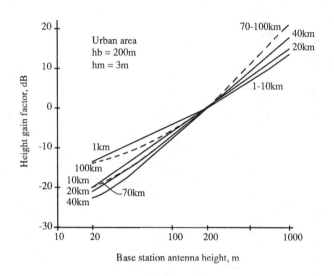

Figure 5.3 Prediction curves for base station height-gain factor referred to base antenna height of 200 m.
After: Okumura, Y., et al., *ibid.*, p. 848.

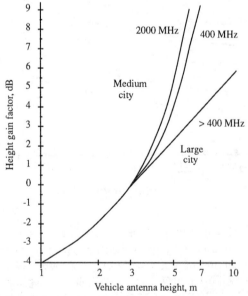

Figure 5.4 Prediction curves for vehicular height gain factor referred to mobile antenna height of 3 m.
After: Okumura, Y., et al., *ibid.*, p. 850.

Sample Radio Link Analyses

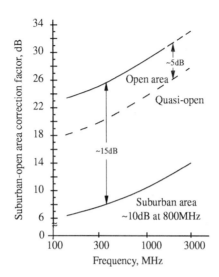

Figure 5.5 Prediction curves for suburban, quasi-open, and open area correction factors.
After: Okumura, Y., et al., *ibid.*, pp. 845-846.

rather modest path lengths common in cellular systems, but they can readily be extended in range as desired. Sample Fortran computer code that includes such an extension is as follows:

```
C
C This program solves for path loss (isotropic to isotropic)
C using Hata's experimental formula fits to Okumura's
C measurements, extended in range courtesy of Al Davidson;
C field strength in dB relative to 1 microvolt per meter
C (dipole to dipole, kW ERP) is also output
C
C MAIN ROUTINE
C
10 CONTINUE
WRITE(6,20)
20 FORMAT(1X,'FREQUENCY IN MHZ=?')
READ(5,30)F
30 FORMAT(E10.3)
WRITE(6,40)
40 FORMAT(1X,'BASE ANTENNA HEIGHT IN FEET=?')
READ(5,30)HBF
```

```
      HBM=HBF/3.281
      WRITE(6,50)
   50 FORMAT(1X,'MOBILE ANTENNA HEIGHT IN FEET=?')
      READ(5,30)HMF
      HMM=HMF/3.281
      WRITE(6,60)
   60 FORMAT(1X,'ENVIRONMENT?: 1=URBAN, 2=SUBURBAN, 3=OPEN
     1, 4=FREE SPACE')
      READ(5,70)IENV
   70 FORMAT(I1)
  100 WRITE(6,110)
  110 FORMAT(1X,'RANGE IN MILES=? (NEG VALUE TO RESTART RUN)')
      READ(5,30)RNGMI
      IF(RNGMI.LT.0.0)GO TO 10
      RNGKM=RNGMI*1.609
      CALL PLOSS(F,HBM,HMM,IENV,RNGKM,PLMED)
      WRITE(6,120)PLMED
  120 FORMAT(1X,'MEDIAN PATH LOSS IN DB = ',E12.4)
      DBU=60.0-PLMED+20.0*ALOG10(F)+79.4
      WRITE(6,130)DBU
  130 FORMAT(1X,'DBU FOR KW ERP/DIPOLE = ',E12.4/)
      GO TO 100
      END
C
C END MAIN ROUTINE, BEGIN SUBROUTINE
C
      SUBROUTINE PLOSS(FMHZ,HBASE,HMOB,IENVIR,RKM,PLMED)
C This routine calculates the isotropic path loss in dB based
C on Hata's empirical fit to Okumura's measured results;
C The fit has been extended past 20 km range courtesy of Al Davidson
C
C INPUT PARAMETERS:
C
C FMHZ=LINK FREQUENCY IN MHZ
C HBASE=BASE(REPEATER) ANTENNA HEIGHT IN METERS
C HMOB=MOBILE ANTENNA HEIGHT IN METERS
C IENVIR=ENVIRONMENTAL TYPE, 1=URBAN, 2=SUBURBAN, 3=OPEN
C RKM=BASE-MOBILE PATH LENGTH IN KM
C
C OUTPUT PARAMETER:
C
C PLMED=MEDIAN PATH LOSS IN DB
```

Sample Radio Link Analyses

```
C
C BEGIN BY CALCULATING REFERENCE PATH LOSS FOR URBAN ENVIRONMENT
PLREF=69.55+26.16*ALOG10(FMHZ)-13.82*ALOG10(HBASE)
PLREF=PLREF+(44.9-6.55*ALOG10(HBASE))*ALOG10(RKM)
C NEXT APPEND CORRECTION FACTOR FOR MOBILE ANTENNA HEIGHT
C USING MEDIUM-SMALL CITY RELATION OF HATA
ADJ1=(1.1*ALOG10(FMHZ)-0.7)*HMOB-(1.56*ALOG10(FMHZ)-0.8)
PLREF=PLREF-ADJ1
C NEXT MODIFY FOR PATH LENGTH IN EXCESS OF 20 KM, IF NECESSARY
ADJ21=0.0
ADJ22=0.0
ADJ23=0.0
IF(RKM.GT.20.0)ADJ22=(RKM-20.0)*(0.3108
1+0.1865*ALOG10(HBASE/100.0))
IF(RKM.GT.64.36)ADJ23=-0.174*(RKM-64.36)
PLREF=PLREF+ADJ21+ADJ22+ADJ23
C FINALLY ADJUST FOR ENVIRONMENT TYPE, IF NECESSARY
PLMED=PLREF
IF(IENVIR.EQ.1)RETURN
IF(IENVIR.EQ.2)GO TO 100
IF(IENVIR.EQ.4)GO TO 200
C OPEN ENVIRONMENT:
ADJ3=-4.78*((ALOG10(FMHZ))**2)+18.33*ALOG10(FMHZ)-40.94
PLMED=PLREF+ADJ3
RETURN
100 CONTINUE
C SUBURBAN ENVIRONMENT:
ADJ4=-2.0*((ALOG10(FMHZ/28.0))**2)-5.4
PLMED=PLREF+ADJ4
RETURN
C FREE-SPACE CALCULATION (4*PI*(DISTANCE/WAVELENGTH))**2, FOR
C ISOTROPIC ANTENNAS)
200 PLMED=20.*ALOG10(FMHZ*RKM)+32.4
RETURN
END
```

Okumura also describes adjustments for terrain undulations that can be applied if topographical data are available. Here undulations are characterized by noting the spread between the 10% highest elevation and the 90% highest elevation along the path. Okumura's nominal results are referenced to a spread of about 20 m for the 800-MHz frequency band. An obvious shortcoming of this approach is its nonuniqueness. There are many terrain profiles that have the same 10%/90% spread, but they are unlikely to have identical shadowing losses. Thus, to achieve more accurate estimates of the effects of terrain, shadowing adjustments are more commonly generated by applying some form of diffraction theory to the terrain data. For example, see Reference [7]; also, refer to Section 10.9.4.

5.2.1 Example 1

Consider a 40-km (25-mi) 850-MHz land-mobile radio link from the Sears Tower in Chicago, Illinois to the northwest suburbs beyond the O'Hare airport. Referring to Figure 5.2, we note a basic median attenuation of about 40 dB. The Sears Tower is roughly 450 m (1500 ft) high; thus, Figure 5.3 indicates a base station height gain of about 9 dB. Antenna heights on passenger cars typically run about 1.5 m (5 ft); thus, Figure 5.4 indicates a vehicular height gain of -3 dB. Finally, although downtown Chicago is definitely an urban environment, the path to the suburbs from such a high site is best characterized as suburban. Hence, Figure 5.5 indicates a correction factor of about 10 dB. Therefore, the prediction is for $40 - 9 + 3 - 10 = 24$ dB of excess path loss (at the median or 50 percentile). The total median loss is obtained by summing in the free-space loss of $20\log(850) + 20\log(40) + 32.4$ to obtain 147 dBi.

FCC rules allow transmit ERPs of 1 kW (relative to a half-wave dipole) at 1000 ft. However, antenna combiner losses and feedline losses make such high ERPs unlikely. A typical buildingtop ERP is more like 200 W relative to a half-wave dipole (or about 300 W EIRP). Such a source would produce a median power level of $54.8 - 147 + 2.15 - 2 = -92$ dBm for a suburban mobile radio with dipole antenna and 2 dB of feedline loss. This perhaps appears quite large since radio sensitivities on the order of 1/4 μV (-119 dBm) are common. Bear in mind, however, that such sensitivity is for static conditions. In the presence of movement, the sensitivity is typically degraded 10 dB by Rayleigh fading. This will be discussed in Chapter 6. Furthermore, average signal powers in decibels tend to vary according to a normal (Gaussian) distribution; hence, the term lognormal shadowing (also discussed in Chapter 6). The standard deviation of this shadowing variation is typically on the order of 8 dB for large areas. Thus, the path loss that is exceeded just 10% of the time (a common design goal for the coverage contour) is $-92 - 1.28(8) = -102$ dBm [4]. Faded sensitivity is 7 dB better than this, so coverage out to 25 miles will be acceptable.

---

[4] The relation between probability and a multiple of the standard deviation, here -1.28, is discussed in Chapter 7.

Sample Radio Link Analyses 45

5.2.2 Example 2

Consider calculation of the separation required between low-power cordless telephone units and public safety land-mobile radio users to avoid talk-in (mobile-to-base direction) interference by the former to the latter. Assume a static sensitivity of 1/4 μV (-119 dBm) for the public safety users. The noise power at sensitivity is typically about 6 dB lower for 25-kHz narrowband FM radios. Hence, the maximum permissible interference power is -119 − 6 = -125 dBm. If the interference source EIRP is 10 W (40 dBm), then a path loss of at least 125 + 40 = 165 dB is required to avoid interference. If the interference source EIRP is just 10 mW, then path loss need only exceed 135 dB to avoid interference.

The distance beyond which sufficient path loss exists to avoid interference can be calculated using Hata's empirical fit to Okumura's measured results. Input parameters needed for this calculation are the antenna heights for the transmitter and receiver, the link frequency, and the environment class. Assume 100 ft, 6 ft, 868 MHz, and open, respectively, for these parameters. One then finds a separation in excess of 33.5 mi is required to ensure noninterference by a 10 W base transmitter. Even with just 10 mW of power a separation of 7.5 mi is required. Had a suburban environment been used instead, the required separations would have been 15.2 and 2.3 mi, respectively.

5.3 Homework Problem

UHF-TV channels can radiate up to 5 MW ERP from antennas up to 2000 ft. Such large powers are necessary to guarantee Grade-B quality television signals at distances on the order of 50 mi. For the definition of this term, and others to follow, the reader is referred to References [8] and [9]. How can it be possible then for amateur radio operators to send nearly snow-free pictures over 50-mi paths with just 100 W transmitters and much lower antennas?

Solution: Commercial television stations generally must radiate omnidirectionally to provide coverage to all users within some coverage radius, regardless of heading. Amateur contacts can be point-to-point in nature and thus higher antenna gains are feasible. Assuming a transmit antenna gain of 17 dBd, the amateur ERP is 37 dBW. This is fully 30 dB below the commercial ERP of 67 dBW. However, even though a substantial antenna gain (15 dBd) is included in the Grade-B planning factors for commercial television, a well-equipped amateur receive setup using four large yagi antennas can provide 22 dBd. This decreases the link shortfall to 23 dB. Planning factors also involve a 10-dB receiver noise figure and 5 dB of loss for the feedline and impedance mismatches. Amateurs often use mast-mounted GaAs *field effect transistor* (FET) LNAs to achieve noise figures of 1 dB and essentially eliminate feedline losses. This gives the amateur a 14-dB advantage, so the shortfall decreases to 9 dB.

Of course few amateurs can afford antennas on commercial buildings like the Sears Tower in Chicago, Illinois. A common amateur antenna height is 70 ft, compared to 1600 ft for the Sears Tower. This causes a shortfall increase from 9 to 31 dB, based on 6-dB antenna height gain per doubling of the height. This situation holds for one end of each case having a 30-ft antenna height. Thirty feet is the planning factor used for commercial television reception and will be taken as the amateur receiving antenna height (not all amateurs are fortunate enough to have a tower; many must be content with a small mast on the roof).

It does not seem that much progress has been made in resolving why the amateur link works. However, temporal and location propagation variability factors have not yet been considered. Grade-B is based on satisfactory reception at half the locations at least 90% of the time. According to the Reference [9], the difference between 90% of the time and 50% of the time is about 9 dB, whereas the difference between the median location and the 10% best location is about 15 dB (for a 12-dB location standard deviation). These factors move the amateur to within 7 dB of the commercial reference.

References

[1] ITT Space Communications Inc. "Domestic Earth Stations: System Designs and Equipment for Multipurpose, Small Aperture Terminals," August 1975.

[2] Miller, J. E., "ATS-6 Television Relay Using Small Terminals Experiment," *IEEE Tr. on Aero. and Elec. Sys.*, Vol. AES-11, No. 6, November 1975, pp. 1038-1047.

[3] Okumura, Y., et al., "Field Strength and Its Variability in VHF and UHF Land-Mobile Radio Service," *Rev. of Elec. Comm. Lab.*, Vol. 16, Nos. 9-10, September-October, 1968, pp. 825-873.

[4] Delisle, G. Y., et al., "Propagation Loss Prediction: A Comparative Study with Application to the Mobile Radio Channel," *IEEE Tr. on Veh. Tech.*, Vol. VT-34, No. 2, May 1985, pp. 86-95.

[5] Jakes, Jr., W. C. (editor), *Microwave Mobile Communications*, New York, NY, John Wiley & Sons, 1974.

[6] Hata, M., "Empirical Formula for Propagation Loss in Land Mobile Radio Services," *IEEE Tr. on Veh. Tech.*, Vol. VT-29, No. 3, August 1980, pp. 317-325.

[7] Bullington, K., "Radio Propagation for Vehicular Communications," *IEEE Tr. on Veh. Tech.*, Vol. VT-26, No. 4, November 1977, pp. 295-308 (see also the correction in August 1978).

[8] Wong, H. K., "A Computer Program for Calculating Effective Interference to TV Service," OST Technical Memorandum, FCC/OST TM 82-2, July 1982.

[9] Damelin, J., et al., "Development of VHF and UHF Propagation Curves for TV and FM Broadcasting," FCC Report R-6602, September 7, 1966 (third printing May 1974).

Chapter 6

Character of Land-Mobile Radio Propagation

This chapter describes the character of land-mobile radio signals in terms of three propagation regimes, differentiated according to the distance scale at which they dominate. The largest scale has the character of power-law propagation. The middle scale involves variations in average power about the median level set by the largest scale. When expressed in decibels such variations tend to be normally (Gaussian) distributed, hence the term lognormal shadowing. Finally, the signal envelope of the smallest scale variations is described by the Rayleigh probability density function, hence the term Rayleigh fading. A simple explanation of such fading is provided, along with an exercise that addresses a generalization of that explanation.

6.1 Power-Law Propagation Regime

Consider an experiment in which a land-mobile receiver is driven from near a base site to far away from that site and the average received power is measured continuously. A plot of received power tends to look like that shown in Figure 6.1. Typically the power rolls off relatively linearly with distance when *both* are expressed in decibels. This is power-law propagation, as discussed in Chapter 3.

6.2 Lognormal-Shadowing Propagation Regime

Substantial variability is also apparent about the power-law trend. That is, for quite small shifts in range, received power variations of many decibels can occur. Experimentally one typically finds that data collected around a circular route (i.e., a route for which path length is kept constant) yield a received power distribution in decibels that plots linearly on normal probability paper (Figure 6.2). Hence, the term lognormal shadowing. The same character is generally found for data collected

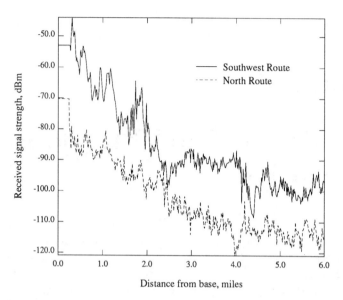

Figure 6.1 Typical average power behavior as a mobile receiver moves radially away from a base transmitter.

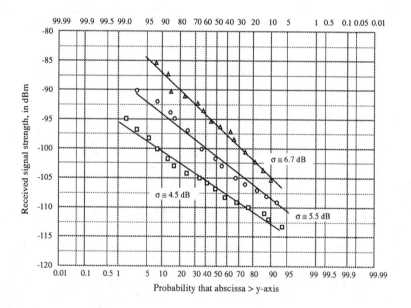

Figure 6.2 Typical distributions of average received power (path length fixed).

Character of Land-Mobile Radio Propagation 51

about any region whose dimensions are modest compared to the distance of the base transmitter. The slope of the probability graph is related to the standard deviation of the lognormal shadowing (half the 16 to 84 percentile spread). Values on the order of 8 to 10 dB are common when the test area covers many miles. However, as the area considered decreases in size the value drops [1].

An explanation for lognormal variation is as follows. Consider the received signal to be the result of the transmitted signal passing through or reflecting off some random number of objects such as buildings, hills, and trees. The individual processes each attenuate the signal to some degree and the final received value is thus the product of many transmission efficiency factors. Therefore, the logarithm of the received signal equals the sum of a large number of factors, each also expressed in decibels. As the number of factors becomes large, the central limit theorem [2] (see also Chapter 8) shows that the distribution of the sum will tend toward Gaussian under quite general circumstances, even if the individual contributors are not Gaussian.

6.3 Rayleigh-Fading Propagation Regime

Finally, if one expands the spatial plot in Figure 6.1 so that variations over just a few wavelengths can be seen (Figure 6.3), a third propagation regime becomes evident. Average power probability distributions at scales under a few tens of wavelengths tend to plot linear on log-log paper (Figure 6.4). This implies an exponential distribution of power; or, equivalently, a Rayleigh distribution of signal envelope (voltage). An explanation of this character has been offered by Clarke [3]. A similar explanation, based on the power spectrum, rather than component waves, has been developed by Gans [4].

Consider the geometry shown in Figure 6.5. Here a mobile receiver is moving in the x-coordinate direction at speed v. The mobile is surrounded by scatterers and assumed to receive signal components from all directions. The electric field intensity for a vertically polarized transmitter can thus be written as:

$$\begin{aligned} E_z &= E_0 \sum_{n=1}^{N} C_n \cos(\omega_c t + \theta_n) \\ \theta_n &= \omega_n t + \phi_n \end{aligned} \quad (6.1)$$

where $E_0 C_n$ is the real amplitude of the nth incoming wave. The C_n random variables are normalized so that their ensemble average sums to unity. Other symbol definitions are: ϕ_n are random phase angles uniformly distributed from 0 to 2π rad, ω_c is the carrier radian frequency of the transmitted signal, and ω_n is the Doppler radian frequency of the nth incoming wave. The latter equals:

$$\omega_n = \beta v \cos(\alpha_n) \quad (6.2)$$

where $\beta = (2\pi/\lambda)$, with λ representing the wavelength.

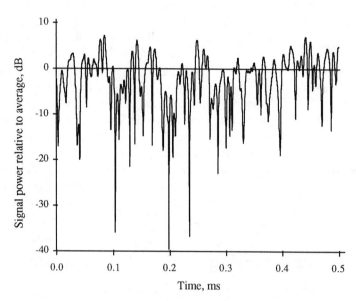

Figure 6.3 Temporal detail of received signal variations at 820 MHz.

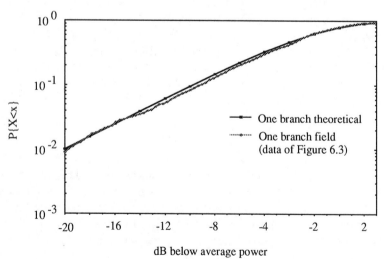

Figure 6.4 Cumulative distribution function for data of Figure 6.3.

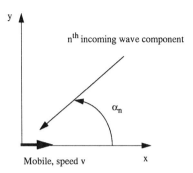

Figure 6.5 A typical component wave incident on a mobile receiver.

It is useful to express E_z in quadrature form [5, 6]:

$$E_z = T_c(t)\cos(\omega_c t) - T_s(t)\sin(\omega_c t) \qquad (6.3)$$

where:

$$\begin{aligned} T_c(t) &= E_0 \sum_{n=1}^{N} C_n \cos(\omega_n t + \phi_n) \\ T_s(t) &= E_0 \sum_{n=1}^{N} C_n \sin(\omega_n t + \phi_n) \end{aligned} \qquad (6.4)$$

As N becomes large, T_c and T_s tend toward Gaussian processes through the central limit theorem, with zero means and standard deviations (average powers) equal to $E_0^2/2$. The rapidity of convergence is discussed in Reference [7] and Section 8.2. The envelope of E_z is distributed as the square root of the sum of the squares of T_c and T_s; i.e., $p(r) = (r/b)\exp(-r^2/2b)$, $r \geq 0$, where b represents the average power. This is the so-called Rayleigh probability density function.

6.4 Summary

In summary, the preceding sections have shown that typically land mobile radio propagation consists of three regimes, differentiated by their spatial scale. The largest scale regime is characterized by power-law propagation such that the average received power decreases with range raised to some exponent. The received power at a fixed range is not constant; rather it tends to be lognormally distributed with a median value set by the power-law propagation. The standard deviation is controlled by the diversity of environments experienced. Values on the order of 8 to 10 dB are common over distances of miles. This second regime can be referred to as the lognormal-shadowing regime. At distance scales under about 100 wavelengths, the average power is essentially constant. Nonetheless, substantial changes in the signal envelope can occur with movements of just one-quarter wavelength. This is due to net cancellation of the various incoming signal components, which often appear to come from all directions because local scatterers commonly surround the mobile receive antenna. This third and smallest scale regime is called the Rayleigh regime after the type of fading that occurs. Separation of the Rayleigh and lognormal-shadowing regimes is rather subjective. The topic is addressed in Reference [8] (also, refer to the final exercise of Chapter 8).

6.5 Homework Problem

In the simple model of Clarke, incoming signal components are assumed to come from all directions. What if instead components only existed for α_n between 0 and 90 deg. Would the short-term probability density function remain Rayleigh? If so, would *any* attribute of the signal differ?

Solution: A generalization of Clarke's model has been analyzed by Aulin [9]. Even with a limited span for angle of arrival, the central limit theorem still applies as the number of components is allowed to increase without bound. Thus, the envelope remains Rayleigh distributed. However, the rate at which fades occur as the mobile moves decreases. The situation posed in the homework problem is equivalent to having a directive antenna on the mobile. Section 3.1.3 in Reference [10] discusses this matter (also, refer to Section 9.5 in this text).

References

[1] Marsan, M., et al., "Shadowing Variability in an Urban Land Mobile Environment at 900 MHz," *Electronics Letters*, Vol. 26, No. 10, May 1990, pp. 646-648.

[2] Papoulis, A., *Probability, Random Variables, and Stochastic Processes*, New York, NY, McGraw-Hill Book Co., 1965.

[3] Clarke, R. H., "A Statistical Theory of Mobile-Radio Reception," *Bell System Technical Journal*, Vol. 47, July-August 1968, pp. 957-1000.

[4] Gans, M. J., "A Power-Spectral Theory of Propagation in the Mobile-Radio Environment," *IEEE Trans. on Veh. Tech.*, Vol. VT-21, No. 1, February 1972, pp. 27-38.

[5] Rice, S. O., "Mathematical Analysis of Random Noise," *Bell System Technical Journal*, Vol. 23, July 1944, pp. 282-332, and Vol. 24, January 1945, pp. 46-156.

[6] Rice, S. O., "Statistical Properties of a Sine Wave Plus Random Noise," *Bell System Technical Journal*, Vol. 27, January 1948, pp. 109-157.

[7] Slack, M., "The Probability Distributions of Sinusoidal Oscillations Combined in Random Phase," *JIEE*, Vol. 93, No. 3, 1946, pp. 76-86.

[8] Lee, W.C.Y., "Estimate of Local Average Power of a Mobile Radio Signal," *IEEE Trans. on Veh. Tech.*, Vol. VT-34, No. 1, February 1985, pp. 22-27.

[9] Aulin, T., "A Modified Model for the Fading Signal at a Mobile Radio Channel," *IEEE Trans. on Veh. Tech.*, Vol. VT-28, No. 3, August 1979, pp. 182-203.

[10] Jakes, Jr., W. C. (editor), *Microwave Mobile Communications*, New York, NY, John Wiley & Sons, 1974.

Chapter 7
Probability Theory Refresher

Land-mobile radio propagation, as already noted in Chapters 5 and 6, is random in nature and thus a probabilistic viewpoint is necessary[1]. In fact, probability is an aspect of nearly all the tasks addressed by land-mobile radio system engineers. A thorough understanding of basic probability theory is therefore essential to such engineers. The reader is presumed to have had previous exposure to probability theory, so this chapter and the two that follow are of a review nature. However, knowledge of the theory and the ability to practically apply it are not synonymous. Consequently, the practical problem of battery energy savings through transmit power control is discussed in detail. Also, the application of probability theory to reliability questions is the subject of two concluding exercises.

The material in Sections 7.1 through 7.5 is drawn primarily from the premier textbook on the subject by Papoulis [2]. Excellent, more elementary sources of information are the textbooks by Walpole and Myers [3] and Brown and Hwang [4].

7.1 What Is Probability?

Probability can be viewed from at least four viewpoints: (1) axiomatic, (2) relative frequency, (3) classical (i.e., the ratio of favorable outcomes to total outcomes possible), and (4) measure of belief.

7.1.1 Axiomatic

For this viewpoint certain terms should first be defined:

---

[1] The suggestion that propagation is not stochastic but rather displays deterministically chaotic behavior has been published by Tannous et al. [1]. This is based on a very limited set of data taken along a hallway with line-of-sight conditions between transmit and receive antennas. Perhaps under such constrained circumstances determinism does hold, but stochastic behavior for mobile and portable propagation links holds in general. This is not an assumption. It is a pragmatic fact based on the success of the probabilistic viewpoint in predicting coverage and interference.

Experiment: Any process that generates raw data.

Raw data: Recorded information in its original collected form (counts or measurements).

Sample space: Set of all possible outcomes in a statistical experiment.

Event: A subset of a sample space.

The probability of an event A is a number, $P(A)$, assigned to this event obeying three postulates:

1. $P(A) \geq 0$.

2. $P(S) = 1$, where S equals the certain event (one that occurs every trial).

3. If A and B are mutually exclusive events, then $P(A+B) = P(A) + P(B)$.

7.1.2 Relative Frequency

Consider an experiment of interest repeated n times. If event A occurs n_A times, then its probability $P(A)$ is defined as the limit of the relative frequency n_A/n of the occurrence of A; i.e., $P(A) = \lim_{n \to \infty} \frac{n_A}{n}$.

7.1.3 Classical

$P(A)$ is determined *a priori* without actual experiment by counting the total number of possible outcomes N. If event A occurs for N_A outcomes, then $P(A)$ is assigned the value (N_A/N). This definition relies on each outcome being equally likely; hence, the definition is circular. It is also difficult to apply when the number of outcomes is not discrete. Even if the alternatives are finite, making them clear may be semantically difficult.

7.1.4 Measure of Belief

Probability is often used as a measure of belief that something is or is not true; for example, "it is probable that X is guilty."

7.2 Operations with Events

The *intersection* of events A and B, denoted as $A \cap B$, is the event containing all elements common to A and B.

The *union* of events A and B, denoted as $A \cup B$, is the event containing all elements that belong to A or B or both (inclusive-OR).

The *complement* of an event A, denoted as A', is the set of all elements of S that are not in A.

7.3 Probability Laws

The probability of the union of events A and B is related to the probabilities of those events and the probability of the intersection of those events through:

$$P(A \cup B) = P(A) + P(B) - P(A \cap B)$$

If A and B are mutually exclusive, then $P(A \cap B) = P(\phi) = 0$. The *conditional probability* of event B, given event A, is defined by:

$$P(B|A) = P(A \cap B)/P(A)$$

where $P(A) > 0$.

Events A and B are *independent* if and only if $P(A \cap B) = P(A)P(B)$. Baye's Rule:

$$\begin{aligned} P(B_k|A) &= \frac{P(B_k \cap A)}{P(A)} = \frac{P(B_k \cap A)}{\sum_{i=1}^n P(B_i \cap A)} \\ &= \frac{P(B_k)P(A|B_k)}{\sum_{i=1}^n P(B_i)P(A|B_i)} \end{aligned}$$

This rule is useful when one is given $P(A|B)$ but wants $P(B|A)$ to find out what happened. For example, one may know the probability that a FET with a defective gate will fail by time t, but want to know the probability that a FET failure by time t is due to a defective gate (when there are numerous other potential failure mechanisms).

7.4 Concept of the Random Variable

7.4.1 Definitions

A *real random variable* X is a real function whose domain is the space S; i.e., a process of assigning a real number $X(\zeta)$ to every outcome ζ of the experiment of interest such that: (1) the set $\{X \leq x\}$ is an event for any real number x and (2) the probability of the events $X = \infty$ and $X = -\infty$ equals zero.

A *complex random variable* Z is a process of assigning to every outcome ζ a complex number $Z(\zeta) = X(\zeta) + jY(\zeta)$, such that the functions X and Y are real random variables.

If a variable can assume only specific values (usually integers), it is called a *discrete* variable.

If a variable can assume any value between certain limits (possibly $-\infty$ and ∞), it is called a *continuous* variable.

7.4.2 Distribution Function

The *distribution* function of the random variable X (*cumulative distribution function* (CDF)) is the function $F_X(x) = P\{X \leq x\}, -\infty \leq x \leq \infty$ equals the probability that $X \leq x$.

A comment on notation is in order here. In general, capitalized symbols will refer to random variables. The real number to which the random variable is being compared is in lower case. When both involve the same symbol, the distribution function subscript is often dropped to simplify the notation; i.e, $F_X(x) = F(x)$.

Properties:

1. $F(-\infty) = 0$ and $F(\infty) = 1$.

2. $F(x_1) \leq F(x_2)$ for $x_1 \leq x_2$; i.e., $F(x)$ is a nondecreasing function of x.

3. $F(x^+) = F(x)$, where $F(x^+) = \lim_{\epsilon \to 0} F(x + \epsilon), \epsilon > 0$; i.e., $F(x)$ is continuous from the right (an aspect that is important for discrete and other discontinuous random variables).

7.4.3 Density Function

The *density* function of the random variable X (*probability density function* (PDF)) is given by the derivative of the distribution function, $f(x) = [dF(x)/dx]$.

Properties:

1. $f(x)$ is non-negative (it may be 0 or ∞ for some x).

2. $\int_{-\infty}^{\infty} f(x) dx = F(\infty) - F(-\infty) = 1$.

3. $P\{x_1 \leq X \leq x_2\} = \int_{x_1}^{x_2} f(x) dx$; note that $P\{X = x\} = 0$ for all x but this is not the same as saying such an event is impossible.

A frequency interpretation of $f(x)$ is as follows: To determine $f(x)$ for a given x, perform the relevant experiment n times and count the number of trials such that $x \leq X(\zeta) \leq x + \delta x$. With $\delta n(x)$ the number of successful trials, one has $f(x)\delta x \cong (\delta n(x)/n)$ for large n and small δx. This is the idea behind the *histogram*, a graphical procedure that approximates the density function (a cumulative histogram or *ogive* approximates the distribution function).

7.5 Mathematical Expectation

7.5.1 Expected Value

The *expected value* (mean) of a real random variable X is the integral (assuming it exists):

$$E(X) = <X> = \mu_x = \int_{-\infty}^{\infty} x f(x) dx$$

The mean places the center of gravity of $f(x)$ and is a measure of location. If X is discrete, taking on values x_n with probability p_n, then:

$$E(X) = \Sigma_n x_n P(X = x_n) = \Sigma_n x_n p_n$$

If $Z = X + iY$, a complex random variable, then $E(Z) = E(X) + iE(Y)$.
Properties:

1. If $f(x)$ is even (i.e., $f(-x) = f(x)$), then $E(X) = 0$.
2. If $f(x)$ is symmetrical about $x = a$ (i.e., $f(a-x) = f(a+x)$), then $E(X) = a$.
3. The expected value of a function of X is given by:

$$E[Y = g(X)] = \int_{-\infty}^{\infty} g(x) f(x) dx$$

The discrete form of this is $\Sigma_k g(x_k) P(X = x_k)$.

4. Additivity:

$$E[g_1(X) + g_2(X) + \cdots + g_n(X)] = E[g_1(X)] + E[g_2(X)] + \cdots + E[g_n(X)]$$

In particular, $E(aX + b) = aE(X) + b$, which shows that expectation is a linear operator.

7.5.2 Conditional Expected Value

The *conditional expected value* is defined by the integral:

$$E(X|M) = \int_{-\infty}^{\infty} x f(x|M) dx$$

For example, consider X as the random variable representing the time of failure of a system. With $M = \{X \geq t\}$, one has:

$$P(M) = P(X \geq t) = F(\infty) - F(t) = 1 - F(t)$$

$$F(x|X \geq t) = \frac{P\{X \leq x, X \geq t\}}{P\{X \geq t\}} = \begin{cases} \frac{F(x) - F(t)}{1 - F(t)} & x \geq t \\ 0 & x < t \end{cases}$$

$$f(x|X \geq t) = \frac{dF(x|X \geq t)}{dx} = \frac{f(x)}{1 - F(t)} = \frac{f(x)}{\int_t^{\infty} f(x) dx}$$

$$x \geq t$$

Thus,
$$E(X|X \geq a) = \frac{\int_a^\infty x f(x) dx}{\int_a^\infty f(x) dx}$$
which represents the average time of failure for all units lasting at least to time a.

7.5.3 Variance

The *variance* of a random variable X with mean μ is given by:
$$\sigma^2 = E[(X - \mu)^2] = \sigma_x^2 = \int_{-\infty}^\infty (x - \mu)^2 f(x) dx$$

It equals the moment of inertia of the probability masses and is a measure of central tendancy. The *standard deviation* is defined as σ.

If X is discrete, then $\sigma^2 = \Sigma_n (x_n - \mu)^2 P(X = x_n)$.

If $Z = X + iY$, then $\sigma^2 = <ZZ^*> - <Z><Z^*>$, where * denotes the complex conjugate. This refinement is necessary to ensure that the variance is positive-real.

7.5.4 General Moments

The k-th *initial moment* is given by:
$$E(X^k) = m_k = \int_{-\infty}^\infty x^k f(x) dx$$

The k-th *central moment* is given by:
$$E[(X - \mu)^k] = \mu_k = \int_{-\infty}^\infty (x - \mu)^k f(x) dx$$

The k,r-th *joint moment* of the random variables X and Y is given by:
$$E(X^k Y^r) = m_{kr} = \int_{-\infty}^\infty \int_{-\infty}^\infty x^k y^r f(x, y) dx dy$$

The *order* of the moment is $k + r$ and the special case of $k = r = 1$ is the *cross-correlation*.

The k,r-th *joint central moment* of the random variables X and Y is given by:
$$E[(X - \mu_x)^k (Y - \mu_y)^r] = \mu_{kr} = \int_{-\infty}^\infty \int_{-\infty}^\infty (x - \mu_x)^k (y - \mu_y)^r f(x, y) dx dy$$

where $\mu_{20} = \sigma_x^2$, $\mu_{02} = \sigma_y^2$, and the *covariance* is:
$$\mu_{11} = E[(X - \mu_x)(Y - \mu_y)] = E(XY) - \mu_x \mu_y$$

Probability Theory Refresher 63

The *correlation coefficient* is $r = \mu_{11}/\sigma_x\sigma_y$.
Two random variables are *uncorrelated* if $E(XY) = E(X)E(Y)$.
They are *orthogonal* if $E(XY) = 0$.
They are *independent* if $f(x,y) = f(x)f(y)$. Independence implies uncorrelation but the reverse is not necessarily true (it does hold for Gaussian random variables). For example, $\sin(X)$ and $\cos(X)$ are uncorrelated and orthogonal but are not independent because $\cos(X) = \sqrt{1 - \sin(X)^2}$.

7.5.5 Characteristic Functions

The *characteristic function* of a random variable X is defined by:

$$\Phi(\nu) = E[\exp(i\nu X)] = \int_{-\infty}^{\infty} \exp(i\nu x) f(x) dx$$

This is the Fourier transform of the PDF. For the discrete case, the characteristic function is given by the sum $\Sigma_k \exp(i\nu x_k) P(X = x_k)$.

Moment theorem:

$$\frac{d^n \Phi(0)}{d\nu^n} = i^n m_n = i^n E(X^n)$$

The proof involves expansion of $\exp(i\nu x)$ in the definition of the characteristic function about $x = 0$ (Maclaurin series):

$$\begin{aligned}
\Phi(\nu) &= \int_{-\infty}^{\infty} f(x)[1 + i\nu x + \cdots + \frac{(i\nu x)^n}{n!} + \cdots] dx \\
&= 1 + i\nu m_1 + \cdots + \frac{(i\nu)^n m_n}{n!} + \cdots
\end{aligned}$$

Given $\Phi(\nu)$, one can obtain $f(x)$ through the inverse Fourier transform:

$$f(x) = \frac{1}{2\pi} \int_{-\infty}^{\infty} \Phi(\nu) \exp(-i\nu x) d\nu$$

The joint characteristic function of the random variables X and Y is given by:

$$\begin{aligned}
\Phi_{XY}(\nu_1 \nu_2) &= E\{\exp[i(\nu_1 X + \nu_2 Y)]\} \\
&= \int_{-\infty}^{\infty} \int_{-\infty}^{\infty} \exp[i(\nu_1 x + \nu_2 y)] f(x,y) dx dy
\end{aligned}$$

In the case of $Z = aX + bY$ one finds:

$$\Phi_Z(\nu) = E[\exp(i\nu z)] = E\{\exp[i(a\nu x + b\nu y)]\} = \Phi_{XY}(a\nu, b\nu)$$

If X and Y are independent:

$$\Phi_{XY}(\nu_1, \nu_2) = \Phi_X(\nu_1) \Phi_Y(\nu_2)$$

7.6 Sample Applications

7.6.1 Gaussian (Normal) Density Function

This is probably the single most important density function. It has the form:

$$f(x) = \frac{1}{\sqrt{2\pi\sigma^2}} \exp[-\frac{(x-\mu)^2}{2\sigma^2}] \tag{7.1}$$

A convenient shorthand notation for a Gaussian-distributed (normal) random variable is $X \sim N(\mu, \sigma)$, where the arguments represent the mean and standard deviation, respectively. The $f(x)$ given above is clearly non-negative. To confirm it is a valid PDF, one must also show that:

$$\frac{1}{\sqrt{2\pi\sigma^2}} \int_{-\infty}^{\infty} \exp[-\frac{(x-\mu)^2}{2\sigma^2}] dx = 1 \tag{7.2}$$

To this end it is useful to make the substitution $y = (x - \mu)/\sigma$. This standardizes the random variable; i.e., transforms it to zero mean and unity variance. The task now becomes verification that:

$$\int_{-\infty}^{\infty} \exp(-\frac{y^2}{2}) dy = \sqrt{2\pi} \tag{7.3}$$

A roundabout way of doing this is to consider the square of the integral

$$\begin{aligned} I^2 &= \int_{-\infty}^{\infty} \exp(-\frac{y_1^2}{2}) dy_1 \int_{-\infty}^{\infty} \exp(-\frac{y_2^2}{2}) dy_2 \\ &= \int_{-\infty}^{\infty} \int_{-\infty}^{\infty} \exp(-\frac{y_1^2 + y_2^2}{2}) dy_1 dy_2 \\ &= 2\pi \end{aligned} \tag{7.4}$$

Transforming to polar coordinates via $y_1 = r\cos(\theta), y_2 = r\sin(\theta)$, and $dy_1 dy_2 = r\, d\theta\, dr$ leads to:

$$\begin{aligned} I^2 &= \int_0^{\infty} [\int_0^{2\pi} d\theta] \exp(-\frac{r^2}{2}) r\, dr \\ &= 2\pi \int_0^{\infty} r \exp(-\frac{r^2}{2}) dr \\ &= -2\pi \int_0^{-\infty} \exp(s) ds \end{aligned} \tag{7.5}$$

The last relation does indeed equate to 2π, hence $f(x)$ is a valid PDF.

7.6.2 Gaussian Distribution Function

The probability that a Gaussian random variable takes on a value less than or equal to an arbitrary value x is found by integrating the density function from $-\infty$ to x:

$$\begin{aligned} F_X(x) &= \text{Prob}(X \leq x) \\ &= \int_{-\infty}^{x} f(x)\, dx \\ &= \frac{1}{\sqrt{2\pi\sigma^2}} \int_{-\infty}^{\infty} \exp[-\frac{(x-\mu)^2}{2\sigma^2}]\, dx \end{aligned} \qquad (7.6)$$

Transforming to the standard variate $y = (x-\mu)/\sigma$ gives:

$$F_X(x) = \frac{1}{\sqrt{\pi}} \int_{-\infty}^{\frac{(x-\mu)}{\sqrt{2}\sigma}} \exp(-y^2)\, dy \qquad (7.7)$$

This integral is not expressible in closed form using elementary functions. It can be expressed in terms of the Laplace function, a special function defined as:

$$L(x) = \frac{1}{\sqrt{2\pi}} \int_0^x \exp(-\frac{t^2}{2})\, dt \qquad (7.8)$$

More commonly, the error function is used. It is defined as:

$$\text{erf}(x) = \frac{2}{\sqrt{\pi}} \int_0^x \exp(-t^2)\, dt \qquad (7.9)$$

The error function arises naturally when the substitution $y^* = (x-\mu)/(\sqrt{2}\sigma)$ is used, rather than the standard variate substitution. Based on these definitions one finds:

$$\begin{aligned} F_X(x) &= \frac{1}{2}[\text{erf}(\frac{x-\mu}{\sqrt{2}\sigma}) + 1] \\ &= \frac{1}{2}L(\frac{x-\mu}{\sigma}) \end{aligned} \qquad (7.10)$$

Tables of the area under the standardized Gaussian curve are common; for example, see Table A.3 in Reference [3]. Graph paper whose probability scale is warped so Gaussian behavior plots as a straight line is also commonly available; for example, Dietzgen graph paper no. 340-ps 90. The scaling is carried out as follows. Let the random variable comparison values correspond to the vertical or y-axis and let the x-axis correspond to the probability values. For some particular comparison value y, the probability is x; i.e., $F_Y(y) = \text{Prob}(Y \leq y) = (1/2)\{1 + \text{erf}[(y-\mu)/(\sqrt{2}\sigma)]\} = x$. Solving this relation for y in terms of x, one obtains $y = \mu + \sqrt{2}\sigma[\text{erf}^{-1}(2x-1)]$. Thus, if x is warped by multiplying it by 2, subtracting 1, and taking the inverse error function, a new coordinate x^* is produced that is linearly related to y. The slope of that relation can be used to deduce the standard deviation and the 50-percentile level can be used to deduce the mean.

7.6.3 Mean of Gaussian Random Variable

$$E(x) = \int x f(x) dx$$
$$= \int_{-\infty}^{\infty} x \frac{1}{\sqrt{2\pi\sigma^2}} \exp[-\frac{(x-\mu)^2}{2\sigma^2}] dx \qquad (7.11)$$

Standardizing the variate by the substitutions $y = (x-\mu)/\sigma$ and $dy = dx/\sigma$ leads to:

$$E(x) = \frac{1}{\sqrt{2\pi}} \int_{-\infty}^{\infty} (\mu + \sigma y) \exp(-\frac{y^2}{2}) dy$$
$$= \mu[\frac{1}{\sqrt{2\pi}} \int_{-\infty}^{\infty} \exp(-\frac{y^2}{2}) dy] + \frac{\sigma}{\sqrt{2\pi}} [\int_{-\infty}^{\infty} y \exp(-\frac{y^2}{2}) dy]$$
$$= \mu \qquad (7.12)$$

The terms in brackets are easily recognized as unity (through the definition of the density function and the fact that such a function must integrate to unity over its full range; the equation offered equals the density function of a zero-mean, unity standard deviation Gaussian random variable) and zero (due to odd symmetry), respectively.

7.6.4 Variance of Gaussian Random Variable

$$Var = E[(x-\mu)^2]$$
$$= \int (x-\mu)^2 f(x) dx$$
$$= \int_{-\infty}^{\infty} \frac{1}{\sqrt{2\pi\sigma^2}} (x-\mu)^2 \exp[-\frac{(x-\mu)^2}{2\sigma^2}] dx \qquad (7.13)$$

Or in terms of the standard variate $y = (x-\mu)/\sigma$:

$$Var = \frac{\sigma^2}{\sqrt{2\pi}} \int_{-\infty}^{\infty} y^2 \exp(-\frac{y^2}{2}) dy \qquad (7.14)$$

The preceding integral can be solved by integration by parts. Letting $u = y$ and $dv = y \exp(-y^2/2) dy$, one obtains $du = dy$ and $v = -\exp(-y^2/2)$. Then, since $uv = \int u dv + \int v du$:

$$Var = \frac{\sigma^2}{\sqrt{2\pi}} [uv|_{-\infty}^{\infty} - \int_{-\infty}^{\infty} v du]$$
$$= \frac{\sigma^2}{\sqrt{2\pi}} [-y \exp(-\frac{y^2}{2})|_{-\infty}^{\infty} + \int_{-\infty}^{\infty} \exp(-\frac{y^2}{2}) dy] \qquad (7.15)$$

Application of l'Hôpital's rule to the first term in brackets shows it to be zero. The second term is recognized as $\sqrt{2\pi}$ times the CDF of a Gaussian standard variate evaluated at ∞; hence, it simply equals $\sqrt{2\pi}$. Thus, $Var = \sigma^2$.

7.6.5 Linear Transformation of a Gaussian Random Variable

Consider the linear equation $Y = \alpha X + \beta$, where $X \sim N(\mu, \sigma)$. The distribution of Y can be found as follows:

$$\begin{aligned}
F_Y(a) &= \text{Prob}(Y \leq a) \\
&= \text{Prob}(\alpha X + \beta \leq a) \\
&= \text{Prob}[X \leq \frac{(a-\beta)}{\alpha}] \\
&= F_X[\frac{(a-\beta)}{\alpha}] \\
&= \int_{-\infty}^{\frac{(a-\beta)}{\alpha}} \frac{1}{\sqrt{2\pi\sigma^2}} \exp[-\frac{(x-\mu)^2}{2\sigma^2}] dx \\
&= \int_{-\infty}^{a} \frac{1}{\sqrt{2\pi(\alpha\sigma)^2}} \exp[-\frac{(y-(\alpha\mu+\beta))^2}{2(\alpha\sigma)^2}] dy \\
&= \int_{-\infty}^{a} f_Y(y) dy
\end{aligned} \qquad (7.16)$$

Hence, one sees that linear transformations of Gaussian random variables are also Gaussian. The new mean equals the sum of the zero-order coefficient plus the old mean multiplied by the first-order coefficient. The new standard deviation equals the old standard deviation times the first-order coefficient.

7.6.6 Moments of a Rayleigh Random Variable

In Chapter 6 it was shown that the smallest scale variation of average received signal envelope tends to follow a Rayleigh distribution for land-mobile radio links. Consequently, this distribution is also of substantial interest. It has the form:

$$f(r) = \frac{2r}{\alpha} \exp(-\frac{r^2}{\alpha}) \qquad (7.17)$$

where $\alpha/2$ represents the average signal power and the envelope (voltage) must of course be greater than or equal to zero. The initial moments for this distribution are found by solving:

$$\begin{aligned}
E(x^n) &= \frac{1}{\frac{\alpha}{2}} \int_0^\infty x^{n+1} \exp(-\frac{x^2}{\alpha}) dx \\
&= \alpha^{\frac{n}{2}} \int_0^\infty y^{\frac{n}{2}} \exp(-y) dy
\end{aligned} \qquad (7.18)$$

where the substitution $y = x^2/\alpha$ has been made. This gives a form that matches the gamma function, a special function defined as:

$$\Gamma(\theta) = \int_0^\infty x^{\theta-1} \exp(-x) dx \qquad (7.19)$$

Some useful properties of the gamma function are the recursion formula, the relation to the factorial function, and the fractional value [6]:

Recursion Formula
$$\Gamma(\theta + 1) = \int_0^\infty x^\theta \exp(-x)dx \qquad (7.20)$$

The substitutions $u = x^\theta$ and $dv = \exp(-x)dx$ allow the integral to be solved by the integration by parts technique. The result is:

$$\Gamma(\theta + 1) = -x^\theta \exp(-x)|_0^\infty + \theta \int_0^\infty x^{\theta-1} \exp(-x)dx \qquad (7.21)$$

The first term can be shown to be zero through l'Hôpital's rule and the second term involves the basic definition of the gamma function. Thus, the recursion relation $\Gamma(\theta + 1) = \theta\Gamma(\theta)$ has been verified.

Relation to the Factorial Function
In the special case that θ equals a positive integer, the recursion relation can be repeatedly applied to yield:

$$\Gamma(n + 1) = n(n - 1)(n - 2)\cdots\Gamma(1) \qquad (7.22)$$

But $\Gamma(1)$ can readily be integrated and shown to equal unity. Thus, $\Gamma(n+1) = n!$.

Fractional Value
The value of $\Gamma(1/2)$ is particularly relevant to Rayleigh moments. To evaluate it, consider the transformation $y = \sqrt{2x}$. This leads to:

$$\begin{aligned}\Gamma(\theta) &= \int_0^\infty (\frac{y^2}{2})^{\theta-1} \exp(-\frac{y^2}{2})y\,dy \\ &= \frac{1}{2^{\theta-1}} \int_0^\infty y^{2\theta-1} \exp(-\frac{y^2}{2})dy \end{aligned} \qquad (7.23)$$

Choosing $\theta = (1/2)$, one obtains:

$$\Gamma(\frac{1}{2}) = \sqrt{2} \int_0^\infty \exp(-\frac{y^2}{2})dy \qquad (7.24)$$

But earlier, this integral evaluated over $-\infty$ to ∞ was shown to equal $\sqrt{2\pi}$. Since the integral is symmetrical about zero, over the range 0 to ∞ the result is $\sqrt{\pi/2}$. Hence, $\Gamma(1/2) = \sqrt{\pi}$.

Applying the preceding information about gamma functions to the moments of a Rayleigh-distributed random variable, one finds:

$$E(x^n) = \alpha^{\frac{n}{2}}\Gamma(1 + \frac{n}{2})$$

$$E(x) = \sqrt{\frac{(\pi\alpha)}{2}}$$
$$E(x^2) = \alpha$$
$$\begin{aligned}Var &= E((x - E(x))^2) \\ &= E(x^2) - [E(x)]^2 \\ &= \alpha(1 - \frac{\pi}{4})\end{aligned} \quad (7.25)$$

7.6.7 Moments of an Exponential Random Variable

Rather than work with signal envelope (voltage), which is typically Rayleigh distributed in land-mobile radio links, it is often simpler to work with power as the random variable. If R represents the Rayleigh envelope, then:

$$f_R(r) = \frac{2r}{\alpha} \exp(-\frac{r^2}{\alpha}), \; r \geq 0 \quad (7.26)$$

Power is proportional to $s = r^2$. Its distribution can be found as follows[2]:

$$\begin{aligned}\int_{r_1}^{r_2} f_R(r) dr &= \int_{s_1 = r_1^2}^{s_2 = r_2^2} f_S(s) ds \\ &= \int_{\sqrt{s_1}}^{\sqrt{s_2}} f_R(\sqrt{s}) ds (\frac{dr}{ds}) \\ f_S(s) &= f_R(\sqrt{s})(\frac{dr}{ds}) \\ &= \frac{1}{\alpha} \exp(-\frac{s}{\alpha}), \; s \geq 0\end{aligned} \quad (7.27)$$

This is the so-called exponential PDF. The mean power can be shown to be $\alpha\Gamma(2) = \alpha$. The mean square power is $\alpha^2\Gamma(3) = 2\alpha^2$. The variance equals α^2, which implies that the standard deviation of the exponential distribution equals the average value.

7.6.8 Conditional Expectation Example

Consider the distribution of call lengths on a trunked land-mobile radio system (traffic issues are discussed further in Chapter 13). Typically this is exponential in form [7], so let:

$$f(x) = \frac{1}{h} \exp(-\frac{x}{h}), \; x \geq 0 \quad (7.28)$$

where h represents the average call length (hold time). What is the average remaining life of a call that is already under way and has thus far lasted kh seconds? To

---

[2]The correct proportionality factor is one-half; thus, when absolute values are required $\alpha/2$ should be used rather than α and load resistances must be taken into account.

answer this question one needs to evaluate:

$$E[x|x \geq kh] = \int x f(x|x \geq kh) dx \quad (7.29)$$

Recalling the analysis of Section 7.5.2, this can be done in two parts:

$$\begin{aligned} I_1 &= \int_{kh}^{\infty} \frac{x}{h} \exp(-\frac{x}{h}) dx \\ I_2 &= 1 - \int_{0}^{kh} \frac{1}{h} \exp(-\frac{x}{h}) dx \\ E[x|x \geq kh] &= I_1/I_2 \\ &= h(k+1) \end{aligned} \quad (7.30)$$

Interestingly, one finds that the expected call duration is equal to the duration already observed plus the average hold time. This means that the average length of a call, regardless of when one begins timing it, is h. Hence, the exponential distribution is considered to be memoryless.

The same viewpoint occurs in reliability calculations when the failure rate is assumed constant. This gives rise to a time of failure PDF that is exponential and implies that regardless of how long a part has already lasted its mean time to failure is h. Of course, this is not really true because real parts eventually age, invalidating the assumption of constant failure.

7.6.9 Analysis of Peak-to-Average Power Ratio for Speech

The peak-to-average power ratio of speech is of particular importance to linear modulation schemes such as *single sideband* (SSB) and *quadrature amplitude modulation* (QAM). Its value sets the *peak envelope power* (PEP) requirement for the transmitter final amplifier so that sufficiently small voice distortions and out-of-channel emissions (splatter) occur. This is often more difficult and costly to meet than the average power requirement; therefore, much work has been done studying means for artificially compressing the speech peaks to reduce the PEP requirement.

The PDF for speech amplitude has been shown by Paez and Glisson to be approximately gamma distributed [8]. A simpler approximation is the Laplacian distribution [9](Figure 7.1). The latter is given by:

$$p(x) = \frac{1}{\sqrt{2}\sigma_x} \exp\left(\frac{-\sqrt{2}|x|}{\sigma_x}\right) \quad (7.31)$$

while the former is given by:

$$p(x) = \left(\frac{\sqrt{3}}{8\pi\sigma_x|x|}\right)^{\frac{1}{2}} \exp\left(\frac{-\sqrt{3}|x|}{2\sigma_x}\right) \quad (7.32)$$

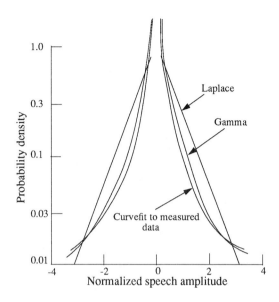

Figure 7.1 Probability distribution of speech power.
After: Paez, M. D. and T. H. Glisson, "Minimum Mean Squared-Error Quantization in Speech," *IEEE Tr. on Comm.*, Vol. COM-20, April 1972, pp. 225-230.

First consider the Laplacian approximation. Transforming to power through $y = x^2$ one has:

$$p(y) = \frac{1}{\sigma_x \sqrt{2y}} \exp\left(\frac{-\sqrt{2y}}{\sigma_x}\right) \tag{7.33}$$

for non-negative y values. The expected value of y is found by integrating this density function multiplied by y over zero to infinity:

$$\begin{aligned} E(y) &= \int_0^\infty y p(y) dy \\ &= \frac{1}{\sqrt{2}\sigma_x} \int_0^\infty \sqrt{y} \exp\left(\frac{-\sqrt{2y}}{\sigma_x}\right) dy \\ &= \frac{\sqrt{2}}{\sigma_x} \int_0^\infty q^2 \exp\left(\frac{-\sqrt{2}q}{\sigma_x}\right) dq \\ &= \sigma_x^2 \end{aligned} \tag{7.34}$$

The CDF for y is given by:

$$F(y) = \int_0^y p(y) dy$$

$$= \frac{\sqrt{2}}{\sigma_x} \int_0^{\sqrt{y}} \exp\left(\frac{-\sqrt{2}q}{\sigma_x}\right) dq$$

$$= 1 - \exp\left(\frac{-\sqrt{2y}}{\sigma_x}\right) \qquad (7.35)$$

Thus, for example, the ratio of the 99-percentile power to the average power is about 10.6 (10.3 dB).

Analysis using the gamma density function is similar. The average power is again found to equal σ_x^2, with the help of relation 3.381 #4 in Reference [10]. The CDF for y can be expressed in terms of the incomplete gamma function using relation 3.381 #1 in Reference [10]:

$$F(y) = \frac{1}{\sqrt{\pi}} \gamma\left(\frac{1}{2}, \frac{\sqrt{3y}}{2\sigma_x}\right) \qquad (7.36)$$

In terms of Tricomi's incomplete gamma function (see p. 260 of Reference [6]) this becomes:

$$F(y) = \sqrt{z}\, \gamma\left(\frac{1}{2}, z\right) \qquad (7.37)$$

where $z = (\sqrt{3y}/2\sigma_x)$. This equation can be solved numerically by computer; for example, by use of the GAMIT subroutine in the special function library of the commercial software offered by IMSL. Doing so one finds that the ratio of the 99-percentile power to the average power is 13.2 (11.2 dB), up somewhat from that obtained using the coarser Laplacian approximation.

7.6.10 Moments of a Rician-Distributed Random Variable

The PDF of a Rician-distributed random variable R can be expressed as (for example, see Section 4.4-7 of Reference [5]):

$$p_R(r) = \frac{2r}{\alpha} \exp\left(-\frac{r^2 + c^2}{\alpha}\right) I_0\left(\frac{2rc}{\alpha}\right), \; r \geq 0 \qquad (7.38)$$

where r equals the resultant amplitude, $(\alpha/2)$ equals the power in the diffuse (Rayleigh) component, $(c^2/2)$ equals the power in the direct (line-of-sight) component and $I_0()$ equals the modified Bessel function of the first kind of order zero. This is a generalization of the Rayleigh-fading distribution typically encountered in land-mobile radio links. As $c \to 0$, the Rician distribution becomes Rayleigh.

The k-th moment of the random variable R is given by:

$$\begin{aligned} m_k &= \int r^k p(r) dr \\ &= \frac{2}{\alpha} \exp\left(-\frac{c^2}{\alpha}\right) \int_0^\infty r^{k+1} \exp\left(-\frac{r^2}{\alpha}\right) I_0\left(\frac{2rc}{\alpha}\right) dr \end{aligned} \qquad (7.39)$$

Probability Theory Refresher

Envelope Mean

Consider the case of $k = 1$; i.e., the mean of the random variable R. Substituting $k = 1$ into the general moment equation gives:

$$m_1 = <r> = \frac{2}{\alpha} \exp(-\frac{c^2}{\alpha}) \bullet INT1 \tag{7.40}$$

where $INT1 = \int_0^\infty r^2 \exp(-r^2/\alpha) I_0(2rc/\alpha) dr$.

Integration by parts can be applied to $INT1$ with the help of relations 9.6.6 and 9.6.28 in Reference [6]. To this end define:

$$u_1 = rI_0(\frac{2rc}{\alpha}), \quad dv_1 = r\exp(-\frac{r^2}{\alpha})dr$$

$$du_1 = \left[I_0(\frac{2rc}{\alpha}) + \frac{2rc}{\alpha}I_1(\frac{2rc}{\alpha})\right]dr, \quad v_1 = -\frac{\alpha}{2}\exp(-\frac{r^2}{\alpha}) \tag{7.41}$$

Integration by parts then gives:

$$INT1 = -\frac{r\alpha}{2}exp(-\frac{r^2}{\alpha})I_0(\frac{2rc}{\alpha})|_0^\infty + \frac{\alpha}{2}INT2 + cINT3 \tag{7.42}$$

where:

$$INT2 = \int_0^\infty \exp(-r^2/\alpha) I_0(2rc/\alpha) dr = (\sqrt{\pi\alpha}/2)\exp(c^2/2\alpha)I_0(c^2/2\alpha)$$

using relation 6.618.4 of Reference [10] and:

$$INT3 = \int_0^\infty r\exp(-r^2/\alpha) I_1(2rc/\alpha) dr$$

This latter quantity can likewise be integrated by parts using (see relation 9.6.26 #3 of Reference [6]):

$$u_2 = I_1(\frac{2rc}{\alpha}), \quad dv_2 = r\exp(-\frac{r^2}{\alpha})dr$$

$$du_2 = \left[I_0(\frac{2rc}{\alpha}) + I_2(\frac{2rc}{\alpha})\right]\frac{c}{\alpha}dr, \quad v_2 = -\frac{\alpha}{2}\exp(-\frac{r^2}{\alpha}) \tag{7.43}$$

Substitution leads to:

$$INT3 = -\frac{\alpha}{2}\exp(-\frac{r^2}{\alpha})I_1(\frac{2rc}{\alpha})|_0^\infty + \frac{c}{2}INT2 + \frac{\alpha}{2}INT4 \tag{7.44}$$

where, using relation 6.618.4 of Reference [10]:

$$INT4 = \int_0^\infty \exp(-r^2/\alpha) I_2(2rc/\alpha) dr = (\sqrt{\pi\alpha}/2)\exp(c^2/2\alpha)I_1(c^2/\alpha)$$

Letting $q = (c^2/2\alpha)$, one obtains:

$$INT3 = \frac{c}{4}\sqrt{\pi\alpha}\exp(q)[I_0(q) + I_1(q)] \qquad (7.45)$$

Substitution of these results then yields:

$$m_1 = <r> = \sqrt{\frac{\pi\alpha}{4}}\exp(-q)[I_0(q)(1 + 2q) + I_1(q)2q] \qquad (7.46)$$

Note that as $c \to 0$, $q \to 0$ and $<r> \to \sqrt{\pi\alpha/4}$. This is as expected since in such a circumstance the Rician distribution degenerates into a Rayleigh distribution (for confirmation see p. 499 of Reference [2]). In the limit as the Rayleigh component disappears, α tends to zero and q tends to infinity. The behavior of both Bessel functions with large arguments is indicated by relation 9.7.1 in Reference [6] as:

$$I_\nu(q)|q \to \infty \sim \exp(q)/\sqrt{2\pi q} \qquad (7.47)$$

Thus, the envelope mean tends toward c as expected.

Envelope Mean Square

The mean square value of R, related to average power, is found by solving the basic moment equation with $k = 2$:

$$\begin{aligned} m_2 = <r^2> &= \frac{2}{\alpha}\exp(-\frac{c^2}{\alpha})\int_0^\infty r^3 \exp(-r^2/\alpha)I_0(2rc/\alpha)dr \\ &= \frac{2}{\alpha}\exp(-\frac{c^2}{\alpha})INT1 \end{aligned} \qquad (7.48)$$

where:

$$INT1 = \int_0^\infty r^3 \exp(-r^2/\alpha)I_0(2rc/\alpha)dr$$

As before, one can integrate by parts, for example using:

$$u_1 = r^2 I_0(\frac{2rc}{\alpha}), \quad dv_1 = r\exp(-\frac{r^2}{\alpha})dr \qquad (7.49)$$

$$du_1 = \left[2rI_0(\frac{2rc}{\alpha}) + \frac{2r^2 c}{\alpha}I_1(\frac{2rc}{\alpha})\right]dr, \quad v_1 = -\frac{\alpha}{2}\exp(-\frac{r^2}{\alpha}) \qquad (7.50)$$

The result is:

$$INT1 = \frac{-r^2\alpha}{2}\exp(-\frac{r^2}{\alpha})I_0(\frac{2rc}{\alpha})|_0^\infty + \alpha INT2 + cINT3 \qquad (7.51)$$

where:

$$INT2 = \int_0^\infty r\exp(-r^2/\alpha)I_0(2rc/\alpha)dr$$

and:
$$INT3 = \int_0^\infty r^2 \exp(-r^2/\alpha) I_1(2rc/\alpha) dr$$

Both integrals can in turn be solved by repeated integration by parts, but a more direct procedure is to return to the original moment expression and use the substitution $s = r^2$ to obtain:

$$m_2 = <r^2> = <s> = (2/\alpha)\exp(-k^2) \int_0^\infty s \exp(-\frac{s}{\alpha}) I_0(2\gamma\sqrt{s}) ds \quad (7.52)$$

where $k^2 = [(c^2/2)/(\alpha/2)]$ equals the ratio of direct to diffuse average powers and $\gamma = (k/\sqrt{\alpha})$. With the help of relation 6.643.2 in Reference [10], we have:

$$m_2 = \frac{\alpha}{k} \exp(-\frac{k^2}{2}) M_{-3/2,0}(k^2) \quad (7.53)$$

where $M_{a,b}(z)$ denotes Whittaker's function. This function can in turn be related to confluent hypergeometric functions via relations 13.1.32, 13.1.27, and 13.8 (example number 4) in Reference [6] as follows:

$$M_{-3/2,0}(k^2) = k\exp(-\frac{k^2}{2}) M(2,1;k^2) \quad (7.54)$$
$$= k\exp(\frac{k^2}{2}) M(-1,1;-k^2) \quad (7.55)$$
$$= k\exp(\frac{k^2}{2})(1+k^2) \quad (7.56)$$

Thus,
$$m_2 = \alpha(1 + k^2) \quad (7.57)$$

Note that as the direct component goes to zero the mean square envelope value approaches α, as it should for Rayleigh-only variation (the average power equals $<s/2> = <r^2/2> = \alpha/2$); also, as α approaches zero the result becomes c^2. Again, we note that the average power is $<s/2> = c^2/2$, as it should be for a direct signal component only. In retrospect, the result that the average power of a Rician signal is the sum of the average powers of the direct and diffuse components is obvious. Chapter 9 will discuss how the PDF of $R^2/2$ can be derived from the PDF of R so average power moments can be obtained directly. The result is a special case of the non-central chi-squared distribution discussed in chapter 28 of Reference [11].

7.6.11 Analysis of Battery Savings Potential Using Power Control

This section contains two analytical approaches for quantifying the average transmit power savings possible through the use of power control. This is important to portable radio design because batteries currently constitute a major portion of the weight and volume of such radios.

Analytical Model 1

The distribution of received power at land-mobile radio trunked base sites can be modeled as the sum of two lognormal distributions: the first representing the bulk of the data and the second representing a high-end contamination, possibly dominated by control station transmissions. Thus the PDF for the received power in dBm can be written as:

$$f(x_{\text{dBm}}) \sim \alpha N(\mu_1, \sigma_1) + (1-\alpha) N(\mu_2, \sigma_2) \tag{7.58}$$

where the notation $N(\mu, \sigma)$ stands for a Gaussian (normal) random variable with mean μ and standard deviation σ, and α denotes the weight of the first lognormal component.

The PDF for a single lognormal component can be found as follows. Let $x_{\text{dBm}} = 10\log(x) = k\ln(x)$, where $k = 10/\ln(10)$. Thus, $x = \exp(x_{\text{dBm}}/k)$ and $dx = (x/k) dx_{\text{dBm}}$. The PDF of X_{dBm}, being normal, can be written as:

$$f(x_{\text{dBm}}) = \frac{1}{\sqrt{2\pi\sigma^2}} \exp[-\frac{(x_{\text{dBm}} - \mu)^2}{2\sigma^2}] \tag{7.59}$$

Substitution of the relations between x, x_{dBm}, dx, and dx_{dBm} leads to:

$$f(x) = \frac{k}{x\sqrt{2\pi\sigma^2}} \exp\{-\frac{[k\ln(x) - \mu]^2}{2\sigma^2}\} \tag{7.60}$$

where x must be greater than or equal to zero.

The r-th moment of X is found by solving the integral:

$$E(x^r) = \int_0^\infty \frac{k x^{r-1}}{\sqrt{2\pi\sigma^2}} \exp\{-\frac{[k\ln(x) - \mu]^2}{2\sigma^2}\} dx \tag{7.61}$$

Substituting $y = [k\ln(x) - \mu]/\sqrt{2}\sigma$, this becomes:

$$E(x^r) = \int_{-\infty}^\infty \frac{1}{\sqrt{\pi}} \exp[-(ay^2 + 2by + c)] dy \tag{7.62}$$

where $a = 1$, $b = -(r\sigma/\sqrt{2}k)$, and $c = -(r\mu/k)$. Upon completing the square of the exponential argument, this integral can be expressed in terms of the error function (for example, see relation #7.4.32 in Reference [6]):

$$E(x^r) = \exp(\frac{r^2 \sigma^2}{2k^2} + \frac{r\mu}{k}) \tag{7.63}$$

Consequently, the average power for our dual lognormal signal strength model is given by:

$$E(x) = \alpha \exp(\frac{\sigma_1^2}{2k^2} + \frac{\mu_1}{k}) + (1-\alpha) \exp(\frac{\sigma_2^2}{2k^2} + \frac{\mu_2}{k}) \tag{7.64}$$

Probability Theory Refresher

Next, consider the impact of perfect power control; i.e., a received power ceiling of p_0(dBm) is maintained. The PDF for a single lognormal component to which such control is applied is given by:

$$f(x) = \frac{k}{x\sqrt{2\pi\sigma^2}} \exp\{-\frac{[k\ln(x) - \mu]^2}{2\sigma^2}\} + \delta(x - p_0)F \tag{7.65}$$

where x ranges between 0 and p_0 (numeric) and F equals the probability that X would have been greater than p_0 in the absence of power control. The r-th moment for the received power under power control is thus:

$$\begin{aligned} E(x^r) &= \int_0^{p_0} \frac{kx^{r-1}}{\sqrt{2\pi\sigma^2}} \exp\{-\frac{[k\ln(x) - \mu]^2}{2\sigma^2}\}dx + p_0 F \\ &= \int_{-\infty}^{\beta} \frac{1}{\sqrt{\pi}} \exp[-(ay^2 + 2by + c)]dy + p_0 F \end{aligned} \tag{7.66}$$

where $\beta = [k\ln(p_0) - \mu]/\sqrt{2}\sigma$. For the case of $r = 1$, the average power is given by:

$$\frac{1}{2}\exp(b^2 - c)[\text{erf}(\beta - \frac{\sigma}{\sqrt{2}k}) + 1] + p_0 F$$

As an example applying this analysis, consider November 1984 measurements made of received signal strength on a trunked system atop the Sears Tower in Chicago, Illinois. The dual lognormal model fitting these measurements is $\alpha = 0.95$, $\mu_1 = $ -75 dBm, $\sigma_1 = 11.7$ dB, $\mu_2 = $ -55 dBm, and $\sigma_2 = 2$ dB. These values result in an average power of 1.3 E-6 (-58.8 dBm). Choosing p_0(dBm) $= -70$ gives $F = 0.37$ (0.335 due to component 1 and essentially 1.0 due to component 2, weighted appropriately). The average power under power control is just 5.0 E-8 (-73.0 dBm), a factor of 26 (14.2 dB) less than without such control. This implies a major battery-life extension potential.

Analytical Model 2

Analytical model 1 is based on actual signal strength measurements. To handle situations where no such observations exist, one can assume power-law propagation holds and units are distributed in range according to a power law as well. Thus,

$$P_r = \frac{K_p}{r^n} \tag{7.67}$$

where K_p depends on the transmit ERP and receive antenna gain, and

$$f(r) = K_r r^m, r_l \leq r \leq r_h \tag{7.68}$$

where the integral of the PDF over the full range of r must be unity. This means $K_r = (m+1)/(r_h^{m+1} - r_l^{m+1}), m \neq -1$ and $K_r = 1/\ln(r_h/r_l), m = -1$.

The average received power is then given by:

$$\begin{aligned} E(P_r) &= \int_{r_l}^{r_h} \frac{K_p}{r^n} K_r r^m dr \\ &= \frac{K_p K_r}{m-n+1}(r_h^{m-n+1} - r_l^{m-n+1}),\ m-n \neq -1 \\ &= K_p K_r \ln(\frac{r_h}{r_l}),\ m-n = -1 \end{aligned} \qquad (7.69)$$

Again assume perfect power control such that P_r cannot exceed some threshold p_0. The probability that the received power exceeds p_0 in the absence of power control is the same as the probability that the range is less than $r_0 = (K_p/p_0)^{1/n}$. Calling this probability F as before, one finds:

$$\begin{aligned} F &= \int_{r_l}^{r_0} K_r r^m dr \\ &= \frac{K_r}{m+1}(r_0^{m+1} - r_l^{m+1}),\ m \neq -1 \\ &= K_r \ln(\frac{r_0}{r_l}),\ m = -1 \end{aligned} \qquad (7.70)$$

Hence, with power control one has an expected received power of:

$$\begin{aligned} E(P_r)_c &= \int_{r_0}^{r_h} \frac{K_p}{r^n} K_r r^m dr + p_0 F \\ &= \frac{K_p K_r}{m-n+1}(r_h^{m-n+1} - r_0^{m-n+1}) + p_0 F,\ m-n \neq -1 \\ &= K_p K_r \ln(\frac{r_h}{r_0}) + p_0 F,\ m-n = -1 \end{aligned} \qquad (7.71)$$

For example, consider fourth-law propagation and signal strengths varying between -110 dBm at 40 mi (implying $K_p = 2.56 \times 10^{-5}$) and -40 dBm (implying a minimum range of 0.71 mi). For a uniform distribution of units with range, $m = 0$ and $K_r = 0.0254$. Limiting received power to -70 dBm implies power shutback for all units closer than 4 mi (8.37% probability) and an average received power of -79.3 dBm. This is 17.1 dB less than the -62.2 dBm obtained without power control. Again, substantial battery-life extension potential is indicated.

7.7 Homework Problems

7.7.1 Problem 1

Are "probability equal to zero" and "impossible" synonymous? How about "probability equal to one" and "certain"?

Probability Theory Refresher 79

Solution: The word synonymous means same, or nearly the same. Because of the latter qualification the answer to both questions is yes. However, in neither case are the two things being compared identical. For example, the probability of a dart landing *exactly* at some point on a dart board is zero (recall property 3 in Section 7.4.3); yet the dart does indeed land somewhere! Similarly, an event that has a probability equal to one only means that the likelihood of it not occurring in a number of trials becomes insignificant as the number of trials increases without bound. A certain event of course occurs every trial, regardless of how few trials are carried out.

7.7.2 Problem 2

The term reliability can be precisely defined mathematically as the probability that a system of interest performs satisfactorily up to time t. The probability of successful operation is assumed unity at time zero; i.e., $R(0) = 1$. Letting $f(t)dt$ represent the probability of failure between times t and $t+dt$, the cumulative probability of failure by time t is $F(t) = \int_0^t f(x)dx$; i.e., $f(t)$ is the failure PDF. Since the probability of failure between time t and $t+dt$ is also equal to the reliability at time t less the reliability at time $t+dt$, one also has $f(t) = [R(t) - R(t+dt)]/dt = -dR(t)/dt$ and $F(t) = 1 - R(t)$.

A second density function of interest is the instantaneous failure rate or hazard function $h(t)$. This represents the conditional probability of failure between times t and $t+dt$, given that the system has functioned properly up to time t. Thus $R(t)h(t)dt = f(t)dt$, which can be rearranged into $-h(t)dt = dR(t)/R(t)$. Integrating from zero to time t then yields $R(t) = \exp[-\int_0^t h(x)dx]$. The hazard function typically varies with time as a "bathtub curve." Early in life the hazard function is relatively high and decreasing, reflecting "infant mortalities"; then a constant failure regime is entered where failures occur randomly in time; and finally, as the system ages and wears out, the hazard function again rises.

What is the *mean time to failure* (MTTF) for "burned in" systems; i.e., those systems for which accelerated testing has removed the infant mortalities? What about two independent systems in series or in parallel?

Solution: The expected life of a system is given by the first moment of the failure probability density:

$$\begin{aligned} E(\text{life}) &= \text{MTTF} \\ &= \int_0^\infty t f(t) dt \\ &= -\int_0^\infty t \frac{dR(t)}{dt} dt \\ &= \int_0^\infty R(t) dt \end{aligned} \qquad (7.72)$$

In the case of constant hazard function, say $h(t) = \lambda$,

$$\begin{aligned}
\text{MTTF} &= \int_0^\infty \exp(-\int_0^t \lambda dx) dt \\
&= \int_0^\infty \exp(-\lambda t) dt \\
&= \frac{1}{\lambda}
\end{aligned} \qquad (7.73)$$

In the case of two systems in series, the overall reliability equals the product of the individual reliabilities. Thus,

$$R(t) = R_1(t) R_2(t) = \exp(-\lambda_1 t) \exp(-\lambda_2 t) = \exp[-(\lambda_1 + \lambda_2) t]$$

Integration of this from zero to infinity yields MTTF $= 1/(\lambda_1 + \lambda_2)$.

For two systems in parallel, the overall reliability is:

$$\begin{aligned}
R(t) &= 1 - \text{Prob(both units fail by time } t) \\
&= 1 - [1 - R_1(t)][1 - R_2(t)] \\
&= \exp(-\lambda_1 t) + \exp(-\lambda_2 t) - \exp[-(\lambda_1 + \lambda_2) t]
\end{aligned} \qquad (7.74)$$

Integration of this from zero to infinity yields MTTF $= (1/\lambda_1) + (1/\lambda_2) - [1/(\lambda_1 + \lambda_2)]$.

Of course, in practice, parallel systems often share some particular burden and the failure of one system puts additional stress on the remaining system. The general notion of handling reliability in terms of a stress distribution is discussed in Chapter 10 of Reference [5]. Changes in stress invalidate the constant hazard assumption and a concept like equivalent time is needed to solve such problems [12].

Mean time to (first) failure is often simply called *mean time between failures* (MTBF). However, unless systems are perfectly repaired, the mean time to the next failure will not be the same as the mean time to the failure that just occurred. Even without this issue, MTBF is a rather misunderstood parameter. In the constant hazard situation, about 63% of the units will fail at times shorter than the MTBF. As Reference [13] points out, failure rate per month is operationally a more meaningful parameter.

7.7.3 Problem 3

Figure 7.2 is a graph of the estimated MTBF performance of an 800-MHz base station as a function of ambient temperature. Here MTBFs are calculated from the reciprocal of the sum of the constant hazard values for each component in the radio. That is, the system is viewed as a series connection of all components and a failure of any single component constitutes a failure of the radio. The hazard values were obtained from MIL-HDBK-217C part failure rate models; however, certain components that interface with the outside world via the antenna feedline, telephone lines

Probability Theory Refresher

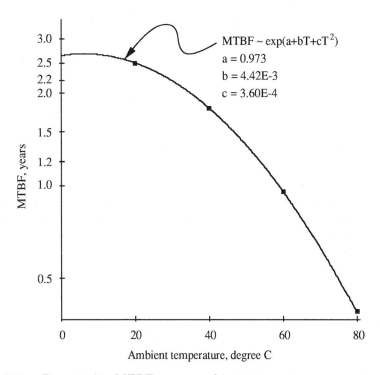

Figure 7.2 Base station MTBF versus ambient temperature.

and ac power lines are adjusted upward in failure rate to account for a "lightning" factor observed in practice.

A least squares error quadratic fit to the data of Figure 7.2 gives

$$\text{MTBF}(T) \approx \exp(a + bT + cT^2) \tag{7.75}$$

where $a = 0.973$, $b = 4.42E-3$, $c = -3.60E-4$, and T is the ambient temperature in Celsius. Assuming the ambient temperature to be Gaussian distributed with mean $T_m = 40$ °C and standard deviation $\sigma = 20$ °C, what is the temperature-averaged MTBF of the base station?

Solution: The general temperature-averaged MTBF is found by averaging the quadratic fit weighted by a Gaussian density function for temperature:

$$\begin{aligned}
\text{MTBF} &= \int_{-\infty}^{\infty} \text{MTBF}(T) f(T) dT \\
&= \int_{-\infty}^{\infty} \exp(a + bT + cT^2) \frac{1}{\sqrt{2\pi\sigma^2}} \exp[-(\frac{T - T_m}{\sqrt{2}\sigma})^2] dT
\end{aligned}$$

$$= \int_{-\infty}^{\infty} \frac{1}{\sqrt{2\pi\sigma^2}} \exp[-(AT^2 + 2BT + C)]dT \tag{7.76}$$

where $A = (1 - 2c\sigma^2)/2\sigma^2$, $B = -(T_m + b\sigma^2)/2\sigma^2$, and $C = (T_m^2 - 2a\sigma^2)/2\sigma^2$. Using relation 7.4.32 in Reference [6] one finds:

$$\text{MTBF} = \frac{1}{\sqrt{1 - 2c\sigma^2}} \exp[(a + \frac{b^2\sigma^2}{2}) + bT_m + cT_m^2]$$

$$\frac{\text{MTBF}}{\text{MTBF}(T_m)} = \frac{1}{\sqrt{1 - 2c\sigma^2}} \exp(\frac{b^2\sigma^2}{2}) \tag{7.77}$$

Substituting the given mean and standard deviation values shows that the MTBF averaged over temperature is about 89% of the MTBF at the average temperature. Thus, the increase in MTBF at low temperatures is fairly well offset by the decrease in MTBF at high temperatures and use of the MTBF at the average temperature may be adequate.

References

[1] Tannous, C., et al., "Strange Attractors in Multipath Propagation," *IEEE Tr. on Comm.*, Vol. 39, No. 5, May 1991, pp. 629-631.

[2] Papoulis, A., *Probability, Random Variables, and Stochastic Processes*, New York, NY, McGraw-Hill Book Co., 1965.

[3] Walpole, R. E. and R. H. Myers, *Probability and Statistics for Engineers and Scientists*, New York, NY, Macmillan Publishing Company, 1985.

[4] Brown, R. G. and P.Y.C. Hwang, *Introduction to Random Signals and Applied Kalman Filtering*, 2nd edition, New York, NY, John Wiley & Sons, 1992.

[5] Beckmann, P., *Probability in Communication Engineering*, New York, NY, Harcourt, Brace & World, Inc., 1967.

[6] Abramowitz, M. and I. Stegun (editors), *Handbook of Mathematical Functions*, New York, NY, Dover Publications, Inc., 1972.

[7] Hess, G. and J. Cohn, "Communication Load and Delay in Mobile Trunked Systems," *31st IEEE Vehicular Technology Conference*, Washington, D.C., April 1981, pp. 269-273.

[8] Paez, M. D. and T. H. Glisson, "Minimum Mean Squared-Error Quantization in Speech," *IEEE Tr. on Comm.*, Vol. Com-20, April 1972, pp. 225-230.

[9] Rabiner, L. R. and R. W. Schafer, *Digital Processing of Speech Signals*, Englewood Cliffs, NJ, Prentice-Hall, Inc., 1978.

[10] Gradshteyn, I. S. and I. M. Ryzhik, *Table of Integrals, Series, and Products*, New York, NY, Academic Press, 1965.

[11] Johnson, N. L. and S. Kotz, *Continuous Univariate Distribution-2*, New York, NY, John Wiley & Sons, 1970.

[12] Yang, W. and C. C. Lin, "A Reliability Model for Dependent Failures in Parallel Redundant Systems," *IEEE Tr. on Reliability*, Vol. R-23, No. 3, October 1974, pp. 286-287.

[13] Editorial, "On MTBF," *IEEE Tr. on Reliability*, Vol. 37, No. 5, December 1988, p. 449.

Chapter 8
Central Limit Theorem

With the basic probability tools of Chapter 7 now in hand, we proceed to apply them to a common situation where some desired output equals the sum of a number of random quantities. For example, the superposition of signal components from many headings as discussed in Chapter 6 or the interference caused by multiple cochannel sources in cellular frequency reuse systems as discussed in Chapter 10. Based on Reference [1], this chapter demonstrates through the central limit theorem why the Gaussian (normal) probability distribution is so pervasive. Hopefully, it will also caution the reader that not all sums tend toward the Gaussian distribution. Furthermore, well out on the distribution tails of even those sums that do, the character may deviate substantially from Gaussian. As radio system coverage and reliability requirements rise toward 100%, it is prudent to keep in mind the caveats of the central limit theorem.

8.1 Sum of Many Independent, Identically Distributed Random Variables

Consider the sum:

$$X = X_1 + X_2 + \cdots + X_n$$

where n is very large and the X_j are all independent and have the same distribution, not necessarily normal, with mean a and variance σ^2. The mean of X is thus na and the variance is $D(X) = n\sigma^2$ (the notation D is natural as variance is a measure of deviation about the mean; it is used interchangeably with Var, the notation introduced in Chapter 7). It is convenient to introduce the normalized variate:

$$Y = \frac{X - na}{\sigma\sqrt{n}}$$

with zero mean and unity variance. The sum can then be written as:

$$Y = \Sigma_{j=1}^{n} Y_j, \quad Y_j = \frac{X_j - a}{\sigma\sqrt{n}}$$

Now consider the deviation of X_j; i.e., $d = X_j - <X_j> = X_j - a$. The characteristic function of d is:

$$\begin{aligned}
\Phi_d(\nu) &= E[\exp(i\nu d)] = E\{\exp[i\nu(X_j - a)]\} \\
&= \int p(x) \exp[i\nu(x-a)] dx \\
&= \int p(x)[1 + i\nu(x-a) + \frac{i^2\nu^2(x-a)^2}{2!} + \cdots] dx \\
&= \Sigma_{k=0}^{\infty} \frac{i^k \nu^k}{k!} \int (x-a)^k p(x) dx \\
&= \Sigma_{k=0}^{\infty} \left(\frac{i^k \nu^k}{k!}\right) M_k
\end{aligned}$$

where $M_k = \mu_k = $ k-th central moment and the exponential function has been expanded in a Taylor series about the origin. The characteristic function of Y_j is thus:

$$\Phi_{Y_j}(\nu) = \Phi_d(\frac{\nu}{\sigma\sqrt{n}}) = \Sigma_{k=0}^{\infty} \left(\frac{i^k \nu^k}{k!}\right) M_k \frac{1}{\sigma^k n^{\frac{k}{2}}}$$

and the characteristic function of Y, because of the independence of its constituents, is:

$$\begin{aligned}
\Phi_Y(\nu) &= [\Phi_{Y_j}(\nu)]^n \\
&= \left(1 - \frac{\nu^2}{2n} - \frac{iM_3 \nu^3}{3!\sigma^3 n^{\frac{3}{2}}} + \frac{M_4 \nu^4}{4!\sigma^4 n^2} + \cdots\right)^n
\end{aligned}$$

Taking the natural logarithm gives:

$$\ln \Phi_Y(\nu) = n \ln\left(1 - \frac{\nu^2}{2n} - \frac{iM_3 \nu^3}{3!\sigma^3 n^{\frac{3}{2}}} + \frac{M_4 \nu^4}{4!\sigma^4 n^2} + \cdots\right)$$

Using the expansion:

$$\ln(1+x) = x - \frac{x^2}{2} + \frac{x^3}{3} - \cdots$$

and retaining terms to order $1/n$ then gives:

$$\ln \Phi_Y(\nu) = -\frac{\nu^2}{2} - \frac{iM_3 \nu^3}{3!\sigma^3 \sqrt{n}} + \frac{M_4 \nu^4}{4!\sigma^4 n} - \frac{\nu^4}{8n} + \cdots$$

As n approaches infinity the characteristic function becomes simply $\exp(-\nu^2/2)$.

Central Limit Theorem

To make sense of this result, consider the characteristic function of a general Gaussian random variable with mean μ and standard deviation σ:

$$\Phi_Z(\nu) = \int_{-\infty}^{\infty} \exp(i\nu z) \frac{1}{\sqrt{2\pi\sigma^2}} \exp[-\frac{(z-\mu)^2}{2\sigma^2}] dz$$

$$= \exp[-\frac{(\mu^2 - \alpha^2)}{2\sigma^2}] \int_{-\infty}^{\infty} \frac{1}{\sqrt{2\pi\sigma^2}} \exp[-\frac{(z-\alpha)^2}{2\sigma^2}] dz$$

where $\alpha = \mu + i\nu\sigma^2$. The required integral can easily be recognized as the PDF of a Gaussian random variable evaluated over the entire range of possible values. Hence, the integral equals unity and the characteristic function is:

$$\Phi_Z(\nu) = \exp[-\frac{\nu^2\sigma^2}{2} + i\nu\mu]$$

Notice that the characteristic function of a standardized Gaussian variate, zero mean ($\mu = 0$) and unity standard deviation ($\sigma = 1$), is just $\exp(-\nu^2/2)$. This is exactly the form found when the sum of a large number of independent, identically distributed random variables was examined. Therefore, one concludes that the density function for such a sum becomes Gaussian in character regardless of the distribution of the individual terms.

8.2 Rapidity of Convergence

Consider the following substitutions in the logarithm of the characteristic function of Y given in Section 8.1: the skew coefficient $s = (M_3/\sqrt{M_2^3}) = (M_3/\sigma^3)$ and the coefficient of excess $\gamma = (M_4/M_2^2) - 3 = (M_4/\sigma^4) - 3$. The expression then becomes:

$$\ln \Phi_Y(\nu) = -\frac{\nu^2}{2} - \frac{is\nu^3}{3!\sqrt{n}} + \frac{\gamma\nu^4}{4!n} + \cdots$$

or, taking the antilogarithm,

$$\Phi_Y(\nu) = \exp(-\frac{\nu^2}{2}) \exp(-\frac{is\nu^3}{3!\sqrt{n}} + \frac{\gamma\nu^4}{4!n} + \cdots)$$

If the latter exponential is expressed as a series and terms to order $(1/n)$ are retained, the result is:

$$\Phi_Y(\nu) = \exp(-\frac{\nu^2}{2})(1 - \frac{is\nu^3}{3!\sqrt{n}} + \frac{\gamma\nu^4}{4!n} - \frac{s^2\nu^6}{2(3!)^2 n} + \cdots)$$

Taking the inverse Fourier transform yields the density function for Y as:

$$p_Y(y) = \frac{1}{2\pi} \int \exp(-\frac{\nu^2}{2} - i\nu y)(1 - \frac{is\nu^3}{3!\sqrt{n}} + \frac{\gamma\nu^4}{4!n} - \frac{s^2\nu^6}{2(3!)^2 n} + \cdots) d\nu$$

The individual terms of this integral are of the form:

$$\int \nu^k \exp(-\frac{\nu^2}{2} - i\nu y) d\nu = I_k$$

Now:

$$\frac{d^k}{dy^k}[\exp(-\frac{\nu^2}{2} - i\nu y)] = (-i\nu)^k \exp(-\frac{\nu^2}{2} - i\nu y)$$
$$= (-i)^k \nu^k \exp(-\frac{\nu^2}{2} - i\nu y)$$

Thus,

$$I_k = (-i)^{-k} \int \frac{d^k}{dy^k}[\exp(-\frac{\nu^2}{2} - i\nu y)]d\nu$$
$$= (i)^k \frac{d^k}{dy^k}\left[\int \exp(-\frac{\nu^2}{2} - i\nu y)d\nu\right]$$
$$= (i)^k \frac{d^k}{dy^k}\left[\sqrt{2\pi}\exp(-\frac{y^2}{2})\right]$$

where the integral comes from the characteristic function of a Gaussian random variable with zero mean and unit variance.

Applying the definition for Hermite polynomials [2]:

$$H_k(y) = (-1)^k \exp(\frac{y^2}{2}) \frac{d^k}{dy^k}\left[\exp(-\frac{y^2}{2})\right]$$

one has:

$$\frac{d^k}{dy^k}\left[\exp(-\frac{y^2}{2})\right] = H_k(y)(-1)^{-k} \exp(-\frac{y^2}{2})$$

and finally:

$$I_k = \sqrt{2\pi}(-i)^k H_k(y)\exp(-\frac{y^2}{2})$$

Using this expression term by term in the density function of Y then gives:

$$p_Y(y) = \frac{\exp(-\frac{y^2}{2})}{\sqrt{2\pi}}\left[1 + \frac{sH_3(y)}{6\sqrt{n}} + \frac{\gamma H_4(y)}{24n} - \frac{s^2 H_6(y)}{72n} + \cdots\right]$$

where:

$$H_3(y) = y^3 - 3y$$
$$H_4(y) = y^4 - 6y^2 + 3$$
$$H_6(y) = y^6 - 15y^4 + 45y^2 - 15$$

Notice that density functions with small higher order moments (in particular small skew coefficients) will sum to Gaussian behavior most rapidly.

8.3 Central Limit Theorem

This theorem states the conditions under which the distribution of the sum of random variables will asymptotically approach the Gaussian distribution as n approaches infinity. The preceding analysis showed that a sufficient condition is that all the terms are independently and identically distributed and that their variance exists. "Identically distributed" implies not only the same kind of distribution, but also the same parameters. Necessary conditions are less stringent. The least stringent conditions for which the theorem holds are the Lindeberg-Feller conditions:

$$\lim_{n \to \infty} \frac{1}{s_n^2} \Sigma_{j=1}^n \int_{s-<X>>\epsilon s_n}^{\infty} (x- <X>)^2 f_j(x) dx = 0 \, , \epsilon > 0$$

where the variance of the j-th term is σ_j^2 and the variance of the sum is s_n^2. For engineering purposes this means that the sum will tend toward a Gaussian distribution provided the variance of each single term is negligible compared to the variance of the sum.

8.4 Homework Problems

8.4.1 Problem 1

Is the sum of a large number of independent draws from an exponentially distributed population Gaussian?
Solution: Consider the exponential PDF to be:

$$f(x) = \frac{1}{h} \exp(-\frac{x}{h}) \tag{8.1}$$

The characteristic function for such a random variable is given by:

$$\begin{aligned}
E[\exp(i\nu x)] &= \int_0^\infty \exp(i\nu x) \frac{1}{h} \exp(-\frac{x}{h}) dx \\
&= \frac{1}{h} \int_0^\infty \exp(-\alpha x) dx \\
&= \frac{1}{h\alpha} \\
&= \frac{1}{1-i\nu h} \\
&= \frac{a}{a-i\nu}
\end{aligned} \tag{8.2}$$

where $\alpha = a - i\nu$ and $a = (1/h)$. The characteristic function for the sum of n such random variables is therefore $[a/(a-i\nu)]^n$. This can be inverse Fourier transformed to obtain the PDF of the sum. However, a simple way to obtain the answer is to begin with the basic relation:

$$\int_0^\infty \exp(i\nu x)\exp(-ax)dx = \frac{1}{a-i\nu} \qquad (8.3)$$

Differentiating this relation m times with respect to a gives:

$$\int_0^\infty \exp(i\nu x)(-1)^m x^m \exp(-ax)dx = \frac{(-1)^m m!}{(a-i\nu)^{m+1}} \qquad (8.4)$$

Choosing $m = n - 1$, multiplying both sides by a^n, and moving the factorial to the left-hand side of the equation leads to:

$$\int_0^\infty \exp(i\nu x)[\frac{a^n}{(n-1)!}x^{n-1}\exp(-ax)]dx = (\frac{a}{a-i\nu})^n \qquad (8.5)$$

The right-hand side is precisely the characteristic function of interest. Therefore, the left-hand side expression in brackets is the PDF for the sum of n exponentially distributed random variables. The sum follows what is called the gamma distribution. This is *not* (exactly) the same as the Gaussian distribution and yet the conditions necessary for the central limit theorem to hold *are* present.

8.4.2 Problem 2

Can a random variable have a PDF of the form $f(x) = k/(a^2 + x^2)$? If so, solve for the mean and variance and comment on the applicability of the central limit theorem.

Solution: For validity as a PDF we require:

$$1 = \int_{-\infty}^\infty \frac{kdx}{a^2+x^2} \qquad (8.6)$$

By symmetry about the origin, this equates to:

$$1 = \frac{2k}{a^2}\int_0^\infty \frac{dx}{1+(\frac{x}{a})^2} \qquad (8.7)$$

Substituting $\tan(\theta) = (x/a)$, the right-hand side becomes:

$$\frac{2k}{a^2}\int_0^{\pi/2} \frac{ad\theta}{\cos^2(\theta)[1+\tan^2(\theta)]} = \frac{2k}{a}\int_0^{\pi/2} d\theta \qquad (8.8)$$

implying the proper value for k is (a/π).

Central Limit Theorem

The median value is readily seen to be zero due to odd symmetry about the origin. However, the mean value is not zero and in fact is not even finite since:

$$\begin{aligned}
E(x) &= \int_{-\infty}^{\infty} x f(x) dx \\
&= k \int_{-\infty}^{\infty} \frac{x dx}{a^2 + x^2} \\
&= \left(\frac{2a}{\pi}\right) \left(\lim_{x_u \to \infty} \int_0^{\frac{x_u}{a}} \frac{y dy}{1 + y^2}\right) \\
&= \left(\frac{a}{\pi}\right) \left(\lim_{x_u \to \infty} \ln(1 + y^2) \Big|_0^{\frac{x_u}{a}}\right) \\
&= \infty
\end{aligned} \quad (8.9)$$

Neither is the mean square value finite since:

$$\begin{aligned}
E(x^2) &= \int_{-\infty}^{\infty} x^2 f(x) dx \\
&= k \int_{-\infty}^{\infty} \frac{x^2 dx}{a^2 + x^2} \\
&= 2k \int_0^{\infty} \frac{x^2 dx}{a^2 + x^2}
\end{aligned} \quad (8.10)$$

Noting that:

$$\begin{aligned}
x_u &= \int_0^{x_u} dx \\
&= \int_0^{x_u} \frac{a^2 + x^2}{a^2 + x^2} dx \\
&= \int_0^{x_u} \frac{a^2 dx}{a^2 + x^2} + \int_0^{x_u} \frac{x^2 dx}{a^2 + x^2}
\end{aligned} \quad (8.11)$$

leads to:

$$\begin{aligned}
\int_0^{x_u} \frac{x^2 dx}{a^2 + x^2} &= x\Big|_0^{x_u} - a^2 \int_0^{x_u} \frac{dx}{a^2 + x^2} \\
&= x_u - a\frac{\pi}{2}
\end{aligned} \quad (8.12)$$

Clearly, as x_u approaches infinity the mean square becomes unbounded. In fact, no moments above order zero are finite. The requirement that the variance exists is thus not met and therefore the central limit theorem does not apply to this distribution (known as the Cauchy distribution).

8.4.3 Problem 3

Does the central limit theorem apply to the sum of a number of lognormally distributed interference sources (for example, like in a cellular frequency reuse system)?

Solution: The answer is yes, but the rate of convergence is so slow that assuming the result also to be lognormally distributed is of practical use [3]. Other useful references on this subject are References [4, 5, 6].

References

[1] Beckmann, P., *Probability in Communication Engineering*, New York, NY, Harcourt, Brace & World, Inc., 1967.

[2] Abramowitz, M. and I. Stegun (editors), *Handbook of Mathematical Functions*, New York, NY, Dover Publications, Inc., 1972.

[3] Fenton, L. F., "The Sum of Log-Normal Probability Distributions in Scatter Transmission Systems," *IRE Tr. on Comm. Sys.*, March 1960, pp. 57-67.

[4] Nasell, I., "Some Properties of Power Sums of Truncated Normal Random Variables," *Bell System Technical Journal*, November 1967, pp. 2091-2110.

[5] Janos, W. A., "Tail of the Distribution of Sums of Log-Normal Variates," *IEEE Tr. on Inform. Th.*, Vol. IT-16, No. 3, May 1970, pp. 299-302.

[6] Schwartz, S. C. and Y. S. Yeh, "On the Distribution Function and Moments of Power Sums with Log-Normal Components," *Bell System Technical Journal*, Vol. 61, No. 7, September 1982, pp. 1441-1462.

Chapter 9

Probability Theory Refresher, Continued

This third and final chapter reviewing probability theory considers functions of random variables and extends the concept to more than a single variable. An important application is the proof that Clarke's simple model of land-mobile propagation introduced in Chapter 6 does indeed produce a Rayleigh-distributed envelope. This is done through consideration of the distribution of the square root of the sum of squares of two independent, zero-mean Gaussian random variables. Other applications involve capture probability in a multiple transmitter simulcast situation (more on this in Chapter 12) and the effect of ideal selection diversity on fade statistics (more on this in Chapter 11).

Next, the estimation of distribution parameters is discussed. In particular, the maximum likelihood method is applied to the problems of estimating the average power of a Rayleigh-faded signal, the median and standard deviation of a lognormally shadowed signal, and the average message length of messages on a trunked radio system.

Rayleigh fading is further scrutinized by considering its level crossing rate and average fade duration. Several ways of checking data to see if they indeed come from a Rayleigh population are discussed.

9.1 Functions of a Single Continuous Random Variable

Consider the task of finding the PDF of some random variable Y, given knowledge of the PDF of some other random variable X to which Y is related. Let $Y = u(X)$ define a one-to-one correspondence between the values of X and Y so that the equation $y = u(x)$ can be uniquely solved for x in terms of y, say $x = w(y)$. Then

the PDF of Y is given by:
$$g_Y(y) = f_X[w(y)]|J|$$
where $J = dx/dy = w'(y)$ is the Jacobian of the transformation [1]. The absolute value operation is necessary so both increasing and decreasing functions are handled properly. In situations where the relationship is not one-to-one, and the interval over which X is defined can be partitioned into k mutually exclusive disjoint sets such that each of the inverse functions $x_1 = w_1(y), x_2 = w_2(y), \cdots x_k = w_k(y)$ of $y = u(x)$ defines a one-to-one correspondence, the PDF of Y is given by
$$g_Y(y) = \Sigma_{i=1}^{k} f_X[w_i(y)]|J_i|$$
where $J_i = w'_i(y)$ for $i = 1, 2, \cdots, k$.

An application of the preceding is the square-law detector. Here $Y = aX^2$ and $a > 0$. If $y < 0$, then $y = ax^2$ has no real solutions and thus $f_Y(y) = 0$. However, for $y \geq 0$ two solutions exist: $x_1 = \sqrt{y/a}$ and $x_2 = -\sqrt{y/a}$. The Jacobians are $(dx_1/dy) = (1/2\sqrt{ay})$ and $(dx_2/dy) = (-1/2\sqrt{ay})$, respectively. Thus, the desired PDF is:
$$g_Y(y) = \frac{1}{2\sqrt{ay}}[f_X(\sqrt{\frac{y}{a}}) + f_X(-\sqrt{\frac{y}{a}})]$$
for $y \geq 0$. The CDF of Y is:
$$\begin{aligned} F_Y(y) &= \text{Prob}(Y \leq y) \\ &= \text{Prob}(-\sqrt{\frac{y}{a}} \leq X \leq \sqrt{\frac{y}{a}}) \\ &= F_X(\sqrt{\frac{y}{a}}) - F_X(-\sqrt{\frac{y}{a}}) \end{aligned}$$

9.2 Extension of Random Variable Concept to Two Variables

The function $f_{XY}(x, y)$ is a joint density function of continuous random variables X and Y if–
1. $f_{XY}(x, y) \geq 0$ for all (x, y);
2. $\int_{-\infty}^{\infty} \int_{-\infty}^{\infty} f_{XY}(x, y) dx dy = 1$;
3. $\text{Prob}[x_1 \leq X \leq x_2, y_1 \leq Y \leq y_2] = \int_{y_1}^{y_2} \int_{x_1}^{x_2} f_{XY}(x, y) dx dy$.

The third relation defines the joint distribution function, which is tied to the joint density function through partial differentiation: $f_{XY}(x, y) = \partial^2 F_{XY}(x, y)/\partial x \partial y$.

The PDFs of X and Y alone, termed the marginal densities, can be obtained by integrations of the joint density function:
$$\begin{aligned} g_X(x) &= \int_{-\infty}^{\infty} f_{XY}(x, y) dy \\ h_Y(y) &= \int_{-\infty}^{\infty} f_{XY}(x, y) dx \end{aligned}$$

If X and Y are independent, the joint density function will simply be the product of the marginal densities.

9.3 Sample Applications

9.3.1 Sum of Two Random Variables

Consider the sum $Z = X + Y$. With the help of Figure 9.1, the distribution function of Z can be written as:

$$\begin{aligned}
F_Z(z) &= \text{Prob}(z \le x + y) \\
&= \int_{-\infty}^{\infty} \int_{-\infty}^{z-y} f_{XY}(x,y) dx dy \\
f_Z(z) &= \frac{dF_Z(z)}{dz} \\
&= \int_{-\infty}^{\infty} f_{XY}(z-y, y) dy
\end{aligned} \quad (9.1)$$

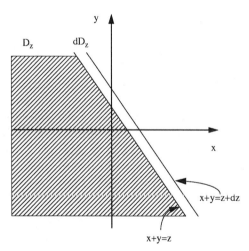

Figure 9.1 Probability space for the distribution function of $Z = X + Y$. After: Papoulis, A., *Probability, Random Variables, and Stochastic Processes*, McGraw-Hill Book Co., New York, NY, 1965, p. 189.

If X and Y are independent this simplifies to:

$$f_Z(z) = \int_{-\infty}^{\infty} f_X(z-y) f_Y(y) dy$$

$$= \int_{-\infty}^{\infty} f_X(x) f_Y(z-x) dx \tag{9.2}$$

so the density function of the sum of independent random variables equals the convolution of the constituent density functions.

Consider, for example, the sum of two zero-mean, equal variance Gaussian random variables:

$$\begin{aligned}
f_Z(z) &= \int_{-\infty}^{\infty} \frac{1}{\sqrt{2\pi\sigma^2}} \exp[-\frac{(z-y)^2}{2\sigma^2}] \frac{1}{\sqrt{2\pi\sigma^2}} \exp[-\frac{y^2}{2\sigma^2}] dy \\
&= \frac{1}{2\pi\sigma^2} \exp(-\frac{z^2}{4\sigma^2}) \int_{-\infty}^{\infty} \exp[-(\frac{1}{\sigma^2})y^2 + (\frac{z}{\sigma^2})y] dy \\
&= \frac{1}{2\pi(2\sigma^2)} \exp[-\frac{z^2}{2(2\sigma^2)}] \cdot \\
&\quad \{\frac{1}{\sqrt{2\pi(\sigma^2/2)}} \int_{-\infty}^{\infty} \exp[-\frac{(y-\frac{z}{2})^2}{2(\sigma^2/2)}] dy\}
\end{aligned} \tag{9.3}$$

The term inside the braces equates to unity since it represents integration of a Gaussian density function over its full range. The remaining factor is seen to be Gaussian also, with zero mean but variance increased twofold. In general, when n independent Gaussian terms are summed, the result remains Gaussian with mean equal to the sum of the means of all constituents and variance equal to the sum of the individual variances. Differences are handled in like manner, with negative signs associated with the means of random variables that involve subtraction.

Consider, for example, a simulcast capture problem. Assume that signals are received from two land-mobile sources. Due to lognormal shadowing, the received powers expressed in decibels are both normally distributed. Let $X \sim N(x_0, \sigma)$ and $Y \sim N(y_0, \sigma)$. The difference between the two received powers is thus $Z \sim N(x_0 - y_0, \sqrt{2}\sigma)$. Acceptable communications can occur when the stronger of the two signals exceeds the weaker by some amount, say C. The probability of failure or noncapture is given by:

$$\begin{aligned}
\text{Prob}(-C \leq Z \leq C) &= \int_{-C}^{C} \frac{1}{2\pi(\sqrt{2}\sigma)^2} \exp\{-\frac{[z-(x_0-y_0)]^2}{2(\sqrt{2}\sigma)^2}\} dz \\
&= \frac{1}{2}[\text{erf}(\frac{C-\mu}{2\sigma}) + \text{erf}(\frac{C+\mu}{2\sigma})]
\end{aligned} \tag{9.4}$$

where $\mu = x_0 - y_0$ and relation 7.1.22 of Reference [2] has been used.

Probability Theory Refresher, Continued

9.3.2 Square Root of Sum of Squares of Two Random Variables

Consider two independent, zero-mean Gaussian random variables with equal variance σ^2. Their joint density function is given by:

$$f_{XY}(x,y) = [\frac{1}{\sqrt{2\pi\sigma^2}}\exp(-\frac{x^2}{2\sigma^2})][\frac{1}{\sqrt{2\pi\sigma^2}}\exp(-\frac{y^2}{2\sigma^2})]$$
$$= \frac{1}{2\pi\sigma^2}\exp(-\frac{x^2+y^2}{2\sigma^2}) \quad (9.5)$$

The Jacobian for a transformation to polar coordinates r and θ (i.e., $x = r\cos(\theta)$ and $y = r\sin(\theta)$) is given by:

$$J = \frac{\partial(x,y)}{\partial(r,\theta)}$$
$$= \begin{vmatrix} \cos(\theta) & -r\sin(\theta) \\ \sin(\theta) & r\cos(\theta) \end{vmatrix}$$
$$= r \quad (9.6)$$

Thus,

$$f_{R,\Theta}(r,\theta) = \frac{1}{2\pi\sigma^2}\exp(-\frac{r^2}{2\sigma^2})r$$
$$F_Z(z) = \int_0^z \int_0^{2\pi} f_{R,\Theta}(r,\theta)d\theta dr$$
$$= \int_0^z \frac{1}{\sigma^2}\exp(-\frac{r^2}{2\sigma^2})r dr$$
$$= 1 - \exp(-\frac{z^2}{2\sigma^2}) \quad (9.7)$$

Differentiation of the distribution function with respect to z yields the density function:

$$f_Z(z) = \frac{z}{\sigma^2}\exp(-\frac{z^2}{2\sigma^2}), \quad z \geq 0 \quad (9.8)$$

which is recognized as Rayleigh in form.

9.3.3 Maximum of Two Random Variables

Consider the random variable obtained by selecting the maximum of two other random variables, $Z = \max(X,Y)$. With the aid of Figure 9.2, the density function of Z can be written as:

$$F_Z(z) = F_{XY}(z,z)$$

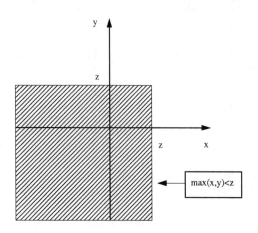

Figure 9.2 Probability space for the distribution function of Z = max(X,Y). After: Papoulis, A., *Probability, Random Variables, and Stochastic Processes*, McGraw-Hill Book Co., New York, NY, 1965, p. 192.

$$f_Z(z) = \frac{\partial F_{XY}(z,z)}{\partial x} + \frac{\partial F_{XY}(z,z)}{\partial y}$$
$$= \int_{-\infty}^{z} f_{XY}(z,y)dy + \int_{-\infty}^{z} f_{XY}(x,z)dx \quad (9.9)$$

If X and Y are independent, then:

$$F_Z(z) = F_X(z)F_Y(z)$$
$$f_Z(z) = f_X(z)F_Y(z) + f_Y(z)F_X(z) \quad (9.10)$$

For example, consider the case of ideal selection diversity to combat Rayleigh fading. Let both Rayleigh-faded random variables have equal average power α. Their envelope density functions are then:

$$f_{R_1}(r_1) = \frac{r_1}{\left(\frac{\alpha}{2}\right)} \exp(-\frac{r_1^2}{\alpha}), \; r_1 \geq 0$$
$$f_{R_2}(r_2) = \frac{r_2}{\left(\frac{\alpha}{2}\right)} \exp(-\frac{r_2^2}{\alpha}), \; r_2 \geq 0 \quad (9.11)$$

Or in terms of power $s = r^2$, the density functions of interest are exponential:

$$f_{S_1=R_1^2}(s_1) = \frac{1}{\alpha} \exp(-\frac{s_1}{\alpha}), \; s_1 \geq 0$$
$$f_{S_2=R_2^2}(s_2) = \frac{1}{\alpha} \exp(-\frac{s_2}{\alpha}), \; s_2 \geq 0 \quad (9.12)$$

Probability Theory Refresher, Continued

Assuming both signals independent, the joint density function in power is:

$$f_{S_1,S_2}(s_1, s_2) = \frac{1}{\alpha^2} \exp(-\frac{s_1 + s_2}{\alpha}) \tag{9.13}$$

The distribution function resulting from ideal selection diversity is:

$$F_Z(z) = [1 - \exp(-\frac{z}{\alpha})]^2 \tag{9.14}$$

which equals the square of the distribution function for either single branch. Thus for deep fades (i.e., reference levels that are small compared to the average power α), ideal selection diversity improvement is in proportion to the depth of fade. For example, the probability of a diversity output being less than one-tenth the average level is about 10 times less than it would be without diversity.

9.4 Estimation of Distribution Parameters

9.4.1 Rayleigh-Distribution Parameter

An important aspect of many laboratory and field tests is the estimation of average signal power in a signal undergoing Rayleigh fading. This task is discussed in detail by Peritsky [3]. The PDF of such a signal is given by (Section 7.6.6):

$$f(r) = \frac{2r}{\alpha} \exp(-\frac{r^2}{\alpha}) \tag{9.15}$$

where $(\alpha/2)$ represents the average signal power and the envelope r must of course be greater than or equal to zero. Transforming via $s = (r^2)^1$ yields an exponential density function for power:

$$f(s) = \frac{1}{\alpha} \exp(-\frac{s}{\alpha}) \tag{9.16}$$

This density is also a special case of the gamma density function discussed in Section 8.4.1. This is important because an obvious potential estimator of average power is the sample average:

$$Y_1 = \frac{1}{n} \sum_{i=1}^{n} s_i \tag{9.17}$$

based on a set of n independent measurements of instantaneous power. Since the sum of n independent, identically distributed exponential random variables is distributed exactly in accord with the gamma distribution, then:

$$f_W(w) = \frac{1}{\Gamma(n)\alpha^n} w^{n-1} \exp(-\frac{w}{\alpha}) \tag{9.18}$$

---

[1] We will ignore the one-half factor needed for power here and assume it is compensated by a load resistance factor of one-half. Otherwise, one can simply replace α by $(\alpha/2)$ in the analysis.

where $W = nY_1 \geq 0$. The expected value of estimator Y_1 is thus:

$$\begin{aligned}
E(Y_1) &= \frac{1}{n}E(W) \\
&= \frac{1}{n}\frac{1}{\Gamma(n)\alpha^n}\int_0^\infty w^n \exp(-\frac{w}{\alpha})dw \\
&= \alpha
\end{aligned} \quad (9.19)$$

implying the estimator is unbiased.

Because of the large variation of signal strengths caused by Rayleigh fading, it is common to collect data using logarithmic detection. For example, spectrum analyzers are often used in the "zero span" mode to act as receivers. This allows very large frequency ranges to be covered by a single piece of equipment and also provides selectable IF bandwidth. Such instruments commonly provide outputs whose levels are proportional to power in dBm. Consequently, it is useful to consider a second average power estimator:

$$Y_2 = \frac{1}{n}\sum_{i=1}^n c\log_d r_i \quad (9.20)$$

where $c = 20$ and $d = 10$ for results in decibels. The expected value of this estimator is:

$$\begin{aligned}
E(Y_2) &= \frac{c}{n}E[\log_d(\prod_{i=1}^n r_i)] \\
&= cE(\log_d r)
\end{aligned} \quad (9.21)$$

where the sample subscript has been dropped because all samples are identically and independently distributed. With the help of relation A(16) in Reference [4], one obtains:

$$\begin{aligned}
E(\log_d r) &= \int_0^\infty (\log_d r)\frac{r}{\frac{\alpha}{2}} \exp(-\frac{r^2}{\alpha})dr \\
&= \frac{1}{2\ln(d)}[-\gamma + \ln(\alpha)] \\
E(Y_2) &= \frac{c}{2\ln(d)}[-\gamma + \ln(\alpha)]
\end{aligned} \quad (9.22)$$

where $\gamma = 0.5772\ldots$ is Euler's constant. Note that for results in decibels the estimator has an underestimation bias of about 2.5 dB. This is sufficiently large to be significant in many applications. Therefore, it must be accounted for or the samples must be converted back to linear form and the earlier unbiased estimator used.

The variance of the estimator can be found by use of relation 4.335 in Reference [5]:

$$E(Y_2^2) = E\{(\frac{c}{n})^2[\log_d(\prod_{i=1}^n r_i)]^2\}$$

Probability Theory Refresher, Continued

$$
\begin{aligned}
&= (\frac{c}{n})^2\{nE[(\log_d r)^2] + (n^2 - n)E^2(\log_d r)\} \\
E[(\log_d r)^2] &= \int_0^\infty (\log_d r)^2 \frac{r}{\frac{\alpha}{2}} \exp(-\frac{r^2}{\alpha}) dr \\
&= \frac{1}{4\alpha(\ln d)^2} \int_0^\infty (\ln z)^2 \exp(-\frac{z}{\alpha}) dz \\
&= \frac{1}{(\ln d)^2}\{[\ln(\sqrt{\alpha}) - \frac{\gamma}{2}]^2 + \frac{\pi^2}{24}\} \\
\sigma_{Y_2}^2 &= E(Y_2^2) - E^2(Y_2) \\
&= \frac{1}{n}(\frac{c}{\ln d})^2 \frac{\pi^2}{24}
\end{aligned}
\qquad (9.23)
$$

The standard deviation of estimator error in decibels is thus found to be about $(5.6/\sqrt{n})$, independent of the average power α.

9.4.2 Lognormal Distribution Parameters

Szajnowski [6] discusses estimation of parameters needed to specify a lognormal population. The density function can be written as:

$$
f_X(x) = \frac{1}{x\sqrt{2\pi\sigma^2}} \exp[-\frac{\ln^2(\frac{x}{m})}{2\sigma^2}] \qquad (9.24)
$$

where x is the value assumed by the lognormal variable X, m is the median value of X, and σ is the standard deviation of $\ln(X/m)$. This is obtained by transforming a Gaussian density via $y = \ln(x)$. The mean and variance of the lognormal random variable are:

$$
\begin{aligned}
E(X) &= m \exp(\frac{\sigma^2}{2}) \\
Var(X) &= m^2 \exp(\sigma^2)[\exp(\sigma^2) - 1]
\end{aligned}
\qquad (9.25)
$$

respectively. It is more convenient, however, to take the mean-to-median ratio ρ as the second variable, rather than the variance. Szajnowski shows that the *maximum likelihood estimators* (MLEs) for a set of n independent observations are given by:

$$
\begin{aligned}
m_{\mathrm{MLE}} &= (\prod_{i=1}^n x_i)^{\frac{1}{n}} \\
\rho_{\mathrm{MLE}} &= [\prod_{i=1}^n (\frac{x_i}{m_{\mathrm{MLE}}})^{\frac{\ln x_i}{m_{\mathrm{MLE}}}}]^{\frac{1}{2n}}
\end{aligned}
\qquad (9.26)
$$

9.4.3 Maximum Likelihood Estimation in General

Assume that the true relationship between random variables X and Y is described by the linear equation $Y(X) = a_0 + b_0 X$ [7]. Then for some observed value of X, say x_i, one can state the probability of observing y_i, assuming the observations are Gaussian distributed with variance σ_i^2 about the true value $y(x_i)$, as:

$$P_i = \frac{1}{\sqrt{2\pi\sigma_i^2}} \exp\{-\frac{[y_i - y(x_i)]^2}{2\sigma_i^2}\} \qquad (9.27)$$

The probability of observing values $y_1, y_2, \ldots, y_n$ for n independent observations with $x = x_1, x_2, \ldots, x_n$ is thus:

$$\begin{aligned} P &= \prod_{i=1}^{n} P_i \\ &= \prod_{i=1}^{n} \frac{1}{\sqrt{2\pi\sigma_i^2}} \exp\{-\frac{1}{2} \sum_{i=1}^{n} [\frac{y_i - y(x_i)}{\sigma_i}]^2\} \end{aligned} \qquad (9.28)$$

Similarly, for any estimated coefficient values (a, b), the probability of observing $y_1, y_2, \ldots, y_n$ is:

$$\hat{P} = \prod_{i=1}^{n} \frac{1}{\sqrt{2\pi\sigma_i^2}} \exp\{-\frac{1}{2} \sum_{i=1}^{n} [\frac{y_i - (a + bx_i)}{\sigma_i}]^2\} \qquad (9.29)$$

The maximum likelihood estimation procedure is based on the assumption that the observed set of measurements is more likely to have come from the parent population (a_0, b_0) than any other distribution with other coefficient values. Thus, P represents the maximum probability obtainable by $\hat{P}$ and the best estimates (a, b) of (a_o, b_o) are those that maximize $\hat{P}$. This is done by minimizing the sum in the exponential argument.

Define

$$\chi^2 = \sum_{i=1}^{n} \frac{(y_i - a - bx_i)^2}{\sigma_i^2} \qquad (9.30)$$

Differentiating with respect to each parameter and setting the results equal to zero gives:

$$\begin{aligned} \frac{\partial \chi^2}{\partial a} &= 0 \\ &= -2 \sum_{i=1}^{n} \frac{(y_i - a - bx_i)}{\sigma_i^2} \\ \frac{\partial \chi^2}{\partial b} &= 0 \\ &= -2 \sum_{i=1}^{n} x_i \frac{(y_i - a - bx_i)}{\sigma_i^2} \end{aligned} \qquad (9.31)$$

Probability Theory Refresher, Continued   103

The solution for these two linear equations in unknowns a and b is:

$$a = \frac{1}{\delta}\left(\sum_{i=1}^{n}\frac{x_i^2}{\sigma_i^2}\sum_{i=1}^{n}\frac{y_i}{\sigma_i^2} - \sum_{i=1}^{n}\frac{x_i}{\sigma_i^2}\sum_{i=1}^{n}\frac{x_i y_i}{\sigma_i^2}\right)$$

$$b = \frac{1}{\delta}\left(\frac{n}{\sigma_i^2}\sum_{i=1}^{n}\frac{x_i y_i}{\sigma_i^2} - \sum_{i=1}^{n}\frac{x_i}{\sigma_i^2}\sum_{i=1}^{n}\frac{y_i}{\sigma_i^2}\right)$$

$$\delta = \frac{n}{\sigma_i^2}\sum_{i=1}^{n}\frac{x_i^2}{\sigma_i^2} - \left(\sum_{i=1}^{n}\frac{x_i}{\sigma_i^2}\right)^2 \qquad (9.32)$$

The procedure can easily be extended to handle polynomial relations of arbitrary order. It can also be used for certain nonlinear relations that can be transformed into polynomial form. For example, if $y = ab^x$ then taking the natural logarithm of both sides of the relation yields the linear relation $y^* = a^* + b^*x$, where $y^* = \ln(y)$, $a^* = \ln(a)$, and $b^* = \ln(b)$. Finally, a linear fit through the origin can be made by only differentiating with respect to b and setting that result alone equal to zero.

The reader is cautioned that while least square error techniques are mathematically attractive, they are not robust. Measurement errors and outliers in the data strongly influence the fit results. Rousseeuw and Leroy [8] discuss a curve-fitting technique, based on the median square error, that is robust.

9.5 Other Aspects of Rayleigh Fading

9.5.1 Level Crossing Rate

The simple model of land-mobile propagation developed by Clarke and Gans explains the observations of Rayleigh variation in the signal envelope over small distances (Chapter 6). Rayleigh variation means that fades of about 10 dB or more occur with a probability of 10%, fades of 20 dB or more occur with a probability of 1%, etc; or, equivalently, 10% of the time the signal will be faded at least 10 dB, 1% of the time the signal will be faded at least 20 dB, etc. To learn more about the character of such fades (i.e., how often they might be expected and for how long), it is instructive to solve for the level crossing rate associated with the Clarke/Gans model. Level crossing rate is defined precisely as the average rate at which the signal envelope crosses some specified level in an increasing, positive sense:

$$N(R) = \int_0^\infty \dot{r} f(R,\dot{r}) d\dot{r} \qquad (9.33)$$

where $f(R,\dot{r})$ represents the joint PDF for the signal envelope in terms of level $r = R$ and rate of change of level $\dot{r}$. A simple derivation of the preceding relation is given in Reference [9]. By considering not only the in-phase (T_c) and quadrature (T_s) components of a large number of randomly arriving signal components, but their time derivatives as well, Rice [10, 11] gives the joint density function in terms of envelope, envelope derivative, phase, and phase derivative as:

$$f(r,\dot{r},\theta,\dot{\theta}) = \frac{r^2}{4\pi^2 b_0 b_2} \exp[-\frac{1}{2}(\frac{r^2}{b_0} + \frac{\dot{r}^2}{b_2} + \frac{r\dot{\theta}^2}{b_2})] \qquad (9.34)$$

where $\tan(\theta) = -(T_s/T_c)$, b_0 is the mean square of T_c, and T_s, b_2 is the mean square of the time derivatives of T_c and T_s. Integrating phase over $(0, 2\pi)$ and phase derivative over $(-\infty, \infty)$ yields:

$$f(r,\dot{r}) = [\frac{r}{b_0}\exp(-\frac{r^2}{2b_0})][\frac{1}{\sqrt{2\pi b_2}}\exp(-\frac{\dot{r}^2}{2b_2})]$$
$$= f_r(r) f_{\dot{r}}(\dot{r}) \qquad (9.35)$$

The separability of this joint density function into a product of density functions for the constituents proves that r and $\dot{r}$ are independent and uncorrelated. The level crossing rate for the Clarke/Gans model can now be solved as:

$$N(R) = \frac{f_r(R)}{\sqrt{2\pi b_2}} \int_0^\infty \dot{r}\exp(-\frac{\dot{r}^2}{2b_2})d\dot{r}$$
$$= \sqrt{\frac{b_2}{\pi b_0}}\rho\exp(-\rho^2) \qquad (9.36)$$

where ρ is the reference level R divided by the rms envelope level. For the case of the vertical electric field, one obtains:

$$N(R) = \sqrt{2\pi} f_m \rho \exp(-\rho^2) \qquad (9.37)$$

where f_m represents the maximum Doppler frequency; i.e., vehicle speed divided by carrier wavelength. Figure 9.3 shows the level crossing rate normalized by f_m for the vertical electric field component [12].

9.5.2 Average Fade Duration

In addition to the rate at which fades occur, their duration is also of importance. The average fade duration below an arbitrary reference $r = R$ is in general given by $E(\tau) = [\text{Prob}(r \leq R)/N(R)]$. For the Rayleigh density function and vertical electric field, this means:

$$E(\tau) = \sqrt{\frac{\pi b_0}{b_2}}\frac{1}{\rho}\exp(\rho^2 - 1)$$
$$= \frac{\exp(\rho^2 - 1)}{\rho f_m \sqrt{2\pi}} \qquad (9.38)$$

Figure 9.4 is a graph of average fade duration normalized by f_m for the vertical electric field component [12].

Specification of complete interfade statistics is not mathematically tractable. However, simulation results indicate that deep fades are Poisson distributed [13].

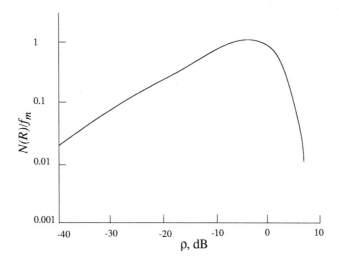

Figure 9.3 Normalized level crossing rate of the vertical electric field component, Rayleigh-fading model.
After: Jakes, Jr., W. C. (editor), *Microwave Mobile Communications*, John Wiley & Sons, New York, NY, 1974, p. 35.

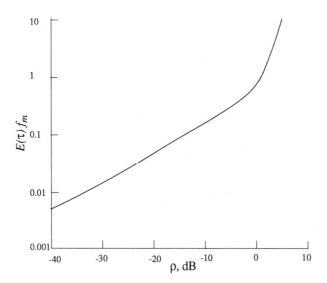

Figure 9.4 Normalized average fade duration level crossing rate of the vertical electric field component, Rayleigh-fading model.
After: Jakes, Jr., W. C. (editor), *ibid*, p. 37.

9.5.3 Test of Distribution

In Chapter 6, field test data for the small-scale power variation of land-mobile links was shown to plot reasonably straightline on a probability scale based on Rayleigh variation. Hence, the conclusion that such variation is Rayleigh. But just how closely must measured data follow some assumed probability distribution to be of significance? This question has been addressed at length for Rayleigh fields by CCIR Technical Committee 12, Radio Communications, Subcommittee 12F, Equipment Used in the Mobile Services, Working Group 1. The working group secretary, C. M. Willyard, ran extensive simulations of Rayleigh fields.

One test of the hypothesis that a particular sample of the field is indeed drawn from a Rayleigh population, as opposed to some other distribution, is based on the chi-square statistic:

$$\chi^2 = \sum_{i=1}^{k} \frac{(f_i - e_i)^2}{e_i} \tag{9.39}$$

where f_i and e_i are, respectively, the observed and expected frequencies of occurrence for events in the k arbitrary intervals. The sampling distribution of this statistic is approximately chi-square with $k - m$ degrees of freedom, where m is the number of quantities, obtained from the observations, used in calculating the expected frequencies (for example, see Table A.5 in Reference [1]). Table 9.1 shows a sample application of the chi-square test to 360 samples equally spaced around a circle of 5 wavelength radius. In the table, an arbitrary choice of 30 intervals has been made. Further, they have been specified such that the expected number of samples to fall in each interval is always 12 (this requires use of the total number of samples, 360; hence, $m = 1$ and the degrees of freedom is 29). The experimental chi-square value of 14.8333 falls well below the value that would be produced by random data 10% of the time. Thus the true distribution for the data can indeed be accepted as Rayleigh.

The *Kolmogorov-Smirnov* (KS) test is a nonparametric way of testing the distribution. The one-sample version concerns agreement between an observed cumulative distribution of samples and some assumed population distribution. It is based simply on the maximum difference between the two distributions. This is generally far easier to compute than the chi-square. Furthermore, the KS test is generally more efficient than the chi-square test for small samples and has no minimum expected frequency requirement for the intervals. Although strictly speaking the KS test applies only to continuous random variables, it can be applied to discrete random variables with only slight error. Table 9.2 shows the relation between sample size, level of significance, and maximum difference between observed and assumed cumulative distribution functions. For more detail see Reference [14] and the KSONE subroutine in the special function library of the commercial software offered by IMSL. Figure 9.5 indicates the region within which the sample cumulative distribution should lie if one is to accept the sampled field as being Rayleigh at a significance level of 10%.

Table 9.1 Chi-Square Test for Rayleigh Field Sample

Cell size in dB[a]	Expected Nr of samples	Actual Nr of samples	Chi-square
100 to 6.36542	12	14	0.33333
6.36542 to 5.37567	12	9	0.75
5.37567 to 4.67126	12	11	0.08333
4.09164 to 3.5819	12	12	0
3.5819 to 3.11584	12	11	0.08333
3.11584 to 2.67859	12	17	2.08333
2.67859 to 2.26061	12	14	0.33333
2.26061 to 1.85527	12	8	1.33333
1.85527 to 1.45755	12	13	0.08333
1.45755 to 1.06342	12	11	0.08333
1.06342 to 0.669434	12	15	0.75
0.669434 to 0.272452	12	16	1.33333
0.272452 to -0.130552	12	13	0.08333
-0.130552 to -.542645	12	14	0.33333
-0.542645 to -.967095	12	9	0.75
-0.967095 to -1.40754	12	14	0.33333
-1.40754 to -1.86817	12	8	1.33333
-1.86817 to -2.35404	12	17	2.08333
-2.35404 to -2.87137	12	10	0.33333
-2.87137 to -3.42817	12	13	0.08333
-3.42817 to -4.03511	12	11	0.08333
-4.03511 to -4.70693	12	9	0.75
-4.70693 to -5.46506	12	13	0.08333
-5.46506 to -6.34252	12	12	0
-6.34252 to -7.39448	12	10	0.33333
-7.39448 to -8.72413	12	12	0
-8.72413 to -10.5629	12	11	0.08333
-10.5629 to -13.6487	12	11	0.083333
-13.6487 to -100	12	13	0.083333

a. Number of samples=360, number of cells=30, radius in wavelengths=5, chi-square value=14.8333, critical value=37.9 for significance level of 10%

Table 9.2 Critical Values for the KS One-Sample Test

Sample size[a]	Level of significance		
	10%	5%	1%
1	0.950	0.975	0.995
2	0.776	0.842	0.929
3	0.642	0.708	0.828
4	0.564	0.624	0.733
5	0.510	0.565	0.669
10	0.368	0.410	0.490
15	0.304	0.338	0.404
20	0.264	0.294	0.356
30	0.220	0.240	0.290

a. For sample sizes greater than 30 approximate the 10% significance value by 1.22 divided by the square root of the number of samples (use 1.36 for 5% and 1.63 for 1%, respectively)

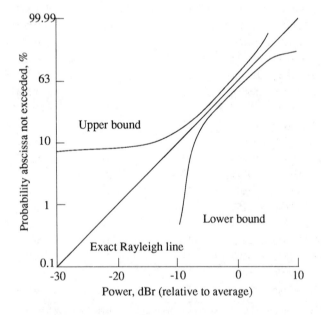

Figure 9.5 KS test bounds for sample of size 300 taken from a presumed Rayleigh field.

A third test of distribution has been proposed by Gotoh and Kubota of the CCIR working group cited earlier in this section. It is based on the idea that counting experiments result in binomial distributions. That is, for any level on the cumulative distribution curve, say p, the number of samples will follow the binomial distribution with mean Np and variance $Np(1-p)$, where N is the total number of samples taken. When the number of samples is large, the binomial form becomes approximately Gaussian. Hence, the actual number of samples should only fall outside the range $[p - 1.65\sqrt{Np(1-p)}, p + 1.65\sqrt{Np(1-p)}]$ 10% of the time. Figure 9.6 indicates the region within which the sample cumulative distribution should lie if one is to accept the sampled field as being Rayleigh at a significance level of 10%. The Gotoh bounds are generally somewhat more stringent than the KS bounds.

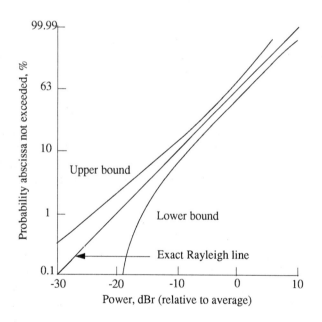

Figure 9.6 Gotoh test bounds for sample of size 360 taken from a presumed Rayleigh field.

9.6 Homework Problems

9.6.1 Problem 1

Traffic load is not generally uniformly distributed over metropolitan areas. This is because user density tends to decrease as the distance from the city center increases. An exponential model is often used to characterize this situation [15]; i.e., $p(r) = d_0 \exp(-\zeta r)$, where $p(r)$ equals the user density per unit area at distance r from the system center, d_0 equals the user density per unit area at the system center, and ζ equals a user density rolloff with distance factor. With such an assumption, what is the PDF for the distribution of users with range?

Solution: The total number of users within some radius r of the system center is obtained by integrating the product of $p(r)$ and $2\pi r\,dr$ (the differential area at radius r) from zero to r. The result is:

$$\begin{aligned} N_{u,r} &= 2\pi d_0 \int_0^r x \exp(-\zeta x)\,dx \\ &= \frac{2\pi d_0}{\zeta^2}[1 - \exp(-\zeta r)(1+\zeta r)] \end{aligned} \quad (9.40)$$

Using this result with $r = R_{\max}$ (the radius corresponding to the total coverage area) allows the constant d_0 to be evaluated if the total number of users is known or vice versa. The fraction of users within some arbitrary radius $r = R$ is given by:

$$F_{R\,|\,\zeta, R_{\max}} = \frac{1 - \exp(-\zeta R)(1+\zeta R)}{1 - \exp(-\zeta R_{\max})(1+\zeta R_{\max})} \quad (9.41)$$

This simply represents the probability that a user is active at a range of at most R. Measurements on a large 800-MHz trunked system in Chicago, Illinois agree well with this equation using $R_{\max} = 70$ mi and $\zeta = 0.13$. Differentiating this CDF function with respect to $r = R$ yields the density function sought:

$$f(r) = \frac{\zeta^2 r \exp(-\zeta r)}{1 - \exp(-\zeta R_{\max})(1+\zeta R_{\max})} \quad (9.42)$$

This has the form of a Gamma distribution with parameters 2 and ζ^{-1} (becoming exactly that as $R_{\max}$ approaches infinity). That distribution is also known as the Erlang distribution and represents the sum of two independent exponentially distributed random variables with parameter ζ^{-1}. Consequently, Monte Carlo simulation draws from this population can be made by summing a pair of exponentially distributed draws. Such draws can be made by taking the natural logarithm of numbers drawn from a uniform distribution over (0,1), then multiplying by $-\zeta^{-1}$ [16].

The denominator of the density function is generally near unity since $R_{\max}$ is usually large, especially compared to ζ. Differentiating the numerator with respect to range and setting the result to zero shows that the most likely range equals the reciprocal of ζ.

Probability Theory Refresher, Continued 111

9.6.2 Problem 2

Message lengths on trunked land-mobile radio systems are approximately exponentially distributed [17]. What is the 90% confidence interval for the average message length based on independent measurement of message length for 100 messages?
    Solution: Consider a sequence of samples of message lengths

$$T = (t_1, t_2, \ldots, t_n)$$

where all t_i are *independent, identically distributed* (IID) observations from an exponential population with average message length $h = (1/\lambda)$. The joint PDF of T can thus be written as:

$$\begin{aligned} f(T) &= \prod_{i=1}^{n} \lambda \exp(-\lambda t_i) \\ &= \lambda^n \exp(-\lambda \sum_{i=1}^{n} t_i) \end{aligned} \quad (9.43)$$

The MLE of λ is found by differentiating with respect to λ and setting the result to zero. For mathematical convenience one should take the natural logarithm first, then differentiate. The result obtained is $\hat{\lambda} = (n/\sum_{i=1}^{n} t_i)$, or equivalently $\hat{h} = (\sum_{i=1}^{n} t_i/n)$, where the "hat" notation denotes an estimate of the true population parameter value.
    To determine the precision of the estimate as a function of the number of samples n, consider the random variable $S = \sum_{i=1}^{n} t_i$. From past analysis it is known that S will be gamma distributed because it represents the sum of IID exponential random variables. Thus, S has the density function:

$$f_S(s) = \frac{1}{\Gamma(n)(\frac{1}{\lambda})^n} s^{n-1} \exp(-\lambda s) \quad (9.44)$$

where $s \geq 0$. From this one can show the expected value of the average message length estimate is $E(\hat{h}) = [E(s)/n] = h$ and the variance of the standard deviation of the estimate is $(h/\sqrt{n})$.
    Confidence intervals for h can be established by noting that the CDF for S can be written as:

$$\begin{aligned} F_S(s) &= \int_0^s f_S(x) dx \\ &= \frac{1}{2^{\frac{\nu}{2}} \Gamma(\frac{\nu}{2})} \int_0^{\mu} y^{\frac{\nu-2}{2}} \exp(-\frac{y}{2}) dy \end{aligned} \quad (9.45)$$

where $\nu = 2n$, $\mu = 2\lambda s$, and $y = 2\lambda x$. But this is the well-tabulated chi-square distribution with $\nu = 2n$ degrees of freedom (for example, see Table A.5 in Reference [1]). A two-sided confidence interval for the estimate of h can be found by noting the probability that the estimate falls below the lower bound with probability ($\alpha/2$) and exceeds the upper bound with that same probability; i.e., the estimate lies within those bounds with probability $1-\alpha$. The bounds are given by:

$$\text{Prob}\left\{\frac{\nu \hat{h}}{\chi^2_{\frac{\alpha}{2}}} < h < \frac{\nu \hat{h}}{\chi^2_{1-\frac{\alpha}{2}}}\right\} = 1-\alpha \quad (9.46)$$

When the degrees of freedom is large, the chi-square value can be approximated as:

$$\chi^2_\gamma \approx \frac{1}{2}[\sqrt{2\nu-1}+\mu_\gamma]^2 \quad (9.47)$$

where μ_γ is the γ percent upper point of the cumulative standard normal probability integral (1.65 for 5%, 1.28 for 10%, -1.28 for 90%, -1.65 for 95%, etc.). Thus, for $n=100$ and $\alpha=5\%$ one obtains the confidence interval $0.86\hat{h} < h < 1.19\hat{h}$.

9.6.3 Problem 3

As noted in Section 9.4.1, estimation of average signal power levels is an important part of many laboratory and field tests. To cope with the wide dynamic range associated with Rayleigh fading, logarithmic detection is often used. Care must be taken during analysis because the average signal power expressed in decibels (i.e., true average signal power) is not the same as the average of the logarithmic power level in decibels; nor is it the same as the average signal envelope expressed in decibels. Compare the average power predictions made using these three quantities, given that the signal is both Rayleigh faded and lognormally shadowed.

Solution: The PDF of the signal envelope for the signal in question can be written as [18] (this is considered further in Section 10.3):

$$f_S(s) = \sqrt{\frac{\pi}{8\sigma^2}} \int_{-\infty}^{\infty} \frac{s}{10^{\frac{\overline{s_d}}{10}}} \exp\left(\frac{-\pi s^2}{4\cdot 10^{\frac{\overline{s_d}}{10}}}\right) \exp\left[-\frac{(\overline{s_d}-m_d)^2}{2\sigma^2}\right] d\overline{s_d} \quad (9.48)$$

where the density function is defined over the range $[0,\infty]$ and $\overline{s_d} = 20\log(\overline{s})$, with $\overline{s}$ equaling the local mean envelope value, characterized by a Gaussian variation (in decibels) with median and mean equal to m_d and standard deviation equal to σ. The expected value for the signal envelope can be found by solving:

$$E(s) = \int_0^\infty s f(s) ds \quad (9.49)$$

A closed-form expression is obtainable by reversing the order of integration and using relation 7.4.4 in Reference [2] and relation 5.(A2) in Reference [4]. The final result, published by Muammar and Gupta [19], is:

$$E(s) = \exp\left(\frac{m_d}{2}k + \frac{\sigma^2 k^2}{8}\right) \quad (9.50)$$

Probability Theory Refresher, Continued

where $k = [\ln(10)/10]$. Relation 7.4.5 in Reference [2] also allows one to solve for the expected value of the envelope squared:

$$E(s^2) = \frac{4}{\pi} \exp(m_d k + \frac{\sigma^2 k^2}{2}) \tag{9.51}$$

One other needed quantity is the expected value of the logarithm of the envelope.

$$E[\ln(s)] = \int_0^\infty \ln(s) f(s) ds \tag{9.52}$$

Substitution of $q = [(\pi/4) \, 10^{(\overline{s_d}/10)}]s^2$ and use of relation 4.331 #1 in Reference [5] allow the integration in s to be carried out first. The remaining integration in $\overline{s_d}$ can then be completed using relations 7.1.22 of Reference [2] and 5.(A4) of Reference [4]. The result is:

$$E[\ln(s)] = \frac{k}{2} m_d - \frac{1}{2}[\gamma + \ln(\frac{\pi}{4})] \tag{9.53}$$

where γ is Euler's constant (0.577...) and dependence on the lognormal-shadowing variability is absent.

Assuming unity load resistance, three candidate expressions for average signal power in decibels are:

$$\begin{aligned} P1 &= 10 \log[\frac{E(s^2)}{2}] &= m_d - \frac{1}{k} \ln(\frac{\pi}{2}) + \frac{k}{2}\sigma^2 \\ P2 &= 20 E[\log(s)] &= m_d - \frac{1}{k}[\gamma + \ln(\frac{\pi}{4})] \\ P3 &= 20 \log[E(s)] &= m_d + \frac{k}{4}\sigma^2 \end{aligned} \tag{9.54}$$

Table 9.3 shows how the results can differ substantially from the correct $P1$ results, depending on σ (values between 6 and 12 dB are typical). Note that when σ approaches zero, the correct value for average power is not simply m_d because that quantity was defined in terms of the logarithm of the mean envelope. Section 9.4.1 has shown that this results in a bias.

Table 9.3 Comparison of Three Average Signal Power Expressions

	dB offset from median		
Form	sigma=0 dB	sigma=6 dB	sigma=12 dB
P1	-2.05	2.19	14.6
P2	-1.46	-1.46	-1.46
P3	0.0	2.07	8.29

9.6.4 Problem 4

Section 9.4.1 addresses the number of independent samples required to pin down the average power of a Rayleigh-faded signal to some desired standard deviation of error. In practice, adjacent samples in neither time nor space are independent, so a practical question still remains: over what time period or over what distance should samples be taken to obtain credible estimates of the local average power of a Rayleigh-faded signal?

Solution: A careful balance between too long a time (distance)² and too short a time (distance) is required. The former causes errors due to nonstationarity of the local average as it allows the larger scale lognormal-shadowing variations to impact the result; the latter causes errors due to statistical fluctuations of the estimate.

We begin by considering a general stationary random process, say $\zeta(z)$. The time and spatial averages of the process can be written, respectively, as:

$$m_T = \frac{1}{T} \int_{t_0}^{t_0+T} \zeta(z) dz$$

$$m_L = \frac{1}{2L} \int_{x_0-L}^{x_0+L} \zeta(z) dz \tag{9.55}$$

where t_0 and x_0 are arbitrary initial time and location coordinates and the arbitrary averaging scales are T and $2L$. When the limits exist, both averages converge to the true process average μ. However, with finite averaging both results are Gaussian distributed about μ with variance [20]:

$$\sigma_T^2 = \frac{2}{T} \int_0^T (1 - \frac{\tau}{T})[R_\zeta(\tau) - \mu^2] d\tau$$

$$\sigma_L^2 = \frac{1}{L} \int_0^{2L} (1 - \frac{\delta}{2L})[R_\zeta(\delta) - \mu^2] d\delta \tag{9.56}$$

where $R_\zeta(\tau) = E[\zeta(t+\tau)\zeta(t)]$ and $R_\zeta(\delta) = E[\zeta(x+\delta)\zeta(x)]$ are the autocorrelation functions for the process in time and location, respectively.

Hata and Nagatsu [21] have shown that for any function of the signal envelope, say $g(r)$, the autocorrelation function can be expressed via an orthogonal expansion in terms of the Rayleigh envelope PDF, $f(r)$, as:

$$R_{g(r)}(\rho) = \int_0^\infty \int_0^\infty g(r_1) g(r_2) f(r_1, r_2) dr_1 dr_2$$

$$= \int_0^\infty \int_0^\infty g(r_1) g(r_2) f(r_1) f(r_2) \sum_0^\infty L_m(\frac{r_1^2}{2\sigma^2}) L_m(\frac{r_2^2}{2\sigma^2}) \rho^{2m}$$

$$= \sum_0^\infty \alpha_m^2 \rho^{2m} \tag{9.57}$$

---

[2]Note that time and space (location) are related through "distance equals the product of speed and time." In place of speed we may substitute the product of maximum Doppler frequency and wavelength.

where $\alpha_m = \int_0^\infty g(r)(r/\sigma^2)\exp(-r^2/2\sigma^2)L_m(r^2/2\sigma^2)dr$, σ^2 represents the average power of the Rayleigh-faded signal, ρ^2 equals the normalized autocorrelation function of a narrowband Gaussian process[3], and L_m is the zero-order Laguerre function used to expand the joint PDF $f(r_1, r_2)$[4].

Let us use this to examine true power; i.e., $g(r) = (r^2/2)$ and $\mu = \sigma^2$. We find:

$$\begin{aligned}\alpha_m &= \int_0^\infty \frac{r^2}{2}\frac{r}{\sigma^2}\exp(-\frac{r^2}{2\sigma^2})L_m(\frac{r^2}{2\sigma^2})dr \\ &= \int_0^\infty \sigma^2 q \exp(-q)L_m(q)dq \\ &= \sigma^2 \frac{\Gamma(m-1)}{m!\Gamma(-1)}\end{aligned} \quad (9.58)$$

where $q = (r^2/2\sigma^2)$ and relation 7.414 #11 in Reference [5] has been used. Only α_0 and α_1 have nonzero values, both equal to σ^2; hence, the autocorrelation function is simply $R_{(r^2/2)}(\rho) = \sigma^4(1+\rho^2)$. The variance of the estimate of power averaged over T seconds is $\sigma_T^2 = 2\sigma^4 INT(\tau)$, where $INT(\tau) = (1/T)\int_0^T [1-(\tau/T)]\rho^2 d\tau$. Similarly, the variance of the estimate of power averaged over a distance of $2L$ is $\sigma_L^2 = 2\sigma^4 INT(\delta)$, where $INT(\delta) = (1/2L)\int_0^{2L}[1-(\delta/2L)]\rho^2 d\delta$.

Another case of interest is the envelope directly; i.e., $g(r) = r$ and $\mu = \sqrt{(\pi/2)}\sigma$. This case is discussed by Lee [22]. Here we have:

$$\begin{aligned}\alpha_m &= \int_0^\infty \frac{r^2}{\sigma^2}\exp(-\frac{r^2}{2\sigma^2})L_m(\frac{r^2}{2\sigma^2})dr \\ &= \int_0^\infty \sqrt{2}\sigma q^{\frac{1}{2}}\exp(-q)L_m(q)dq \\ &= \frac{\sigma}{2\sqrt{2}}\frac{\Gamma(m-\frac{1}{2})}{m!}\end{aligned} \quad (9.59)$$

so all terms in the autocorrelation function are nonzero. Nonetheless, because the magnitude of ρ is at most unity, the series converges rapidly and the function can be reasonably approximated by just the first two terms; i.e., $R_r(\rho) \approx (\pi\sigma^2/2)[1+(\rho^2/4)]$. The variance of the estimate of average envelope averaged over T seconds is thus approximately $\sigma_T^2 \approx (\pi\sigma^2/4)INT(\tau)$ and averaged over the distance $2L$ is approximately $\sigma_L^2 \approx (\pi\sigma^2/4)INT(\delta)$.

---

[3] ρ is itself a function of either time shift τ or location shift δ. Section 1.3 of Reference [12] develops the expressions $\rho = J_0(\omega_d \tau) = J_0(\beta\delta)$, where ω_d equals the maximum radian Doppler frequency and $\beta = 2\pi$ divided by the carrier frequency wavelength.

[4] This involves a zero-order modified Bessel function of the first kind [12]. It can be obtained by integrating over phase the more general joint density function involving both envelope and phase of two signals separated in time (or space). The development of that function is along the lines of the analysis in problem 2, Section 15.7.2.

A third case of interest is $g(r) = 20\log(r) = 2K\ln(r)$, where $K = [10/\ln(10)]$, the reciprocal of k defined for problem 3. Here:

$$\alpha_m = \int_0^\infty 2K\ln(r)\frac{r}{\sigma^2}\exp(-\frac{r^2}{2\sigma^2})L_m(\frac{r^2}{2\sigma^2})dr \qquad (9.60)$$

For $m = 0$ we have (Section 9.4.1):

$$\begin{aligned}\alpha_0 &= 2K\int_0^\infty \ln(r)\frac{r}{\sigma^2}\exp(-\frac{r^2}{2\sigma^2})L_m(\frac{r^2}{2\sigma^2})dr \\ &= K[-\gamma + \ln(2\sigma^2)]\end{aligned} \qquad (9.61)$$

For $m = 1$ we have:

$$\begin{aligned}\alpha_1 &= 2K\int_0^\infty \ln(r)\frac{r}{\sigma^2}\exp(-\frac{r^2}{2\sigma^2})L_1(\frac{r^2}{2\sigma^2})dr \\ &= 2K\int_0^\infty \ln(r)\frac{r}{\sigma^2}\exp(-\frac{r^2}{2\sigma^2})[1-(\frac{r^2}{2\sigma^2})]dr \\ &= 2K(I_1 + I_2)\end{aligned} \qquad (9.62)$$

where relation 8.973 #5 of Reference [5] has been used to express the Laguerre polynomial in terms of its argument. The needed integrations are:

$$\begin{aligned}I_1 &= \int_0^\infty \ln(\sqrt{q})\exp(-q)dq \\ &= -\frac{\gamma}{2} \\ I_2 &= -\frac{1}{2}\int_0^\infty q\ln(q)\exp(-q)dq \\ &= -\frac{1}{2}(1-\gamma)\end{aligned} \qquad (9.63)$$

where relations 4.331 #1 and 4.352 #1 and 2 in Reference [5] have been used. The end result is $\alpha_1 = -K$.

Similarly, one can show $\alpha_2 = -(K/2)$, $\alpha_3 = -(K/3)$, etc. Thus, the autocorrelation function for this third case can be expressed as $R_{2K\ln(r)}(\rho) = K^2[-\gamma+\ln(2\sigma^2)]^2 + K^2\rho^2[1 + (\rho^2/4) + \cdots] \approx K^2[-\gamma + \ln(2\sigma^2)]^2 + K^2\rho^2$. Therefore, the variance of the estimate of $20\log(r)$ averaged over T seconds is approximately $\sigma_T^2 \approx 2K^2 INT(\tau)$ and averaged over the distance $2L$ is approximately $\sigma_L^2 \approx 2K^2 INT(\delta)$.

The so-called one-sigma spread is useful for comparing the three cases just analyzed. This quantity is defined as the decibel value of the quantity average plus the standard deviation of estimation error divided by the quantity average minus the standard deviation of estimation error. Thus, for the three situations examined:

$$\text{Spread}_{1,1} = 10\log(\frac{\sigma^2 + \sqrt{2\sigma^4 INT}}{\sigma^2 - \sqrt{2\sigma^4 INT}})$$

… Probability Theory Refresher, Continued

$$= 10\log(\frac{1+\sqrt{2INT}}{1-\sqrt{2INT}})$$

$$\text{Spread}_{1,2} = 20\log(\frac{\sqrt{\frac{\pi\sigma^2}{2}}+\sqrt{\frac{\pi\sigma^2 INT}{4}}}{\sqrt{\frac{\pi\sigma^2}{2}}-\sqrt{\frac{\pi\sigma^2 INT}{4}}})$$

$$= 20\log(\frac{1+\sqrt{(INT/2)}}{1-\sqrt{(INT/2)}})$$

$$\text{Spread}_{1,3} = \{K[-\gamma+\ln(2\sigma^2)]+\sqrt{2K^2 INT}\}$$
$$-\{K[-\gamma+\ln(2\sigma^2)]-\sqrt{2K^2 INT}\}$$
$$= 2K\sqrt{2INT} \tag{9.64}$$

where $INT = INT(\tau) = INT(\delta)$. Interestingly, all three cases yield one-sigma spreads of approximately $2K\sqrt{2\,INT}$ when INT is small compared to unity and that is the practical case of interest. For example, numerical solutions of the integral show that averaging over $\beta\delta = 20$ (3.18 wavelengths) yields one-sigma spreads of 3.53, 3.38, and 3.34 dB for situations 1, 2, and 3, respectively. Increasing the integration length (or time, with that distance traversed at constant speed) threefold decreases the spreads to 2.19, 2.16, and 2.15 dB, respectively. Lee [22] suggests averaging over distances of 20 to 40 wavelengths to achieve both good statistical accuracy and minimal risk of lognormal variation contamination.

The preceding analysis answers the question of how long (in distance or time) to average to achieve good estimates of local average power in a Rayleigh field. However, the question of how many samples to take over that distance or time has been skirted because the analysis assumes analog (continuous) averaging. If instead we take N uniformly spaced samples and average them, an additional statistical fluctuation occurs. For large N the central limit theorm applies and the sample average will have a standard deviation equal to the standard deviation of a single sample divided by the square root of N. Section 10.3.1 shows that the standard deviation of the power of a Rayleigh-faded signal expressed in decibels is 5.57 dB. Thus, the standard deviation of an N sample average power estimate is $(5.57/\sqrt{N})$. This cannot be made arbitrarily small because in practice samples must be separated in space by roughly 0.8 wavelengths to be uncorrelated [22]. Averaging over 20 (40) wavelengths therefore limits the effective number of samples to about 25 (50). Of course, a somewhat larger number might still be desirable to filter away noise effects under low signal-to-noise conditions.

References

[1] Walpole, R. E. and R. H. Myers, *Probability and Statistics for Engineers and Scientists*, New York, NY, Macmillan Publishing Company, 1985.

[2] Abramowitz, M. and I. Stegun (editors), *Handbook of Mathematical Functions*, New York, NY, Dover Publications, Inc., 1972.

[3] Peritsky, M. M., "Statistical Estimation of Mean Signal Strength in a Rayleigh-Fading Environment," *IEEE Tr. on Veh. Tech.*, Vol. VT-22, No. 4, November 1973, pp. 123-130.

[4] Ng, E. and M. Geller, "A Table of Integrals of the Error Function," *J. of Res. of the NBS*, B. Mathematical Sciences, Vol. 73B, No. 1, January-March 1969, pp. 1-20.

[5] Gradshteyn, I. S. and I. M. Ryzhik, *Table of Integrals, Series, and Products*, New York, NY, Academic Press, 1965.

[6] Szajnowski, W. J., "Estimators of Log-Normal Distribution Parameters," *IEEE Tr. on Aero. and Elec. Sys.*, Vol. AES-13, No. 5, September 1977, pp. 533-536.

[7] Bevington, P. R., *Data Reduction and Error Analysis for the Physical Sciences*, New York, NY, McGraw-Hill Book Co., 1969.

[8] Rousseeuw, P. J. and A. M. Leroy, *Robust Regression and Outlier Detection*, New York, NY, John Wiley & Sons, 1987.

[9] Lee, W.C.Y., "Statistical Analysis of the Level Crossings and Duration of Fades of the Signal from an Energy Density Mobile Radio Antenna," *Bell System Technical Journal*, Vol. 46, February 1967, pp. 417-448.

[10] Rice, S. O., "Mathematical Analysis of Random Noise," *Bell System Technical Journal*, Vol. 23, July 1944, pp. 282-332, and Vol. 24, January 1945, pp. 46-156.

[11] Rice, S. O., "Statistical Properties of a Sine Wave Plus Random Noise," *Bell System Technical Journal*, Vol. 27, January 1948, pp. 109-157.

[12] Jakes, Jr., W. C. (editor), *Microwave Mobile Communications*, New York, NY, John Wiley & Sons, 1974.

[13] Arnold, H. W. and W. F. Bodtmann, "Interfade Interval Statistics of a Rayleigh-Distributed Wave," *IEEE Tr. on Comm.*, Vol. COM-31, No. 9, September 1983, pp. 1114-1116.

[14] McCuen, R. H., *Statistical Methods for Engineers*, Englewood Cliffs, NJ, Prentice-Hall, Inc., 1985.

[15] Sakamoto, M. and M. Hata, "Efficient Frequency Utilization Techniques for High-Capacity Land Mobile Communications System," *Rev. of Elec. Comm. Labs*, Vol. 35, No. 2, March 1987, pp. 89-94.

[16] Rubinstein, R. Y., *Simulation and the Monte Carlo Method*, New York, NY, John Wiley & Sons, 1981.

[17] Hess, G. and J. Cohn, "Communication Load and Delay in Mobile Trunked Systems," *31st IEEE Vehicular Technology Conference*, April 1981, Washington, D.C., pp. 269-273.

[18] French, R., "The Effect of Fading and Shadowing on Channel Reuse in Mobile Radio," *IEEE Tr. on Veh. Tech.*, Vol. VT-28, No. 3, August 1979, pp. 171-181.

[19] Muammar, R. and S. C. Gupta, "Co-channel Interference in High Capacity Mobile Radio Systems," *IEEE Tr. on Comm.*, Vol. COM-30, No. 8, August 1982, pp. 1973-1978.

[20] Papoulis, A., *Probability, Random Variables, and Stochastic Processes*, New York, NY, McGraw-Hill Book Co., 1965.

[21] Hata, M. and T. Nagatsu, "Mobile Location in a Cellular System," *IEEE Tr. on Veh. Tech.*, Vol. VT-29, No. 2, May 1980, pp. 245-252.

[22] Lee, W.C.Y., "Estimate of Local Average Power of a Mobile Radio Signal," *IEEE Tr. on Veh. Tech.*, Vol. VT-34, No. 1, February 1985, pp. 22-27.

Chapter 10
Coverage Analysis and Simulation

Coverage prediction is probably by far the dominant activity of land mobile radio system design. The length of this chapter reflects that. The chapter begins by quantifying coverage versus thermal noise analytically under the assumption of power-law propagation and lognormal shadowing. This stimulates a discussion of how to include the effect of Rayleigh fading and an approximate solution is generated that agrees well with Monte Carlo simulation results. Section 10.4 then describes how information on certain signal characteristics, needed by the sections that follow, can be experimentally gathered.

Coverage is addressed not only in terms of the desired signal against thermal noise (Sections 10.1, 10.2, and 10.3), but also against adjacent and offset channel interference (Sections 10.5, 10.6, and 10.8.2), against cochannel interference (Sections 10.7, 10.8.1, 10.9, and 10.10 (which introduces the cellular concept)), and against intermodulation interference (Section 10.11). Several of these sections pertain directly to matters recently considered by the FCC.

10.1 Area Coverage with Power-Law Propagation and Lognormal Shadowing

Under the assumptions of power-law propagation and lognormal shadowing, it is possible to develop an analytical expression for the percentage of locations within some radius R of a base station for which the received signal strength exceeds some threshold value. Section 2.5.3 in Reference [1] solves this problem; unfortunately, numerous typographical errors are present.

Defining s_0 as the reference signal level in decibels that must be exceeded and P_{s_0} as the probability that the signal received in some differential area does indeed exceed the reference, the area coverage probability equates to:

$$\text{ACP} = \frac{1}{\pi R^2} \int P_{s_0} dx dy$$

$$= \frac{1}{\pi R^2} \int P_{s_0} r\, dr\, d\theta \qquad (10.1)$$

The power-law assumption implies that the mean signal strength at range r is proportional to that range raised to some negative power, such as:

$$E(s) = \alpha - 10n \log\left(\frac{r}{R}\right) \qquad (10.2)$$

where α accounts for the transmitter ERP, receive antenna gain, feedline losses, etc. and, for convenience, the range at which the mean signal strength equals α has been set to R. The lognormal-shadowing assumption implies that the received power is not constant for fixed range, but rather is normally distributed (in decibels) with mean value $E(s)$ and some standard deviation, say σ. Consequently (refer to Section 7.6.2):

$$P_{s_0} = \frac{1}{2} - \frac{1}{2}\text{erf}\left[\frac{s_0 - \alpha + 10n \log(\frac{r}{R})}{\sigma\sqrt{2}}\right] \qquad (10.3)$$

Substituting $a = (s_0 - \alpha)/(\sigma\sqrt{2})$ and $b = [10n \log(e)]/(\sigma\sqrt{2})$ then leads to:

$$\text{ACP} = \frac{1}{2} - \frac{1}{R^2}\int_0^R r\, \text{erf}\left[a + b \ln\left(\frac{r}{R}\right)\right] dr \qquad (10.4)$$

A further substitution of $t = a + b\ln(r/R)$ gives:

$$\text{ACP} = \frac{1}{2} - \frac{1}{b}\exp\left(-\frac{2a}{b}\right)\int_{-\infty}^a \exp\left(\frac{2t}{b}\right)\text{erf}(t)\, dt \qquad (10.5)$$

The required integration can be carried out via integration by parts using $u = \text{erf}(t)$ and $dv = \exp(2t/b)dt$, which imply $du = (2/\sqrt{\pi})\exp(-t^2)dt$ and $v = (b/2)\exp(2t/b)$, respectively. The answer obtained is:

$$\text{ACP} = \frac{1}{2}\left\{1 - \text{erf}(a) + \exp\left(\frac{1-2ab}{b^2}\right)\left[1 - \text{erf}\left(\frac{1-ab}{b}\right)\right]\right\} \qquad (10.6)$$

This equation indicates how contour coverage probability, P_{s_0} when $r = R$, relates to area coverage probability. Figure 10.1 [1] shows the relation in terms of the parameter (σ/n).

10.2 Impact of Rayleigh Fading on Coverage

The simple model in the preceding section did not account for Rayleigh fading, a common circumstance of land-mobile radio links. Subjective voice quality measurements of standard 25-kHz analog FM radios typically indicate that sensitivity in fading drops about 10 dB relative to static conditions. Measurements of digital voice systems tend to show similar sensitivity decreases between faded and nonfaded circumstances. The 10-dB number is hardly sacred. It can be decreased in analog FM

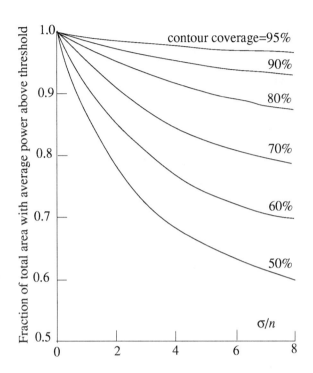

Figure 10.1 Fraction of total area with average power above threshold. After: Jakes, Jr., W. C. (editor), *Microwave Mobile Communications*, John Wiley & Sons, New York, NY, 1974, p. 127.

radios by sophisticated compandoring. Similarly, error coding and error mitigation strategies affect the value of digital modulation schemes. However, the important point is that operationally the impact of Rayleigh fading can be viewed as some fixed loss in sensivitity. Hence, to achieve some desired coverage probability, one simply needs to provide a link margin equal to that fixed loss plus the appropriate shadowing factor (for example, 1.28σ for 90% coverage and 1.65σ for 95% coverage). Let this methodology be termed the separate effects method of coverage prediction.

A different methodology, hinted at in the classic paper by Okumura [2], might be termed the combined standard deviations method. It will be shown shortly that the standard deviation of a composite Rayleigh-lognormal distribution expressed in decibels equals the square root of the sum of the squares of the standard deviations for each effect alone. It is tempting therefore to use this standard deviation multiplied by the usual Gaussian factors to obtain coverage probabilities. Depending on the magnitude of shadowing and the percentile sought, this can produce poor coverage designs.

10.3 Analysis of Composite Rayleigh-Lognormal Distribution

The PDF for the instantaneous power of a random variable whose envelope is Rayleigh distributed is exponential in form and given by:

$$f(s|\alpha) = \frac{1}{\alpha} \exp(-\frac{s}{\alpha}) \qquad (10.7)$$

where s equals the instantaneous power (greater than or equal to zero) and α equals the average signal power. The moments of this distribution are well known (Section 7.6.7). The moments of power expressed in decibels are not particularly well known so derivations for the first two moments follow.

10.3.1 Mean and Standard Deviation of Power in Decibels for Rayleigh-Faded Signal

Consider $s_{dB} = 10\log(s) = K\ln(s)$, where $K = 10/\ln(10)$. The PDF for s_{dB} can be obtained from the density function for s via:

$$\begin{aligned} f_{s_{dB}}(s_{dB}|\alpha) &= f_s(s_{dB}|\alpha) / \left|\frac{ds_{dB}}{ds}\right| \\ &= \frac{1}{K\alpha} \exp\left[\frac{s_{dB}}{K} - \frac{1}{\alpha}\exp(\frac{s_{dB}}{K})\right] \end{aligned} \qquad (10.8)$$

Using this PDF one evaluates the mean value as follows:

$$\begin{aligned} <s_{dB}|\alpha> &= <10\log(s)> \\ &= K<\ln(s)> \\ &= K\int_0^\infty \frac{\ln(s)}{\alpha}\exp(-\frac{s}{\alpha})ds \\ &= K[\ln(\alpha) - C] \end{aligned} \qquad (10.9)$$

where relation #4.331.1 in Reference [3] has been used and $C \approx 0.5772$ is Euler's constant. The mean square power in decibels can similarly be evaluated with the help of relation #4.335.1 in Reference [3]:

$$\begin{aligned} <s_{dB}^2|\alpha> &= K^2 <[\ln(s)]^2> \\ &= K^2 \int_0^\infty \frac{[\ln(s)]^2}{\alpha}\exp(-\frac{s}{\alpha})ds \\ &= K^2\left\{\frac{\pi^2}{6} + [C - \ln(\alpha)]^2\right\} \end{aligned} \qquad (10.10)$$

Thus, the standard deviation is 5.57 dB, independent of α.

10.3.2 Mean and Standard Deviation of Power in Decibels for Rayleigh plus Lognormal-Faded Signal

In land-mobile radio propagation, α itself is actually a random variable whose form is generally lognormal in character. Defining $\beta = 10\log(\alpha)$, one thus has:

$$f(\beta) = \frac{1}{\sqrt{2\pi\sigma^2}}\exp\left[-\frac{(\beta-\mu)^2}{2\sigma^2}\right] \quad (10.11)$$

for the PDF of the average signal power in decibels, where μ equals the median average power level in decibels and σ equals the standard deviation of average power in decibels. The average instantaneous power expressed in decibels is now given by:

$$\begin{aligned}<s_{dB}> &= \int_0^\infty <s_{dB}|\alpha> f(\alpha)d\alpha \\ &= \int_{-\infty}^\infty <s_{dB}|\beta> f(\beta)d\beta \end{aligned} \quad (10.12)$$

Because $<s_{dB}|\beta> = K\ln[\exp(\beta/K)] - KC = \beta - KC$, one finds:

$$<s_{dB}> = \int_{-\infty}^\infty \frac{\beta}{\sqrt{2\pi\sigma^2}}\exp\left[-\frac{(\beta-\mu)^2}{2\sigma^2}\right]d\beta - KC\int_{-\infty}^\infty p(\beta)d\beta \quad (10.13)$$

The latter integral is unity and the former integral can be evaluated with the help of relation 5.(A4) in Reference [4] as simply μ. Thus $<s_{dB}> = \mu - KC$.

The mean square value is found by using:

$$\begin{aligned}<s_{dB}^2> &= \int_0^\infty <s_{dB}^2|\alpha> p(\alpha)d\alpha \\ &= \int_{-\infty}^\infty <s_{dB}^2|\beta> p(\beta)d\beta \\ &= \int_{-\infty}^\infty (C_1 - 2KC\beta + \beta^2)p(\beta)d\beta \end{aligned} \quad (10.14)$$

where $C_1 = (K^2\pi^2/6) + K^2C^2$. Of the three parts in this integral, the first integrates to just C_1 and the second integrates to $-2KC\mu$ because it is like the integral already solved for the mean value. The third integral can be solved with the help of relations 5.(A7) and 5.(A8) in Reference [4]. The answer is $\mu^2 + \sigma^2$. Combining all three parts and subtracting the square of the mean shows that the overall variance equals the sum of the lognormal variance σ^2 and the Rayleigh variance 5.57^2.

10.3.3 Composite Distribution Function

The unconditional power PDF is given by:

$$\begin{aligned}f(s) &= \int_{-\infty}^\infty 10^{\frac{-\beta}{10}}\exp(-s/10^{\beta/10})\frac{1}{\sqrt{2\pi\sigma^2}}\exp\left[-\frac{(\beta-\mu)^2}{2\sigma^2}\right]d\beta \\ &= \frac{1}{\sqrt{2\pi\sigma^2}}\int_{-\infty}^\infty \exp\left[-\beta k - s\exp(-\beta k) - \frac{(\beta-\mu)^2}{2\sigma^2}\right]d\beta \end{aligned} \quad (10.15)$$

where $k = \ln(10)/10 = K^{-1}$. The probability that the signal power is less than or equal to some reference level s_r is thus given by:

$$\text{Prob}(s \leq s_r) = \int_0^{s_r} f(s)ds$$

$$= \frac{1}{\sqrt{2\pi\sigma_e^2}} \int_0^{s_r} \int_{-\infty}^{\infty} \exp\left[-\gamma - s\exp(-\gamma) - \frac{(\gamma - \mu_e)^2}{2\sigma_e^2}\right] d\gamma\, ds \quad (10.16)$$

where $\mu_e = \mu k$, $\sigma_e = \sigma k$, and $\gamma = \beta k = \ln(\alpha)$.

To evaluate the preceding integral, reverse the order of integration and consider:

$$I_1 = \int_0^{s_r} \exp(-A - sB)ds \quad (10.17)$$

where $A = \gamma + [(\gamma - \mu_e)^2/2\sigma_e^2]$ and $B = \exp(-\gamma)$. The straightforward result is:

$$I_1 = \exp\left[-\frac{(\gamma - \mu_e)^2}{2\sigma_e^2}\right] \{1 - \exp[-s_r \exp(-\gamma)]\} \quad (10.18)$$

Next, one must evaluate:

$$I_2 = \frac{1}{\sqrt{2\pi\sigma_e^2}} \int_{-\infty}^{\infty} I_1 d\gamma$$

$$= \frac{1}{\sqrt{2\pi\sigma_e^2}} \left\{ \int_{-\infty}^{\infty} \exp[-\frac{(\gamma-\mu_e)^2}{2\sigma_e^2}]d\gamma \right.$$

$$\left. - \int_{-\infty}^{\infty} \exp[-\frac{(\gamma-\mu_e)^2}{2\sigma_e^2} - s_r \exp(-\gamma)]d\gamma \right\}$$

$$= I_3 - I_4 \quad (10.19)$$

where:

$$I_3 = \frac{1}{\sqrt{2\pi\sigma_e^2}} \int_{-\infty}^{\infty} \exp[-\frac{(\gamma-\mu_e)^2}{2\sigma_e^2}]d\gamma \quad (10.20)$$

$$= 1$$

and

$$I_4 = \frac{1}{\sqrt{2\pi\sigma_e^2}} \int_{-\infty}^{\infty} \exp[-\frac{(\gamma-\mu_e)^2}{2\sigma_e^2} - s_r \exp(-\gamma)]d\gamma \quad (10.21)$$

Equation (10.21) is intractable, but can be approximated via Laplace's method (see p. 127 of Reference [5]). The idea is to express the exponential argument in a Taylor series expanded about the value of γ for which the first derivative is zero.

Coverage Analysis and Simulation

This minimizes the argument and thus maximizes the contribution to the integral.
Let:

$$A(\gamma) = -\frac{(\gamma - \mu_e)^2}{2\sigma_e^2} - s_r \exp(-\gamma) \qquad (10.22)$$

The first and second derivatives with respect to the variable of integration γ are:

$$\begin{aligned} A'(\gamma) &= -\frac{(\gamma - \mu_e)}{\sigma_e^2} + s_r \exp(-\gamma) \\ A''(\gamma) &= -\frac{1}{\sigma_e^2} - s_r \exp(-\gamma) \end{aligned} \qquad (10.23)$$

Using a Taylor series expansion about reference point γ_0 one has:

$$A(\gamma) = A(\gamma_0) + A'(\gamma_0)(\gamma - \gamma_0) + A''(\gamma_0)\frac{(\gamma - \gamma_0)^2}{2} + \cdots \qquad (10.24)$$

Choosing γ_0 so that $A'(\gamma_0) = 0$ yields $\gamma_0 = \mu_e + s_r\sigma_e^2 \exp(-\gamma_0)$. If γ_0 is small, the transcendental equation reduces to $\gamma_0 \approx (\mu_e + s_r\sigma_e^2)/(1 + s_r\sigma_e^2)$. Otherwise γ_0 must be obtained by nonlinear numerical solution techniques such as the secant method [6].

Integral I_4 can now be written as:

$$\begin{aligned} I_4 &= \frac{1}{\sqrt{2\pi\sigma_e^2}} \exp\left[-\frac{(\gamma_0 - \mu_e)^2}{2\sigma_e^2} - s_r \exp(-\gamma_0)\right] \bullet \\ &\quad \int_{-\infty}^{\infty} \exp[-\frac{(\gamma - \gamma_0)^2}{2\phi^2} + \cdots]d\gamma \end{aligned} \qquad (10.25)$$

where $\phi = \sigma_e/\sqrt{1 + s_r\sigma_e^2 \exp(-\gamma_0)}$. Dropping the higher order terms and using $(\gamma_0 - \mu_e)/\sigma_e^2 = s_r \exp(-\gamma_0)$ leads to:

$$\begin{aligned} I_4 &\approx \frac{1}{\sqrt{2\pi\sigma_e^2}} \exp\left[-\frac{x(x+2)}{2\sigma_e^2}\right] \int_{-\infty}^{\infty} \exp[-\frac{(\gamma-\gamma_0)^2}{2\phi^2}]d\gamma \\ &\approx \frac{1}{\sqrt{x+1}} \exp\left[-\frac{x(x+2)}{2\sigma_e^2}\right] \end{aligned} \qquad (10.26)$$

where $x = \gamma_0 - \mu_e = s_r\sigma_e^2 \exp(-\gamma_0)$. Consequently:

$$\text{Prob}(s \leq s_r) \approx 1 - \frac{1}{\sqrt{x+1}} \exp\left[-\frac{x(x+2)}{2\sigma_e^2}\right] \qquad (10.27)$$

10.3.4 Monte Carlo Simulation Results

The composite distribution function that proved intractable can of course be evaluated for specific values of μ, σ, and s_r by numerical integration techniques. Alternatively, the solution can be obtained from the cumulative histogram for random draws taken from a composite Rayleigh-lognormal population with specified mean and standard deviation. This is the so-called Monte Carlo method [7]. Figure 10.2 shows 10,000 trial results for standard deviations spanning 2 to 10 dB in steps of 2 dB relative to the median average power level (taken as 0 dB).

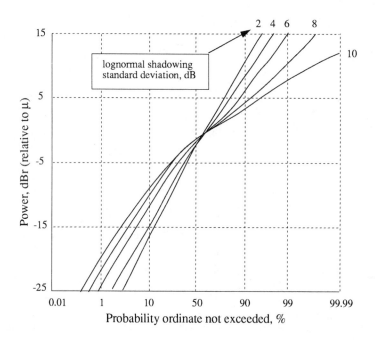

Figure 10.2 Cumulative distribution function for combined Rayleigh-lognormal variation.

The approximate cumulative probability derived in the preceding section has been solved for 8-dB standard deviation and is given in Table 10.1 (method 4). Also noted are: (1) the Monte Carlo results, which can be taken as the correct answers; (2) the results using the combined standard deviation method and a Rayleigh standard deviation of 7.5 dB[1]; and (3) the results using the combined standard deviation method with the true Rayleigh standard deviation of 5.57 dB.

---

[1]Inferrable from Okumura's measured data [2]. Interestingly this excess is not a real-world influence caused by deviations from Rayleigh small-sector character. Though such deviations are

Table 10.1 Cumulative Probability Values for Composite Rayleigh-Lognormal Distribution

Method 1=Monte Carlo results
Method 2="Okumura" with 7.5 dB Rayleigh std dev
Method 3="Okumura" with 5.57 dB Rayleigh std dev
Method 4=Approximate equation

Level/method	Part I: arranged by level			
	1	2	3	4
0 dBr	58.5%	50%	50%	58.7%
-3	45.8	39	38	45.5
+3	70.5	61	62	70.9
-6	33.9	29.2	27.2	32.7
+6	80.4	70.8	72.8	81.1
-10	20.9	18.1	15.6	18.4
+10	90.5	81.9	84.4	90.7
-20	3.8	3.4	2.2	2.6
+20	99.3	96.6	97.8	99.3

%-tile/method	Part II: arranged by percentile			
	1	2	3	4
50%	-2.0 dBr	0	0	-2.0
10	-14.9	-14.0	-12.5	-13.4
90	9.7	14.0	12.5	9.6
5	-18.6	-18.1	-16.1	-16.9
95	13.0	18.1	16.1	12.9
1	-26.6	-25.6	-22.7	-24.2
99	19.0	25.6	22.7	19.0

Parameter values: median average power=0 dBr, standard deviation=8 dB

From Table 10.1 one sees that the approximate solution is reasonably accurate for levels ±20 dB about the median. Performance of the combined standard deviation method is good for deep fades, but the error in level grows as fade depth decreases. Performance is sensitive to the lognormal standard deviation since as that parameter shrinks, the cumulative distribution becomes more curved on a normal probability scale.

This sensitivity is important. Field tests in Chicago [8] support a shadowing standard deviation of only about 5.5 dB for the urban small-area variability of constant radius routes. Variability around the entire circumference of a constant radius route is of course much larger, typically 8 to 10 dB; however, for coverage analyses where path-specific terrain effects are included in the estimation of median signal power, the small-area variability is appropriate. The same field tests produced even lower values for radial routes (3.7 dB being most representative). For random routes both values would probably apply equally, which implies an effective standard deviation of just 4.6 dB. Comparing this to Figure 10.2 with a 4-dB standard deviation (the nearest curve to 4.6 dB) note that method 2 very accurately predicts fade levels in the 10% to 1% range, but is off by more than 5 dB at the 90% peak level and 8.5 dB at the 99% peak level.

It is interesting to note that the median of the composite distribution is not equal to the lognormal median. Rather it is somewhat less; the amount being a weak function of the lognormal standard deviation. There does, however, appear to be a common crossover point independent of the lognormal standard deviation occurring at about -1 dBr and 53%.

10.3.5 Regression Analysis

In the preceding section it was pointed out that due to the curved nature of the CDF, the combined standard deviation method can not be accurate over the entire probability range. Of course, for a sufficiently large lognormal standard deviation, the method may be accurate enough for the range of interest.

Table 10.2 shows the worst case errors for three types of Gaussian least mean square error regression fits to the Monte Carlo results [9]: (1) unconstrained, (2) fit forced through -1.75 dBr (a representative value for the range of lognormal standard deviations considered), and (3) fit forced through 0 dBr. A fourth case is also listed: namely, the method of (3) using the standard deviation for the solution equal to the square root of the sum of 7.5 dB squared (for the Rayleigh factor) and the lognormal standard deviation squared. The error performance is much improved, especially for case four, if only fades are considered. Table 10.3 shows results over the range of -25 to 0 dBr.

---

expected due to occasional direct-path circumstances, they would lead to a smaller standard deviation than 5.57 dB, not a larger one. Rather, the excess stems from force fitting Gaussian behavior to a process that is not truly Gaussian.

Table 10.2 Summary of Least Mean Square Error Regression Analysis of Combined Standard Deviation Method

Lognormal std dev	Abs max level error	Fit parameters Median	Std dev
Case I: unconstrained			
10 dB	0.6 dB	-2.44 dBr	11.4 dB
8	1.3	-2.72	9.69
6	2.24	-3.29	7.99
4	3.34	-4.05	6.82
2	3.73	-4.78	6.00
Case II: forced through -1.75 dBr			
10 dB	1.3 dB	-1.75 dBr	11.4 dB
8	2.3	-1.75	9.66
6	3.8	-1.75	7.88
4	3.34	-1.75	6.82
2	3.73	-1.75	6.00
Case III: forced through 0 dBr			
10 dB	3.0 dB	0.0 dBr	11.4 dB
8	4.1	0.0	9.64
6	5.7	0.0	7.83
4	8.2	0.0	6.51
2	9.0	0.0	5.83
Case IV: like III, but Rayleigh std dev=7.5 dB			
10 dB	5.0 dB	0.0 dBr	12.5 dB
8	6.8	0.0	10.97
6	10.6	0.0	9.6
4	>11	0.0	8.5
2	>13	0.0	7.76

Range of curvefit = -25 dBr to +20 dBr

Table 10.3 Summary of Least Mean Square Error Regression Analysis of Combined Standard Deviation Method, Fades Only

Lognormal std dev	Abs max level error	Fit parameters Median	Std dev
Case I: unconstrained			
10 dB	0.2 dB	-1.97 dBr	11.84 dB
8	0.5	-1.61	10.44
6	0.7	-1.65	9.20
4	1.5	-1.20	8.65
2	1.8	-0.89	8.23
Case II: forced through -1.75 dBr			
10 dB	0.4 dB	-1.75 dBr	11.99 dB
8	0.55	-1.75	10.46
6	0.7	-1.75	9.13
4	1.7	-1.75	8.33
2	1.9	-1.75	7.77
Case III: forced through 0 dBr			
10 dB	2.4 dB	0.0 dBr	13.18 dB
8	2.45	0.0	11.50
6	2.5	0.0	10.05
4	2.7	0.0	9.18
2	2.8	0.0	8.57
Case IV: like III, but Rayleigh std dev=7.5 dB			
10 dB	2.3 dB	0.0 dBr	12.50 dB
8	2.4	0.0	10.97
6	2.4	0.0	9.6
4	3.1	0.0	8.5
2	3.7	0.0	7.76

Range of curvefit = -25 dBr to 0 dBr

Coverage Analysis and Simulation 133

10.4 Measurement of Trunked System Subscriber Unit Signal Characteristics

Precise information on the distributions of talk-in[2] signal strength, frequency error, and range allows coverage and interference probabilities to be estimated with high confidence. This section discusses a means of gathering such information for a Motorola-type trunked system, though the basic ideas suggested can be applied to a degree on other system types.

10.4.1 Acquiring Information about the ISW

On Motorola trunked systems, subscriber radios initiate calls by transmission of *inbound signaling words* (ISWs) on a control channel. Such transmissions begin between 49- and 52-bit times (277 μs/bit) after the positive transition of the central control *word frame interrupt* (WFI) clock, which is a transition that all units lock to while listening to the outbound control channel link. The 3-bit period is included to account for two-way propagation delay (10.73 μs/mi), group delay variations in the transmitter response, and timing variations in the subscriber unit microprocessor controllers. The basic ISW information encompasses 78 bits and the 81-bit period during which the central site might receive an ISW will be termed the decoding window.

ISW-related information is collected by an outboard microprocessor with a pair of *analog-to-digital* (A/D) converter interfaces as follows. During each decoding window, the discriminator output of the base control channel receiver is sampled at a rate of 36000 samples per second (10 times the signaling bit rate) and stored in memory. Concurrently, the second A/D converter samples an analog voltage whose level is related to the logarithm of received signal strength. If an ISW is successfully received and decoded during the decoding window, information regarding the fleet, unit *identification code* (ID), and call type will appear on an RS-232 data link during the following decoding window. The outboard microprocessor is arranged to monitor and intercept such traffic.

10.4.2 Estimating the Subscriber Unit Transmit Frequency

To estimate the transmitted frequency of the ISW, a special receiver is used whose first local oscillator is locked to a precision 5-MHz source. Second local oscillator drift is of minor importance; that oscillator is simply tuned precisely to 33.3 MHz so the discriminator dc output is proportional to the frequency error of the transmitting source. The particular ISW bit pattern must also be taken into account because

---

[2] By this we mean the subscriber unit to base direction. The base to subscriber path is termed talk out. Other common terms are uplink and downlink, respectively.

when the number of "true" bits does not equal the number of "false" bits, a dc shift is also produced.

10.4.3 Estimating the Subscriber Unit Range

Ranging is done by precisely determining the arrival time of the ISW relative to the central time standard. Two-way propagation delay is about 10.73 μs/mi. However, additional delay is also present due to synchronizing error, filtering, and processing at the subscriber unit. Such delay is radio-type dependent, so a mapping of unit ID to radio type is done in the final processing stage to select the most accurate value (based on *a priori* calibration measurements of known radio types at known physical distances).

Precise determination of arrival time is nontrivial because of the relatively slow signaling rate of 3600 bps (277 μs/bit; hence, during 1 bit a two-way propagation distance of about 26 mi is covered). By tenfold oversampling and correlating a hard-limited version of the received signal against what ideally should have been received (this is obtained by recoding the decoded ISW information), a resolution on the order of 2.6 mi is possible. Figure 10.3 shows examples of discriminator outputs. Figure 10.4 shows the peaky correlation that results (Figure 10.5 confirms that the impact of real receivers is minor). Arrival time is associated with the correlation function peak, which can be conveniently quantified using a least squares parabolic fit.

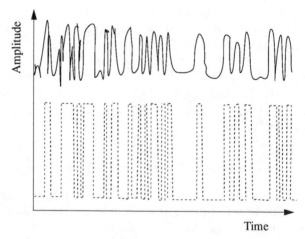

Figure 10.3 Plot of actual discriminator samples and corresponding ISW bit pattern.

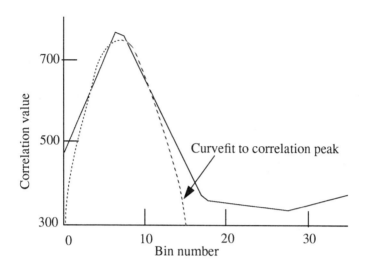

Figure 10.4 Plot of correlation results and least mean square error parabola curvefit.

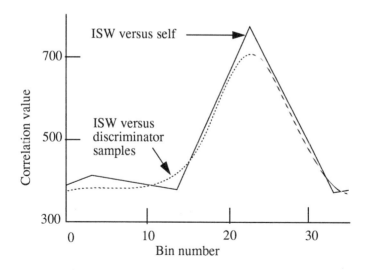

Figure 10.5 Plot of correlation results of discriminator samples with ISW pattern and ISW pattern self-correlation.

10.4.4 Measured Results

Figure 10.6 shows the distribution of signal strength as a function of range for a trunked, *specialized mobile radio* (SMR)[3] provider located atop the Sears Tower in Chicago, Illinois. The lines correspond to the median, the median plus one standard deviation, and the median minus one standard deviation signal strengths. Note that the standard deviation is not particularly sensitive to range and is on the order of 8 dB, as one would expect for lognormal shadowing over a large area. Also note that signal strength initially increases with increasing range before falling off. This is a result of the receive antenna elevation pattern, which discriminates against close-in signals. Because the Sears Tower is such a high site, one must travel several miles to reach the minimum path loss condition. The overall distribution of subscriber unit range is shown in Figure 10.7.

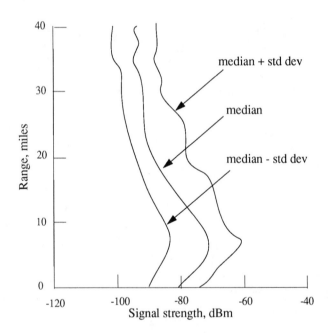

Figure 10.6 Signal strength versus range plot for Chicago, Illinois 800-MHz trunked SMR.

The overall distribution of signal strengths is shown in Figure 10.8. The bulk of this distribution in decibels follows one normal trend; however, probably due to con-

---

[3]The specialized mobile radio service was introduced in the early 1980s to the United States by FCC Docket 18262.

Coverage Analysis and Simulation

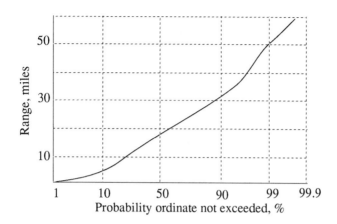

Figure 10.7 Cumulative range distribution for Chicago, Illinois 800-MHz trunked SMR.

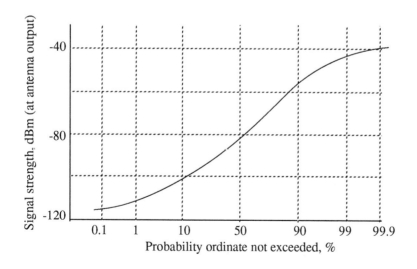

Figure 10.8 Cumulative signal strength distribution for Chicago, Illinois 800-MHz trunked SMR.

trol stations with unnecessarily high ERPs, inclusion of a second normal population is helpful for modeling the signal strength behavior.

Figure 10.9 shows subscriber unit frequency error for the same SMR system. The character is Gaussian with nearly zero mean and a standard devation of about 800 Hz. Few units appear outside the applicable FCC specification of 2.5 ppm stability.

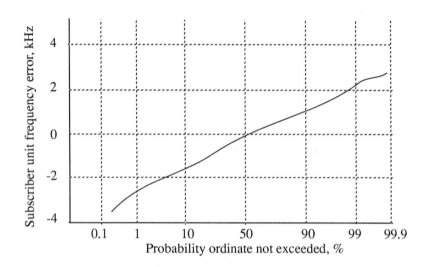

Figure 10.9 Cumulative distribution of subscriber unit frequency error for Chicago, Illinois 800-MHz trunked SMR.

Similar behaviors of signal strength and frequency error were noted for trunked systems in New York City and Los Angeles. The range behavior differed somewhat. As expected, distances were compressed for New York City and stretched out for Los Angeles (where a distinct lack of ranges under 10 mi stems from the use of mountaintop sites that are quite isolated from the population).

10.5 Adjacent Channel Interference Considerations in the Land-Mobile Radio Service

The origin of adjacent channel interference is shown in Figure 10.10 [11]. The figure portrays two transmissions occurring on adjacent channels. Inevitably some signal

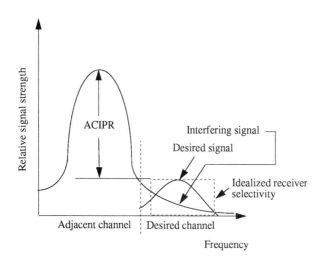

Figure 10.10 Origin of adjacent channel interference.

components spread beyond the channel boundaries and can be intercepted by receivers tuned to the adjacent channel. When the signal strength of the adjacent channel transmission becomes so large that the power intercepted by an on-channel receiver approaches that of the desired on-channel signal source, interference occurs. The ratio of the signal strengths of the two transmissions at the point at which interference is first noted is called the *adjacent channel interference protection ratio* (ACIPR).

Figure 10.11 shows how finite ACIPR can cause a coverage hole in the talk-out performance of a system. Figure 10.12 shows how finite ACIPR can result in overall reduced talk-in range. This is because mobile A must be moved closer to the base receiver to produce sufficient signal relative to the signal inadvertently coupled from close in mobile B on an adjacent channel.

ACIPR can be measured several ways. The simplest procedure, geared toward the frequency-modulated equipment that dominates current land-mobile radio operation, is the EIA Standard RS-204-D adjacent channel selectivity and desensitization measurement [12]. This procedure uses 1-kHz and 400-Hz tones at 60% rated peak deviation on the desired and interfering channels, respectively. Unfortunately, tones are generally poor substitutes for real voice, although a dual-tone technique that does reasonably reflect voice performance has been developed for 12.5-kHz channel measurements [13]. Therefore, a preferable procedure is to have a number of test subjects listen to voice transmissions over the radio and adjust the signal strength of the interferer until some stated interference level (for example, "just noticeable," "hurts intelligibility," and "minimum acceptable quality") is obtained. The subjec-

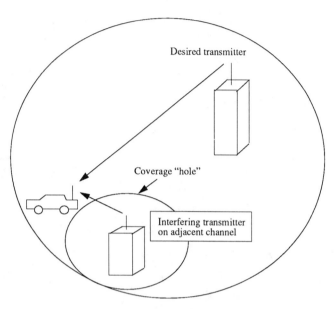

Figure 10.11 Impact of adjacent channel interference on talk-out coverage.

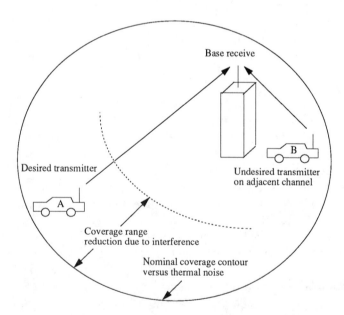

Figure 10.12 Impact of adjacent channel interference on talk-in coverage.

tive difficulty of such tests can be minimized by instead using a 1-kHz tone at 60% rated deviation on the desired channel and noting when the SINAD degrades from 12 to 6 dB on interfering voice peaks [14].

ACIPR values of 70 dB or greater are generally necessary to combat the near-far problems possible in uncontrolled land-mobile environments; i.e., in situations where the undesired source can be much closer than the desired source [15]. Figure 10.13 shows a history of ACIPR and sensitivity performance as channel spacings have been decreased from over 100 kHz to today's best of 12.5 kHz. The graph has been adjusted to reflect the technology used in modern 800-MHz synthesized radios so that trends can be noted.

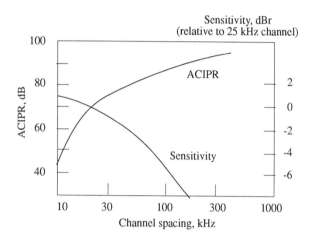

Figure 10.13 History of land-mobile sensitivity and adjacent channel interference performance.

In systems with channel spacings greater than about 20 kHz, ACIPR is limited by internally generated oscillator sideband noise. As the channel size decreases to 12.5 kHz, ACIPR becomes set by transmitter splatter and further decreases in channel bandwidth cause ACIPR to fall rapidly. Sensitivity increases as channel spacing decreases because narrower noise bandwidths can be used. But this improvement has flattened out by 12.5-kHz spacing.

Another problem with such narrow channels is frequency offset. For example, even with 4-ppm frequency tolerance (3.5-kHz error at 850 MHz), ACIPR degrades only about 2 dB for 25-kHz channels. A similar frequency error in a 12.5-kHz channel system degrades ACIPR by about 28 dB. Hence, the far more stringent frequency stability requirements noted in the FCC rules for the 900-MHz band (1.5 ppm for mobiles and 0.1 ppm for bases).

10.6 Spectrum Efficiency Potential of 25-kHz Offset Channel Assignments in the 821- to 824- and 866- to 869-MHz Public Safety Bands

The FCC recently allocated the paired bands of 821 to 824 MHz and 866 to 869 MHz to the Public Safety Radio Service and solicited comment on service rules and technical standards (Docket 87-112). Spectrum efficiency demands ostensibly would favor a 12.5-kHz channelization scheme; however, timeliness, expansion of existing 800-MHz systems, high-speed data and encrypted voice, and mutual aid communication are factors in favor of maintaining 25-kHz channelization. A third possibility is to use 25-kHz channels with 12.5-kHz offset assignments. This would satisfy the issues of timeliness, expansion, etc. and also have the potential to improve spectrum efficiency because public safety users tend to operate within well-defined modest-sized boundaries. Hence, spectrum efficiency via geographic reuse is a reasonable possibility.

This section estimates the spectrum efficiency potential of the offset channelization scheme and compares it to that which might be attained with 12.5-kHz split channels [16]. This is done by breaking the problem into two parts. First, it is established how close systems with offset assignments can be located without unacceptable interference. Second, using these separation requirements, the actual needs of a major metropolitan area (Dallas/Fort Worth, Texas) are considered and users are assigned to frequencies that satisfy the constraints. The spectrum required with the offset strategy can thus be directly compared to that required by the split-channel strategy.

10.6.1 Geographic Separation Requirements

Consider the case of two users requiring a service area of 10 mi and assigned to 25-kHz offset channels. Focus on talkout as the limiting case, assuming that talk-in problems can be handled by remote receiver sites as necessary. In particular focus on the worst square mile location; i.e., the location 10 mi from the desired transmitter in the direction of the undesired transmitter. One can estimate the interference potential of such a worst square mile location as follows.

Consider both desired and undesired signals to be independently faded lognormal processes. The conditional probability that the difference between undesired and desired received signal powers exceeds some splatter value s is then given by:

$$\text{Prob}[(u-d) > s | s] = \frac{1}{2}\text{erfc}[\frac{s - (u_0 - d_0)}{2\sigma_{ln}}] \qquad (10.28)$$

where d equals the desired signal power in dBm, d_0 equals the median desired signal

Coverage Analysis and Simulation

power in dBm, erfc is the complementary error function (1.0 minus the error function [10]), u equals the undesired signal power in dBm, u_0 equals the median undesired signal power in dBm, s equals the splatter protection required to avoid interference in decibels, and σ_{ln} equals the standard deviation of the lognormal shadowing in decibels.

Measurements of splatter protection versus frequency spacing indicate a linear relationship in the region of interest; i.e., one-half to full channel spacing. For example, Figures 10.14 and 10.15 show typical splatter performance versus frequency offset for 25- and 12.5-kHz radios, respectively. Observations of frequency errors for 800-MHz radios in commercial service show them to be reasonably Gaussian in character (Figure 10.9). Together, these findings imply a Gaussian character for the parameter s in Equation (10.28). Averaging over a Gaussian distribution for s leads to:

$$\begin{aligned}
\text{Prob}[(u-d) > s] &= \frac{1}{2\sqrt{2\pi\sigma_f^2}} \int_{-\infty}^{\infty} \exp[\frac{-(s-s_0)^2}{2\sigma_f^2}] \cdot \\
&\quad \text{erfc}[\frac{s-(u_0+d_0)}{2s_{ln}}]ds \\
&= \frac{1}{2}\text{erfc}[\frac{s_0-(u_0-d_0)}{\sqrt{2(\sigma_f^2+2\sigma_{ln}^2)}}]
\end{aligned} \quad (10.29)$$

where s_0 equals the median splatter protection required to avoid interference (in decibels) and σ_f equals the standard deviation of splatter protection in decibels.

To use the preceding analysis, information about the receiving radio's susceptibility to interference when two closely spaced signals are received is required. Interference measurements made using today's 25-kHz radios typically indicate a protection on the order of 10 dB for just noticeable interference at a spacing of 12.5 kHz (Figure 10.14). That is, the undesired signal must be at least 10 dB stronger than the desired signal before interference is just noticeable. This protection number typically improves at a rate of 8 dB/kHz as the spacing increases. The nominal protection can be increased from 10 to 20 dB and the sensitivity decreased from 8 to 5 dB/kHz if three improvements are made [17]: (1) the peak deviation is reduced to 4 kHz, (2) the receiver selectivity is improved by the addition of more poles to the IF filtering, and (3) frequency stabilities are improved to 1.5 ppm for mobiles, portables, and control stations and 0.1 ppm for base stations. This third improvement is necessary to make the second improvement practical.

A 90% thermal noise talk-out coverage contour of 10 mi is consistent with the following parameter assumptions: (1) 150 ft base antenna height, (2) 200 W EIRP, (3) suburban Okumura propagation with 8-dB lognormal shadowing standard deviation, (4) -109-dBm receiver sensitivity (1/4 μV with a 10-dB Rayleigh-fading allowance), (5) 0.8-kHz standard deviation for frequency error (Figure 10.9), and (6) 6-ft-high mobile antenna with 0-dBd antenna gain. For these assumptions and

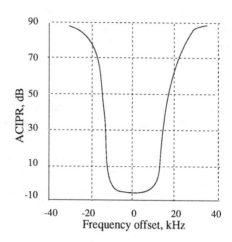

Figure 10.14 Adjacent channel interference protection ratio performance of a representative 25-kHz radio.
After: Hess, G., "Spectrum Efficiency Potential of 25 kHz Offset Channel Assignments in the 821-824/866-869 MHz Public Safety Bands," *38th IEEE Vehicular Technology Conference*, Philadelphia, PA, June 1988, p. 362.

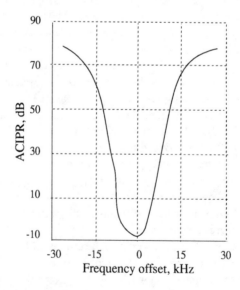

Figure 10.15 Adjacent channel interference protection ratio performance of a representative 12.5-kHz radio.
After: Hess, G., *ibid.*, p. 362.

base transmitters separated by 20 mi, the probability of interference without regard to coverage is 4.8%; this value drops to just 2% conditioned on the desired signal level being above sensitivity. This is relatively small compared to the typical commercial system design criterion of 10% thermal outage at the coverage contour.

For different sized service areas scaling holds reasonably well. For example, two users requiring just 5-mi service areas separated by 10 mi would also experience about 2% interference outage at the worst location. Similarly, users requiring 15-mi coverage should be at least 30 mi apart. In the case of different coverage requirements, the larger value should be assumed for both.

10.6.2 Channel Assignment Procedure

The preceding results point to the possibility of offset assignments with 20-mi base-to-base spacings for public safety use with 10-mi coverage requirements using improved 25-kHz radios. To gauge the impact of this on spectrum efficiency, a channel assignment program is used. This program takes information on channel requests such as location, coverage requirement, and number of channels required and proceeds to assign channels so that all constraints are met. If this is possible with N channels, the program is rerun assuming some lesser number of channels is available. The smallest value N that satisfies all the constraints is the most spectrally efficient assignment scheme.

A key item in the channel assignment program is the channel separation matrix. This is developed from the site separation matrix, whose entries denote the base-to-base spacings between all pairings of sites, and reuse rules such as: (1) cochannel operation requires at least 70-mi spacing, (2) offset operation requires at least 20-mi spacing with 10-mi coverage requirements (proportioned for other coverage requirements), and (3) antenna combining requires at least 10-channel separation (0.25 MHz) among all channels to be combined. The channel separation matrix contains the constraints that our assignments must satisfy.

An algorithm to make assignments is described by F. Box [18]. It addresses requests in order of difficulty. A request that could not be satisfied on iteration i is considered difficult and given a random high weight so that it is one of the first cases considered on the next iteration. For each request under consideration, channels are tried alternately from the low and high ends of the band. The first channel (if any) that satisfies the constraints of the channel separation matrix is used for the assignment. For convenience channels are counted on an offset basis; i.e., the first full 25 kHz is channel 1, 12.5 to 37.5 kHz is channel 2, the second full 25 kHz (25 to 50 kHz) is channel 3, etc.

A summary of the channel assignment procedure is as follows:

1. Obtain site information for assignment under consideration:

    a. latitude/longitude of base sites

b. coverage requirement

2. Determine starting point:

   If the number of assignments already made is even, then the starting point is channel 1, the lowest frequency channel in the contiguous frequency range available for assignments; else, start at channel N, where N is the highest frequency channel available.

3. Determine relevant site spacings for channel I, the channel under consideration for assignment:

   a. Cochannel case - let $D1$ equal the distance between the site under consideration and the site, if any, already assigned to channel I[4].

   b. Lower offset case - let $D2$ equal the distance between the site under consideration and the site, if any, already assigned to channel $I - 1$.

   c. Upper offset case - let $D3$ equal the distance between the site under consideration and the site, if any, already assigned to channel $I + 1$.

4. Determine the relevant coverage requirements for the assignment under consideration:

   a. Cochannel case - let $C1$ equal the greater of the coverage requirement for the assignment under consideration and the coverage requirement of the closest cochannel user already assigned, if any; if no such assignment has been made then so indicate by setting $C1$ to zero.

   b. Lowside offset case - let $C2$ equal the greater of the coverage requirement for the assignment under consideration and the coverage requirement of the closest lowside offset user already assigned, if any; if no such assignment has been made then so indicate by setting $C2$ to zero.

   c. Highside offset case - let $C3$ equal the greater of the coverage requirement for the assignment under consideration and the coverage requirement of the closest highside offset user already assigned, if any; if no such assignment has been made then so indicate by setting $C3$ to zero.

5. Check the assignment viability[5,6]: If $(D1/C1)$ is greater than or equal to $K1$,

---

[4] In assignments where multiple sites are involved, specify the minimum distance of all pairwise site spacings

[5] The $K2$ and $K3$ values depend on the splatter performance of the radios and the amount of interference that can be tolerated. Measurements indicate that values on the order of 2.0 might be viable with improved 25-kHz analog FM radios; i.e., radios with 20-dB nominal offset splatter protection. A cochannel criterion of about $K1 = 5.0$ is suggested for such radios.

[6] Antenna combining of a block of cosited channels may impose additional frequency separation requirements.

and if $(D2/C2)$ is greater than or equal to $K2$, and if $(D3/C3)$ is greater than or equal to $K3$, then the assignment can be made (Go to step 8); else, continue.

6. Assignment attempt in step 5 was unsuccessful so one must try again using a different channel value for I: If starting point was the lowest channel then increment I by 1; else decrement by 1.

7. Test to make sure I is still within the permissible range of 1 to N: If I is in range then go to step 3; otherwise, the process must be repeated with reduced criteria for $K1$, $K2$, and $K3$ to allow an assignment or a random difficulty factor must be assigned and more assignment attempts made.

8. If further assignments are to be made then go to step 1.

In the Dallas/Fort Worth metropolitan region, 155 assignment needs involving 33 sites were identified to satisfy immediate needs and allow for suburban growth through the year 2000. In accordance with the national plan [17], five channels must be set aside to allow for mutual aid communication among users outside their normal jurisdictional boundaries. Figure 10.16 indicates the bandwidth required to satisfy these needs as a function of the fraction of channels that preclude offset assignments (for example, due to the use of encrypted voice or high-speed data, or because statewide coverage is required).

Figure 10.16 involves three curves: one for offset operation with just 10-dB nominal interference protection as with today's radios, one for offset operation with the 20-dB nominal interference protection possible with improved radios, and one for 12.5-kHz channel operation. When all assignments must use 25-kHz channels and offsets are not permitted, Figure 10.16 indicates that 140 equivalent channels are required (a number that exceeds the 120 available through the new allocation). The value is slightly less than the sum of 155 and 5 because of cochannel reuse. Under the more reasonable circumstance of 20% of the channels being incompatible with offset assignments, one finds that improved radios could satisfy the Dallas/Fort Worth needs with just 2.325 MHz of paired spectrum (93 times 25 kHz). Even with 12.5-kHz split-channel operation, 2.15 MHz of paired spectrum would be required (86 times 25 kHz). Thus, the gain in spectrum efficiency is relatively modest in contrast to the problems such a scheme would cause in terms of timeliness, expansion of existing systems, and mutual aid communications.

10.7 Interference Potential of Cordless Telephone Sharing with Public Safety Band Users

Section 5.2.2 addressed the interference potential of a single 10-mW cordless telephone unit to public safety subscriber units listening on the base talk-out frequency,

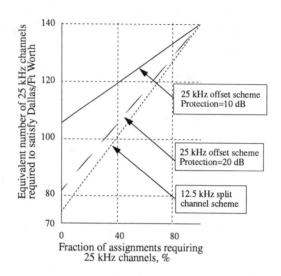

Figure 10.16 Comparison of 25-kHz offset and 12.5-kHz split-channel assignment schemes.
After: Hess, G., *ibid.*, p. 363.

and located at the edge of coverage by that base transmitter. The purpose of this section is to quantify the interference potential of such cordless telephone sharing operation if it becomes commonplace. That would imply a large number of units, perhaps a million, spread over metropolitan areas.

To that end a simulation was written that generates 2500 trials, carried out in five groups of 500 trials each. In each trial, a location is established for the public safety subscriber unit by randomly placing it within a circle of specified radius (typically 5 mi). The center of that circle is taken as the base transmitter location, which can be offset a user-specified distance from the center of population (typically 15 mi). The desired signal strength is determined using Hata's empirical fit to Okumura's measured path loss data and user-specified values for base EIRP, base antenna height, subscriber antenna height and gain, and link frequency (respectively, 100 W, 200 ft, 6 ft, 0 dBd, and 868 MHz). Signal strength is also adjusted for shadowing via draws from a normal population with zero mean and 8-dB standard deviation when on the street, −14.2-dB mean and 12.3-dB standard deviation when inbuilding [19]. Typically, the in-building probability is set to 0.2.

In each trial, 10,000 possible interferers are typically considered. This number represents 1 million units spread over the metropolitan area, transmitting on average one 3-minute call per hour (duty cycle equals 0.05), and overlapping the public safety user frequency 20% of the time. The latter is the result of 20 100-kHz channels being available to the cordless telephone units, whereas the public safety user is on some specific 25-kHz channel. Interferer locations are established by drawing ranges from a gamma probability density (refer to homework problem 1, Section 9.6.1; typical parameter values are 7.5 mi and 2) and angles from a uniform $(0, 2\pi)$ distribution. Of course most of the interferers are too weak to be of significance. The program ignores all cordless telephone units farther than some specified distance away from the public safety subscriber unit (typically 5 mi).

Cordless telephone units are assumed to transmit 10-mW EIRP (2.5 mW into a 25-kHz bandwidth) from an antenna height of 6 ft. The local environment can be set independently from that of the desired link, though generally both are taken to be suburban. For cordless telephone units, the in-building probability is typically set at 0.8 (inbuilding but near windows is like not being inbuilding).

The desired signal power for each trial is compared to the noise floor and if the difference exceeds some capture requirement (typically 17 dB; i.e., static 20-dB quieting plus a 10-dB pad for Rayleigh fading) coverage versus noise is established. If that is true, a test for sufficient desired-to-undesired power difference to avoid interference outage is made. Percent coverage versus noise only and percent coverage versus interference given that coverage versus noise exists are the two key program outputs for each 500 trial group. When all groups have been completed, group means and standard deviations are calculated.

The results are as follows. A public safety system located 15 mi from the metropolitan center and attempting to cover 5 mi has an area coverage probability of 95.0% ($\sigma = 0.9\%$) versus thermal noise only. The conditional coverage in the presence of cordless telephone interference drops to just 76.6% ($\sigma = 2.3\%$) when interferers within 5 mi of the receiving subscriber unit are considered. There are on average 232 such interference sources. However, even when the separation cutoff distance is decreased to 1 mi, the conditional coverage is basically unchanged (77.7% with $\sigma = 1.4\%$; caused by an average of nine interferers). Worse yet, the conditional interference only slowly improves as the desired path length is decreased. For example, if one only considers receiving units within 1.25 mi of the desired base station, the coverage rises to 99.8% ($\sigma = 0.1\%$) versus thermal noise but is still only 93.8% ($\sigma = 0.6\%$) in the presence of interference. This is because it is just too likely that a cordless telephone transmitter will be extremely close (i.e., a few hundred feet or less) to the receiving unit.

10.8 Further Sharing of UHF Television by Private Land-Mobile Radio Services

General Docket 85-172 proposes rules for further sharing of the UHF television band by private land-mobile radio services. This section, based on Reference [20], describes methodologies for evaluating the interference potential to television reception based on the proposed rules. Two general situations are considered: (1) cochannel interference caused by land-mobile operation outside the television coverage area and (2) noncochannel interference caused by land-mobile operation inside (or nearly so) the television coverage area. The interference probabilities are contrasted with those related to self-imposed UHF television interference.

10.8.1 Cochannel Sharing

First consider cochannel sharing, a situation whose sharing rules differ in Docket 85-172 from those already implemented by an earlier Docket, 18261. The proposal would allow maximum facility land-mobile base stations (1kW at 500 ft) to produce a 10% maximum field strength level 40 dB below the median Grade-B television field strength of 64 dBμ. The sharing already in existence requires a 50-dB protection. In terms of separation, Docket 85-172 would only require 75 mi from the Grade-B contour, while Docket 18261 requires 108 mi. Typical Grade-B coverage ranges are on the order of 50 mi; this implies site-to-site separations of about 125 mi for Docket 85-172 and 158 mi for Docket 18261.

This cochannel sharing section consists of seven subsections. The first describes mathematically how both the location and time variability of field strength level can be handled. Next, the service probability/range relationship is examined assuming the absence of interference. This allows one to establish the extent of the Grade-B service area. Then, the probability that the difference between desired and undesired field strengths is less than some value is examined. Of course there can be no interference unless there is service to begin with, so the fourth subsection relates loss of service to the joint probability of the service and interference probabilities obtained via subsections two and three. The analysis to this point follows that given by Wong [21]. Subsection five, however, elaborates on how to account for the receiver-to-receiver variability of interference susceptability. Subsection six outlines the solution for two specific situations, one for a land-mobile interference source and one for a UHF television interference source. This is followed by results for numerous cases of interest.

Field Strength Level in Decibels

The field strength level in decibels produced at an arbitrary range R from a source is taken to be the result of two uncorrelated normal processes. The first describes the level variability among locations, all of which are at range R. It has a median value based on Part 73 of the FCC rules and is calculated via interpolation of a table of values based on FCC report R-6602 [22]. The standard deviation of this process is assumed independent of range. The second process describes the temporal variability. It is a zero-mean normal variation with standard deviation dependent on range.

Coverage Analysis and Simulation

Thus, the field strength exceeded for $L\%$ of the locations at a range R from the source, $T\%$ of the time is given by:

$$F(L, T; R) = F(50, 50; R) - k(T)s_T - k(L)s_L \qquad (10.30)$$

where $F(50, 50; R)$ is the median field strength at range R from the source (i.e., the level exceeded at 50% of the locations 50% of the time); $k(X)$ is the standard variate of a normal distribution at the X percentile (for example, see Table I, p. 326, of Reference [23]); s_T is the standard deviation of field strength variation with time (values are based on FCC report R-6602 [22]); and s_L is the standard deviation of field strength variation with location (values are based on the work of Egli [24] and Fine [25]).

Probability of Service in the Absence of Interference

The probability of service in the absence of interference is based on exceeding some minimum field strength at some percentage of the locations for some percentage of time. The time percentage is fixed at 90% (the normal television planning factor) and the location percentage is output as P_s, the probability of service, as a function of range R.

$$F(50, 50; R) - k(90)s_T - k(P_s)s_L = F\text{min} \qquad (10.31)$$

Fmin can be inferred from Part 73 of the FCC rules, which defines the relationship of the median field strength to each of three service contours:

Service contour	(L,T)	F(50,50) at UHF (dBµ)
Principal Community	(90,90)	80
Grade-A	(70,90)	74
Grade-B	(50,90)	64

For example, the Grade-A service contour refers to the contour on which a standard television receiver is expected to produce acceptable picture quality at least 90% of the time for at least 70% of the locations. The range R of each service contour can be found by using the graphs in FCC report R-6602 (Figures 10.17 and 10.18). It is a function of the frequency (low VHF, high VHF, or UHF), the transmit power, and the antenna height.

Probability of Interference

In the presence of an undesired signal u, interference to a desired signal d is presumed when the desired field strength level minus the interference field strength level is less than some required protection factor. Assuming that the fields (expressed in

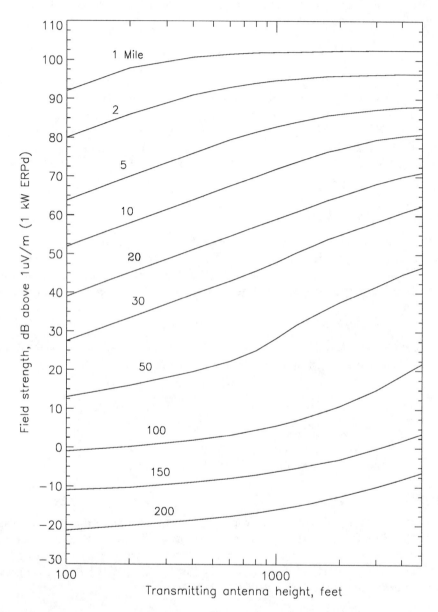

Figure 10.17 Television channels 14-83, estimated field strength exceeded at 50% of the potential receiver locations for at least 50% of the time at a receiving antenna height of 30 ft.
After: Damelin, J. et al., "Development of VHF and UHF Propagation Curves for TV and FM Broadcasting," FCC Report R-6602, September 7, 1966 (third printing May 1974), Figure 29.

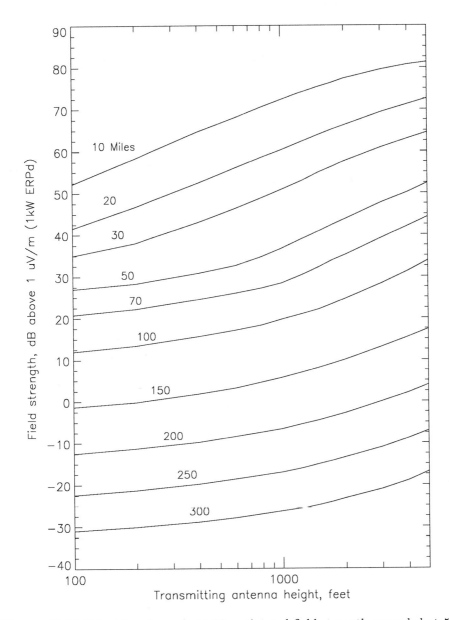

Figure 10.18 Television channels 14-83, estimated field strength exceeded at 50% of the potential receiver locations for at least 10% of the time at a receiving antenna height of 30 ft.
After: Damelin, J. et al., "Development of VHF and UHF Propagation Curves for TV and FM Broadcasting," FCC Report R-6602, September 7, 1966 (third printing May 1974), Figure 30.

decibels) are uncorrelated and distributed normally, the desired-to-undesired field strength difference *not* reached for $L\%$ of the locations $T\%$ of the time is:

$$Fd - Fu = Fd(50, 50; Rd) - Fu(50, 50; Ru) + k(T)\sqrt{s_{Td}^2 + s_{Tu}^2}$$
$$+ k(L)\sqrt{s_{Ld}^2 + s_{Lu}^2} \qquad (10.32)$$

Letting $R(Q)$ equal the field strength protection ratio in decibels required to provide service quality Q and taking 10% as the time criterion, one can solve for the interference contour at spatial level Pi. This contour is the locus of points such that the probability of receiving interference more than 10% of the time is $L = Pi$. Or, given a range of interest, Pi can be found via solution of:

$$k(Pi) = \frac{[Fu(50, 50; Ru) - Fd(50, 50; Rd) - k(10)\sqrt{s_{Td}^2 + s_{Tu}^2} + R(Q)]}{\sqrt{s_{Ld}^2 + s_{Lu}^2}} \qquad (10.33)$$

Probability of Interference to Service

From subsections two and three, the probability of service without regard to interference is Ps and the probability of interference without regard to service is Pi. The interference to service can therefore be judged by the joint probability of Ps and Pi. Assuming Fd and Fu to be uncorrelated normal random variables, this joint probability can be expressed in terms of tabulated cumulative bivariate normal probability functions (for example, see Section 26.3 of Reference [10]).

Receiver Variability of Susceptibility

Implicit in the analysis so far is a fixed field strength protection ratio $R(Q)$ for all receivers. $R(Q)$ equals the difference between the protection required at the television receiver input (referred to as receiver susceptibility) and the protection afforded by the receiving antenna through directivity and polarization discrimination. When solutions for a range of receiver susceptibility values are examined, one finds that the fraction of area at risk due to interference (denoted as R) is approximately related to the receiver susceptibility in decibels (denoted as x) via:

$$R \approx \exp[A(x - C) + D]$$
$$\approx \exp(Ax + B) \qquad (10.34)$$

where A and D are regression coefficients for the particular interference situation being examined, $B = -AC + D$, and C equals the antenna contribution to protection in decibels due to both directivity and polarization discrimination. Consequently, the unconditional area fraction at risk (i.e., the value averaged over the distribution

Coverage Analysis and Simulation 155

of receiver susceptibility) is given by:

$$E(R) = \int \frac{1}{\sqrt{2\pi\sigma_x^2}} \exp(Ax + B) \exp[-\frac{(x-x_0)^2}{2\sigma_x^2}]dx \tag{10.35}$$

This equation is based on the fact that receiver susceptibility in decibels is normally distributed with mean x_0 and standard deviation σ_x ([26]). Technically the upper limit of the equation is finite; however, the bulk of the contribution to the integral occurs when x is within a few standard deviations of x_0. Consequently, it is acceptable to treat the limits as $-\infty$ to ∞. With the help of relation 5.(A2) in Reference [4], the equation can be expressed in closed form as:

$$E(R) = \exp[Ax_0 + B + (\frac{A\sigma_x}{\sqrt{2}})^2] \tag{10.36}$$

Sample Applications

#1 Land-mobile interference sources. Consider, for example, the worst-case situation proposed in Docket 85-172; i.e., UHF television reception at the Grade-B contour directly between a desired television source (equivalent to channel 22, WMPT, Annapolis, MD, 5 MW at 870 ft) and multitransmitter land-mobile sources of maximum permissible ERP and height (1 kW at 500 ft) 75 mi distant. For this so-called worst square mile, the following parameter values apply:

1. $Fd(50,50;49.9 \text{ mi}) = 64 \text{ dB}\mu$
2. $Fu(50,50;75 \text{ mi}) = 10 \text{ dB}\mu + 5 \text{ dB}\mu$ (multisource factor) $= 15 \text{ dB}\mu$ [7]
3. $k(10) = -1.282$
4. $s_{Ld} = s_{Lu} = 12 \text{ dB}$
5. $s_{Td} = 6.9 \text{ dB}$
6. $s_{Tu} = 10.9 \text{ dB}$
7. $R(Q) = 40 \text{ dB}$ (at the set requirement for the median set) - 20 dB (antenna protection) = 20 dB [8]

Substitution into Equation (10.33) gives $k(Pi) = -0.7343$, which implies $Pi = 23.1\%$. The joint probability of service and interference is thus $Pj(50\%,23.1\%) = 2.8\%$.

---

[7]Interference is commonly regulated by specifying an allowable D/U value. Here D refers to a (50,50) field strength value but U refers to a (50,10) value. For D/U = 40 dB as proposed in Docket 85-172, U is 24 dBμ at the Grade-B contour. Figure 26 of FCC report R-6602 shows that the (50,50) field strength value under such conditions is 10 dBμ; A 5-dB field strength increase due to multiple transmitters is cited on page 6, Section III.B. of Reference [27].

[8]Paragraph 19.(corrected) of Docket 85-172 uses a receiver susceptibility value of 45 dB. This is the program passable value for the 90 percentile set found by the Canadian study [26]. The same source cites a median value of 40 dB. The 20-dB antenna protection value is that cited on page 14, Section III.A. of Reference [27].

Applying this analysis to each square mile within the Grade-B contour allows one to quantify the loss of service on an area basis. The loss of service factor overall is 0.40%. However, that value assumes in essence all receivers are as susceptible to interference as the median set and therefore is an underestimation. Regression coefficients describing the relation between loss of service and susceptibility are $A = 0.115$ and $B = -7.83$. Using these values in Equation (10.36), along with a median susceptibility of $x_0 = 40$ dB and a standard deviation of $\sigma_x = 3.9$ dB [26], gives an actual area at risk of 0.44%.

#2 UHF-TV interference source. Consider, for example, the cochannel UHF television sharing of channel 22, WMPT in Annapolis, MD with WDAU in Scranton, PA, located 173 mi away. The latter station produces 3 MW at 1560 ft. For the so-called worst square mile, the following parameters apply:

1. $Fd(50,50;49.9 \text{ mi}) = 64$ dBμ
2. $Fu(50,50;123.1 \text{ mi}) = 36.8$ dBμ
3. $k(10) = -1.282$
4. $s_{Ld} = s_{Lu} = 12$ dB
5. $s_{Td} = 7$ dB
6. $s_{Tu} = 10.2$ dB
7. $R(Q) = 28$ dB (at the set requirement for the median set) - 10 dB (antenna protection) = 18 dB [9]

Substitution into Equation (10.33) gives $k(Pi) = 0.392$, which implies $Pi = 65.2\%$ and $Pj(50\%,65.2\%) = 21.4\%$. Repeating such analyses for each small region within the Grade-B contour leads to the result that 4.3% of the WMPT Grade-B coverage area is at risk to interference from WDAU. This is for typical receivers. The regression coefficients that describe the interference likelihood for other fixed receiver susceptibilities are $A = 0.112$ and $B = -6.29$. Using $x_0 = 28$ dB and $\sigma_x = 3.9$ dB in Equation (10.36) then leads to an overall area at risk of 4.7%.

Summary of Results

Table 10.4 shows cochannel potential interference results for three general situations of interest: (1) land-mobile interference sources at 158 mi spacing, corresponding to conditions of the present sharing Docket 18261; (2) land-mobile interference sources at 125 mi spacing, as proposed in Docket 85-172; and (3) existing UHF-TV to UHF-TV sharing. As expected, the relaxation of the separation requirement in Docket 85-172 relative to Docket 18261 leads to increased interference potential. However,

---

[9] A receiver susceptibility value of about 28 dB for cochannel television interference has been reported by Fredendall [28]. The quality criteria used there approximates the program passable criteria used for land-mobile receiver susceptibility numbers. Also, imprecise frequency offset was used as is the case for WMPT and WDAU. In addition, the spacing criterion for offset UHF-TV listed in the 6th (Final) Report on TV Allocations is 28 dB.

Table 10.4 Results of Cochannel Potential Interference Analysis

After: Hess, G. "Further Sharing of UHF Television by Private Land Mobile Radio Services," *37th IEEE Vehicular Technology Conference*, Tampa, FL, June 1987, p. 329.

Separation (miles)[a]	Receiver susceptibility (median/std dev in dB)	Antenna protection (dB)	Grade-B area at risk (%)
Case 1: Land mobile (20 1 kW sources at 500 ft) sharing with a UHF-TV source equivalent to ch 22, WMPT, Annapolis, MD			
158	40/3.9	20	0.2
125	40/3.9	20	0.4
Case 2: UHF-TV sharing with UHF-TV			
(a) Desired source = ch 22, WMPT, Annapolis, MD (5 MW at 870 ft, 49.9 mi Grade-B contour); undesired source = ch 22, WDAU, Scranton, PA (3 MW at 1560 ft)			
173	28/3.9	10	4.7
(b) Desired source = ch 24, WKJL, Baltimore, MD (1.18 MW at 1073 ft, 49.9 mi Grade-B contour); undesired source = ch 24, WNPB, Morgantown, WV (2.25 MW at 770 ft)			
162.4	28/3.9	10	3.1
(c) Desired source = ch 21, WLIW, Garden City, NY (1.2 MW at 1220 ft, 49.9 mi Grade-B contour); undesired source #1 = ch 21, WHP, Harrisburg, PA (1.2 MW at 1220 ft)			
181.8	28/3.9	10	1.8
undesired source #2 = ch 21, WNHT, Concord, NH (1.85 MW at 1133 ft)			
198.5	28/3.9	10	0.9
Case (c) result of 2 interferers			2.7
Averages for cases (a), (b), and (c)			
175.2	28/3.9	10	3.5

a. 158-mi separation (50-dB *D/U*) pertains to Docket 18261; 125-mi separation (40-dB *D/U*) is proposed in Docket 85-172

the interference likelihood remains small on an absolute basis and also small on a relative basis compared to interference expected from existing UHF-TV cochannel assignments.

10.8.2 Noncochannel Sharing

This section discusses evaluation of interference caused by mechanisms other than cochannel. The methodology already outlined can be applied to situations where the interference source is located sufficiently outside the Grade-B contour. Examples are UHF-TV to UHF-TV interference analysis and first adjacent channel (upper and lower) analysis of land-mobile interference to UHF-TV reception. In the latter situation, the D/U requirement of 0 dB suggested in Docket 85-172 (and in practice via Docket 18261) implies that the land-mobile base station must be at least 12 miles from the Grade-B contour.

For other mechanisms, where the interference source can lie within the Grade-B television coverage contour, a different type of analysis is desirable. This is so the precise transmit and receive antenna directivities can be used for the particular geometry of each location considered. In addition, some short path lengths involved between the undesired land-mobile source and the television receiver are beyond the scope of FCC report R-6602. In its place propagation models that have been verified by the land-mobile community can be used.

Monte Carlo Simulation Description

A representative base station scenario will be examined where 20 undesired sources of 1-kW ERP at an antenna height of 500 ft are located 30 mi from a desired television source (for mechanisms where nonlinearity is involved, 21 such interference sources are included). The desired television source is assumed to radiate 5 MW from 1000 ft. FCC report R-6602 figures are used to establish the desired field strength at points randomly chosen in a circle of user-specified radius about the undesired site. A dense land-mobile site option is used that adjusts the actual ERP in accordance to that which would be seen around a site with 90 antennas on a 90- by 80-ft grid.

Propagation losses between the base station and nearby television receivers are modeled by Hata's empirical formula based on Okumura's measurements [29]. The loss is checked to make sure it at least equals free-space loss. Also, a lognormal spatial variation loss factor is added to the median value that Hata's model produces.

A receiver antenna height of 30 ft is used with a mainlobe antenna gain of 15 dBd at the Grade-B contour, dropping to -1 dBd at the Principal Community contour. A net loss of 5 dB is attributed to feedline and mismatch. A receiver noise figure of 10 dB is used, with additional noise coming in from the antenna, which is assumed to be at a noise temperature of 290 K. A 10-dB polarization protection is assumed for reception of the desired signal over the undesired signal. The protection due

Coverage Analysis and Simulation 159

to antenna directivity is determined in accordance with the pointing angle between the desired source and the undesired source. The amount of protection is based on averaging the results reported for UHF antennas in NTIA report 79-22 [30].

Each simulation consists of 100,000 trials for which the following are noted: (1) whether or not the desired-to-noise power ratio at the receiver exceeds that required for adequate service (taken here as 28 dB; i.e., passable picture quality), (2) whether or not the desired-to-undesired power ratio at the receiver exceeds the receiver susceptability value, and (3) whether or not items (1) and (2) are simultaneously satisfied.

Monte Carlo Analysis Parameters

A rather large number of parameters must be specified to conduct the interference simulation. Monte Carlo results that follow consider interference from land-mobile repeaters located at the maximum proposed distance from city center (30 mi). Television receivers out to some radius away from the repeaters are apt to be affected by various noncochannel interference mechanisms. The particular parameter values used are listed in Table 10.5.

Simulation Results

Table 10.6 lists results for 15 noncochannel interference situations with land-mobile repeater sources and two examples of existing UHF-TV to UHF-TV sharing. The greatest interference potential for the land-mobile cases appears to be only a few hundredths of a percent of the Grade-B service area. Even if there are a number of such land-mobile sites, the interference potential will be substantially less than that predicted for existing UHF-TV assignments examined.

10.9 800-MHz SMR Cochannel Spacing

10.9.1 Summary

This section reviews current FCC rules and methods used to establish cochannel separations between 800-MHz SMRs. An analysis of talk-out interference probability is given, also based on FCC methods. Shortcomings of the FCC report R-6602 propagation data are discussed and an alternative interference analysis using Okumura/Hata propagation and the concept of faded sensitivity is offered. Regardless of approach, however, it is clear that the protection afforded by current rules is not synonymous with interference-free operation. The lack of major interference problems so far can be explained by a number of reasons; unfortunately, the mitigating factors are becoming less common as the spectrum becomes more crowded. The interest in licensing so-called short-spaced sites will further aggravate the problem

Table 10.5 Parameter Values for Monte Carlo Noncochannel Interference Simulations

After: Hess, G. "Further Sharing of UHF Television by Private Land Mobile Radio Services," *37th IEEE Vehicular Technology Conference*, Tampa, FL, June 1987, p. 329.

Attribute	Value
Desired frequency	591.25 MHz
Undesired source	Base station receiver
Undesired/desired spacing	30 mi (48.3 km)
Radius of interest about undesired site	5 km
Median desired field strength	87 dBµ at min range
(based on 5 MW at 1000 ft)	81 dBµ at max range
Desired signal temporal standard deviation	1.9 dB at min range
	2.7 dB at max range
Spatial standard deviation	12 dB
S/N required for TASO "passable" picture quality	28 dB
Undesired ERP	30 dBW + 3 dBW if nonlinear mechanism involved + 5 dBW multisource factor
Undesired antenna height	500 ft (152 m)
Site character	"dense"
Receive antenna height	30 ft (9.14 m)
Undesired propagation model	Hata, suburban environment
Undesired spatial standard deviation	6 dB
Receive antenna gain	Based on desired field strength, ranges from -1 dBd for principal community to 15 dBd at Grade-B contour
Receive antenna polarization discrimination	10 dB
Feedline plus mismatch loss	5 dB
Receiver noise figure	10 dB
Antenna noise temperature	290 K
Receiver noise bandwidth	4 MHz
Receiver input impedance	300 ohms

Table 10.6 Results of Noncochannel Potential Interference Simulations

After: Hess, G. "Further Sharing of UHF Television by Private Land Mobile Radio Services," *37th IEEE Vehicular Technology Conference*, Tampa, FL, June 1987, p. 329.

Interference mechanism[a]	Land mobile interference sources Receiver susceptibility[b]	Fraction of Grade-B service area at risk due to interference
First adjacent situations		
1. Upper 1st, 0-3 MHz	-23 dB	<0.001%
2. Upper 1st, 3-6 MHz	-40 dB	<0.001%
3. Lower 1st, 3-6 MHz	-7 dB	0.017%
4. Lower 1st, 0-3 MHz (Two interferers[c])	-36 dB	<0.001%
5. Lower 1st	-18 dB	0.0035%
6. Upper 1st	-23 dB	0.001%
Other situations		
7. Upper 4th (1/2 IF)	-19 dB	0.015%
8. IF (+7 channel)	-26 dB	0.0055%
9. Image (+14 and +15 channels)	-20 dB	0.012%
10. Lower 2nd	-29 dB	0.0082%
11. Lower 3rd	-33 dB	0.0043%
12. Lower 4th	-36 dB	0.0027%
13. Upper 2nd	-27 dB	0.011%
14. Upper 3rd	-28 dB	0.009%
15. Upper 4th	-38 dB	0.0018%
UHF-TV interference sources		
16. Desired = channel 21 WLIW, Garden City, NY, 1.15 MW at 400 ft, 35.7 mi Grade-B contour	Undesired = channel 20, WTXX, Waterbury, CT, 2.24 MW at 1200 ft, 55.2 mi separation -16 dB[d]	1.9%
17. Desired = channel 21, WLIW, as above	Undesired = channel 25 WNYE, New York, NY, 0.658 MW at 583 ft, 28.3 mi separation -25 dB	2.4%
Net effect of two interferers		4.2%

a. Results for situations 1 through 6, 16, and 17 are based on area interference; other results are based on a Monte Carlo analysis

b. Land mobile interference receiver susceptibility numbers are based on median program passable criteria as discussed in Appendix 3 of reference [31]

c. A nominal 8 dB improvement in receiver susceptibility is possible for cochannel and cross-modulation interference mechanisms if channel spacings near 15 kHz are avoided

d. The receiver susceptibility numbers are those cited by land mobile interests in reference [27], adjusted downward 6 dB to account for the peak/average power difference between multiple carrier measurements and television signals; an antenna protection of just 10 dB is used because polarization discrimination does not apply if the rules for doing so are not sufficiently stringent.

Various methodologies for controlling short-spacing license grants are noted. The methodology perhaps most in line with FCC rules and practice is a 40/20 rule. This means that the 20-dBμ contour of the proposed station, as obtained from FCC report R-6602 $F(50,10)$ UHF curves, must fall outside the 40-dBμ contour of the existing station, as obtained from the $F(50,50)$ UHF curves. Existing SMRs are treated as equivalent to at least 500 W at 500 ft and 70 mi is retained as the maximum required spacing. An additional separation of 5 mi is recommended for multisite low-power, low-height operation.

10.9.2 FCC Spacing Rules and Engineering Justification

FCC rules normally provide an SMR operator with 70 mi of protection, in that no other cochannel operator can be licensed within 70 mi of the existing base site (47 C.F.R. §90.621(b)[10]). Certain exceptions to this rule, such as high mountain sites in California, are afforded additional mileage protection. Shorter spacings than normal have been allowed under waiver, provided that both parties involved consented. More recently, the FCC has considered short spacings justified solely by engineering showings, the precise form of which was the subject of a notice of inquiry early in 1991.

SMRs are allowed up to 1-kW ERPd[11] at 1000-ft *height above average terrain* (HAAT) in general, whereas conventional users are allowed such parameters only in urban areas and are limited to 500 W ERPd at 500-ft HAAT in suburban areas (47 C.F.R. §90.635, Table 2). The rationale behind these numbers is a 20-mi service area, which in turn is related to a 40-dBμ contour (47 C.F.R. §90.621(c)). The service area is defined as the region within the 40-dBμ field strength contour as determined from the $F(50,50)$ UHF-TV curves in FCC report R-6602 [22]. Unstated are two additional factors needed to use the R-6602 curves: (1) a 9-dB adjustment is required to reflect operation at realistic mobile antenna heights (report R-6602 is referenced to a subscriber antenna height of 30 ft), and (2) a 10-dB urban clutter factor is associated with urban operation.

With this in mind, one finds that for a 1-kW urban station the curve should be entered at $40 + 9 + 10 = 59$ dBμ and for a 500 W suburban station the curve entry value is $40 + 9 + 3 = 52$ dBμ (the 3-dB factor is necessary because the curve is presented in terms of field strength per kilowatt ERPd). To standardize results read from report R-6602 curves, the Mass Media Bureau of the FCC has tabulated the various curves and written software to interpolate where necessary among the specified values [32]. Using a carrier frequency of 850 MHz and 164 ft

---

[10]Various FCC rules will be referred to by citing Title 47, Code of Federal Regulations, and the relevant numbered parts as per October 1, 1988.

[11]The "d" is appended to ERP to remind us that the reference antenna is a dipole rather than an isotropic source (for which EIRP is the usual shorthand).

Coverage Analysis and Simulation   163

(50 m) for terrain roughness (the nominal value specified in report R-6602), the computer program indicates a coverage radius of 19.25 mi for a 1-kW urban station at 1000 ft (28.22 mi under suburban propagation conditions) and 20.05 mi for a 500-W suburban station at 500 ft.

The 70 mi normal base-to-base spacing for interference protection is ostensibly tied to the desire to prevent the 30-dBμ $F(50,10)$ contour from impinging on the service area (47 C.F.R. §90.621(c)). However, computer solutions indicate that for a 1-kW station at 1000 ft, $F(50,10) = 30$ dBμ at just 30.95 mi over an urban path, and a 500-W suburban station at 500 ft produces 30 dBμ at 32.05 mi. The base-to-base spacings implied by these computer solutions are thus only 19.25 + 30.95 = 50.2 mi (1 kW, 1000 ft, urban) and 20.05 + 32.05 = 52.1 mi (500 W, 500 ft, suburban). Both of these distances are far shorter than the rules require. However, as will soon be shown, this is fortunate because "protection" is hardly synonymous with "interference-free" operation if such short spacings were in fact to occur. By using 70 mi, the rules imply that at the edge of service the field strength difference between the desired $F(50,50)$ value and the undesired $F(50,10)$ value should be 59 - 35.4 = 23.6 dB (1 kW, 1000 ft, urban) and 52 - 30.8 = 21.2 dB (500 W, 500 ft, suburban). Both those differentials are substantially greater than the 10 dB implied by a simple 40/30 comparison.

10.9.3 Probability of Interference Analysis

The view of interference used by the broadcast arm of the FCC and tied to the propagation curves of report R-6602 is as follows [21]. In the presence of an undesired signal u, interference to a desired signal d is presumed when the desired field strength level minus the interference field strength level is less than some required protection factor. Assuming that the fields (expressed in decibels) are uncorrelated and Gaussian distributed, the desired-to-undesired field strength difference not reached for $L\%$ of the locations $T\%$ of the time is (recall Section 10.8.1):

$$Fd - Fu = Fd(50,50;Rd) - Fu(50,50;Ru) + k(T)\sqrt{s_{Td}^2 + s_{Tu}^2}$$
$$+ k(L)\sqrt{s_{Ld}^2 + s_{Lu}^2} \qquad (10.37)$$

where Rd equals the desired path length, Ru equals the undesired path length, s_{Td} equals the standard deviation of time variability for the desired signal, s_{Tu} equals the standard deviation of time variability for the undesired signal, s_{Ld} equals the standard deviation of location variability for the desired signal, and s_{Lu} equals the standard deviation of location variability for the undesired signal. The k factors represent the multipliers necessary to obtain the T percentile and L percentile for Gaussian random variables (i.e., $k = 1.28$ for 90%, -1.28 for 10%, etc.). The temporal standard deviations can be inferred from the differences between report R-6602 $F(50,50)$ and $F(50,10)$ curves. The location standard deviation is assumed independent of range and equals 10.9 dB for land-mobile applications, according to Reference [33].

Letting $R(Q)$ equal the field strength protection ratio (expressed in decibels) required to provide service quality Q, and taking 10% as the time criterion, one can solve for the interference contour at spatial level Pi. This contour is the locus of points such that the probability of receiving interference more than 10% of the time is $L = Pi$. Or, given a range of interest, Pi can be found by solving:

$$k(Pi) = \frac{[Fu(50,50; Ru) - Fd(50,50; Rd) - k(10)\sqrt{s_{Td}^2 + s_{Tu}^2} + R(Q)]}{\sqrt{s_{Ld}^2 + s_{Lu}^2}} \quad (10.38)$$

Reference [33] assumes that $R(Q)$ equals 6 dB. This is in line with the *static* performance of typical 25-kHz land-mobile radio equipment for circuit merit 3 quality (readable speech with only a few syllables missed and occasional repetition required; this equates well to 20-dB quieting sensitivity, for which the signal-to-thermal noise power ratio is typically 7 dB). At the worst square mile (i.e., at the coverage radius away from the desired station and directly toward the undesired station) the probability of interference for 500 W, 500 ft suburban SMRs spaced 70 mi is given by:

$$\begin{aligned} k(Pi) &= \frac{1}{\sqrt{10.9^2 + 10.9^2}}[Fu(50,50; 49.95) - Fd(50,50; 20.05) \\ &\quad - (-1.28)\sqrt{1.02^2 + 8.44^2} + 6.0] \\ &= \frac{(17.0 - 49.0) + 16.88}{15.41} \\ &= -0.988 \end{aligned} \quad (10.39)$$

implying an interference probability of about 16%, which is even greater than the thermal outage probability in the absence of interference (which through the specification of 40 dBµ is about 10%). Other locations in the coverage area are of course impacted less, but it certainly seems unreasonable to view this situation as providing protection. And yet the 40/30 methodology implies even closer spacings (as low as 52.1 mi) would be allowable. Clearly something is amiss[12].

---

[12]The spacing rules do not address talk-in interference, presumably because with equal power subscriber units uniformly distributed about the service areas of two comparable systems, talkin and talkout are balanced. However, in unbalanced cases talk-in interference probability can exceed talk-out interference probability. Differences in base transmit power and antenna height will cause unbalance. Also, the fact that there is a sizeable spread in mobile transmit powers and increasing use of portables further aggravates the imbalance between talkin and talkout.

10.9.4 Discussion of the Shortcomings

Temporal Variability

One problem with the preceding methodology is its reliance on report R-6602 propagation data. Those data were collected with 30-ft-high receive antennas for broadcast design purposes. Translations of those data to the 5-ft antenna heights common to land-mobile radio operation are of dubious value at percentiles other than the median. For example, the difference between $F(50,50)$ and $F(50,10)$ at 20.05 mi is only about 1.3 dB for 500-ft base antenna height; hence, the standard deviation of $s_{Td} = 1.02$ used in Equation (10.39). Such small variation is understandable for data obtained with a 30-ft receive antenna height; a height well within the radio horizon of a 500-ft transmit antenna and unlikely to be surrounded by nearby scatterers. Yet a mobile antenna at 5 ft *is* normally surrounded by many nearby scatterers, leading to a Rayleigh field and substantial variability.

For longer paths, where the 30-ft-high antenna is unlikely to have a direct view of the transmit antenna, the report R-6602 data do indicate substantial fading, more in line with a Rayleigh field. For example, the difference between $F(50,50)$ and $F(50,10)$ at 49.95 mi is 10.8 dB for 500-ft base antenna height; this implies a standard deviation of 8.44 dB, which was previously used for s_{Tu}. Of course, even if both desired and undesired temporal variations are in line with those associated with Rayleigh fading, the theoretical analysis incorrectly treats those variations as Gaussian in nature.

Spatial Variability

The predicted 16% probability of interference is inflated due to the use of overly high values for both s_{Ld} and s_{Lu}[13]. Location variability of average power (i..e., the Rayleigh effect is filtered away) when expressed in decibels is indeed generally Gaussian in nature for land-mobile paths; hence, the term lognormal shadowing. The standard deviation is a function of the size of the area considered, decreasing in value as the area considered decreases in size [8]. For the large areas associated with high-power, high-site SMRs, a nominal value of at most 8 dB is in order. This should decrease to about 6 dB for low-power, low-site SMRs.

Median Field Strength Accuracy

Numerous organizations have conducted field tests to measure the path loss between base station antenna sites (which are relatively high) and mobile radio units (whose antennas are relatively near ground level). Such data allow the generation of empir-

---

[13]Whereas the probability is decreased by the overly low accounting of temporal (Rayleigh) fading and by the use of a capture requirement 1 dB less than that typically needed for static circuit merit 3 quality.

ical models for predicting path loss. One of the most widely accepted models is that of Okumura [2]. It offers relatively good accuracy with relatively little complexity [34].

The basic procedure has been nicely summarized by four figures presented in Section 5.2. A reference level of median attenuation beyond the free-space path loss (referred to as excess path loss) is first obtained. The reference assumes an urban environment with base antenna height of 200 m and mobile antenna height of 3 m. Prediction curves are noted for path lengths between 1 and 100 km and frequencies between 100 MHz and 3 GHz. The reference loss is then adjusted according to the actual base antenna height, the actual mobile antenna height, and the actual environment type (the choices being urban, suburban, quasi-open, and open). Note that doubling the base antenna height typically decreases path loss by 6 dB (in accord with plane earth propagation), whereas doubling the mobile antenna height typically decreases path loss by just 3 dB. Okumura's results have been fitted to simple mathematical formulas by Hata [29]. As published, the formulas are limited to the rather modest path lengths common in cellular systems, but they can readily be extended in range as desired (refer to the Fortran code included in Section 5.2).

Okumura also describes adjustments for terrain undulations that can be applied if topographical data are available. However, terrain-shadowing adjustments are more commonly generated by applying some form of diffraction theory to the terrain data. The basics of knife-edge diffraction loss for a single obstruction are given in Reference [1]. An approximate formula for the loss in decibels is $10\log(\lambda r_1/4\pi^2 h^2)$, where λ is the wavelength, r_1 is the distance from source to obstacle, and h is the effective height of the obstacle. To be valid, the distance from obstacle to receiver must be large relative to r_1, which in turn must be large relative to h, which in turn must be large relative to λ. In practice, of course, several obstacles often occur along paths of interest. See, for example, Giovaneli [35] for how multiple knife-edge problems can be handled. Also in practice, obstacles seldom behave like knife edges. See de Assis [36] for an example of how rounded obstacles can be treated.

A comparison of report R-6602 and Okumura/Hata predictions yields Table 10.7. Interestingly, it appears that report R-6602 with 50 m terrain undulations yields predictions fairly well in line with the Okumura/Hata empirical model with 20 m undulations. Although the terrain adjustment relation given in the original R-6602 report is ADJ = $-0.03\delta h\{1 - [\text{Freq}(\text{MHz})/300.0]\}$, the curves presented in fact involve a constant term of 4.8 dB so that the adjustment is 0 dB for 650 MHz and δh equal to 50 m. This constant term is explicitly noted in later work [37] and in the FCC rules (47 C.F.R. §73.684(1)). Thus, it is clear why methods 1 and 2 in Table 10.7 differ by about 5.8 dB: 4.8 dB is the constant term and 1 dB is the result of using 850 MHz, rather than the reference value of 650 MHz. The 1-dB impact of the latter seems unimportant but ignoring it can lead to spacing differences of several miles.

Compared to Okumura, report R-6602 predictions tend to overestimate signal

Coverage Analysis and Simulation

Table 10.7 Comparison of FCC report R-6602 and Okumura/Hata Propagation Models

Method / distance (miles)[a]	$F(50,50)$ in dBu[b]						
	5.0	10.0	15.0	20.0	30.0	40.0	50.0
1	78.7	67.8	61.0	54.9	43.8	33.1	24.1
2	73.0	62.1	55.2	49.1	38.0	27.3	18.4
3	70.1	61.5	54.8	48.0	36.5	26.4	20.0

a. Method 1 is R-6602 using 9-dB antenna adjustment for 30 ft to 5 ft and 0-ft roughness factor; method 2 is like method 1 but uses the standard 164-ft (50 m) roughness factor; method 3 uses Okumura suburban propagation with nominal terrain adjustment corresponding to a roughness factor of about 20 m

b. Parameters assumed are 1 kW ERPd source at 1000 ft to dipole receive antenna at 5 ft, 850 MHz

strength at short ranges and underestimate signal strength at long ranges. This behavior becomes more pronounced as the base antenna height is lowered. Thus, report R-6602 interference predictions will be more optimistic (lower in probability) than those made using Okumura propagation.

10.9.5 Alternative Interference Analysis

Operationally, the impact of Rayleigh fading can be viewed as some fixed loss in sensitivity. For standard 25-kHz analog FM radios (as well as many digital voice systems), the sensitivity loss is about 10 dB. Therefore, to achieve circuit merit 3 quality in fading, a protection ratio of 17 dB is suggested. The probability of interference is now simply related to the lognormal location variability (see [16] or Section 10.6.1):

$$\text{Prob}[(u-d) > -R(Q)] = \frac{1}{2}\text{erfc}[\frac{(Fd(50,50) - Fu(50,50)) - R(Q)}{2s_L}] \quad (10.40)$$

where $s_{Ld} = s_{Lu} = s_L$. Solving this equation for the case previously addressed gives an interference probability of (report R-6602 propagation [14]):

$$\begin{aligned}\text{Prob}[(u-d) > -R(Q)] &= \frac{1}{2}\text{erfc}[\frac{((49-9)-(17-9)-17)}{16}] \\ &= \frac{1}{2}\text{erfc}(0.9375)\end{aligned}$$

[14] The 9-dB antenna factors are now explicitly included to facilitate comparison with Okumura/Hata results.

$$= 0.0925 \qquad (10.41)$$

which is comparable to the thermal noise outage of 10%, or (Okumura/Hata propagation):

$$\begin{aligned}\text{Prob}[(u-d) > -R(Q)] &= \frac{1}{2}\text{erfc}[\frac{(38.44-12.51-17)}{16}] \\ &= \frac{1}{2}\text{erfc}(0.5581) \\ &= 0.215 \qquad (10.42)\end{aligned}$$

a substantially worse value because, relative to Okumura/Hata, report R-6602 tend to overestimate signal strength at the shorter ranges and underestimate signal streng at longer ranges.

Of course it is unfair to count as interference situations where the capture crite rion is not met but the desired signal level itself is too weak relative to thermal nois to provide acceptable communication. Thus, the interference probability should b treated as a joint probability of meeting the thermal noise criterion but not meet ing the interference capture criterion. Wong [21] addresses this issue and show how the bivariate Gaussian distribution function can be used to express the desire probability (this was noted earlier in Section 10.8.1). With 90% service probabilit against thermal noise and 9.25% (21.5%) interference probability, the true outag is 14.86% (24.35%). Thus, if report R-6602 propagation holds, the outage rate i increased by about 50% at the worst square mile due to the presence of interference If Okumura/Hata propagation holds, the outage rate is boosted almost 150%.

10.9.6 Equivalent Short-Spacing Methodologies

The preceding analyses make it clear that even with 70 mi spacings substantial inter ference is possible between high-power, high-site SMRs. Yet even shorter spacing apparently are successfully in operation. How can this be? Several possibilities com to mind: (1) intervening terrain provides additional isolation between the systems (2) the coverage areas of the systems tend to be away from, rather than toward each other, (3) the new cochannel site is not yet operational, or fully loaded, an (4) interference is in fact present but the trunking nature of the systems allows call in trouble to move to clear(er) channels. One should also bear in mind that to dat few channels in few markets actually have a cochannel user within 70 mi. Also cases that have arisen to date can generally be handled by careful frequency choice looking for frequencies with lower power, lower site users or users with directiona antennas. Unfortunately, as time passes and radio use expands, such interferenc mitigation possibilities will diminish.

The focus now turns to methodologies that provide the equivalent of what i allowed by existing rules. The analysis of 500 W, 500-ft stations indicates that th current report R-6602 methodology and 70-mi spacing allowance act like a 40/18.8 rule. Analysis of 1000 W, 1000-ft stations produces a similar result (40/16.4 rule). Rounding the lower numbers off, it would not seem unreasonable to call for a 40/20 rule using past procedures. This means that the 20-dBμ contour of the proposed station, as obtained from the $F(50,10)$ report R-6602 UHF curve, must fall outside the 40-dBμ contour of the existing station, as obtained from the $F(50,50)$ R-6602 UHF curve. Crediting existing SMRs for being equivalent to at least 500 W at 500 ft and limiting spacing to at most 70 mi leads to the spacings shown in Table 10.8. For comparison purposes Table 10.9, based on the ostensibly current rule of 40/30,

Table 10.8 40/20 Rule Cochannel Spacing

	Base-to-base separation (miles)[a]					
Existing station ERP(W)/HAAT(ft)[b]	Proposed station ERP(W)/HAAT(ft)					
	1000/1000	750/750	500/500	250/250	100/200	50/100
1000/1000	70	70	70.0	63.0	55.6	46.6
750/750	70	70	70.0	59.4	52.1	43.1
500/500	70	70	66.7	54.8	47.5	38.4
250/250	70	70	66.7	54.7	47.4	38.4
100/200	70	70	66.7	54.7	47.4	38.4
50/100	70	70	66.7	54.7	47.4	38.4

a. Add 5 miles to spacing if proposed station is an ESMR, i.e. a low power, low antenna height station with geographic reuse

b. Conditions: (1) A 9-dB antenna factor for translating the 30-ft antenna height of report R-6602 to the 5-ft antenna height of typical mobiles is used, (2) no urban clutter reduction factor is included, (3) a delta h factor of 164 ft (50 m) is used, and (4) the frequency is specified as 850 MHz

is also given.

Unfortunately, as has earlier been pointed out, the equivalency of this methodology is lost as the spacing drops significantly below 70 mi due to the error in using report R-6602 (50,10) data. A 40/10 rule using only report R-6602 (50,50) data would avoid this error. Another way suggested for avoiding the error involves equalizing desired and undesired signal levels (using report R-6602 (50,50) values) 35 mi from existing sites. The current filing consensus involves a 40/22 rule, which does not avoid the error. The logic behind the number 22 involves the well-known 17-dB cellular design criterion for desired carrier-to-interference power ratio, see Section 10.10. Subtracting that value from 40 gives 23. An additional 1-dB adjustment is included so that report R-6602 tables can be used directly (the impact of operation at 850 MHz rather than 650 MHz is about 1 dB).

Table 10.9 40/30 Rule Cochannel Spacing

	Base-to-base separation (miles)[a]					
Existing station ERP(W)/HAAT(ft)			Proposed station ERP(W)/HAAT(ft)			
	1000/1000	750/750	500/500	250/250	100/200	50/100
1000/1000	70.0	65.9	60.3	50.6	45.2	37.5
750/750	68.5	62.4	56.7	47.1	41.6	34.0
500/500	63.9	57.1	52.1	42.5	37.0	29.4
250/250	63.9	57.1	52.1	42.5	36.9	29.3
100/200	63.9	57.1	52.1	42.5	36.9	29.3
50/100	63.9	57.1	52.1	42.5	36.9	29.3

a. Add 5 miles to spacing if proposed station is an ESMR, i.e. a low power, low antenna height station with geographic reuse

10.10 Cellular System Operation

10.10.1 The Cellular Concept

Today most people are familiar with car telephones, but probably not many really know what cellular radio means. Cellular is a system concept that has come into being because spectrum (the frequencies that carry the radio messages) is a limited resource assigned by the FCC to all types of radio users, including television broadcasters. The key elements of the cellular concept, frequency reuse and cell splitting, are explored in detail in Reference [38]. Substantial details of commercial analog cellular systems around the world are contained in Reference [39].

To provide telephone-quality audio, car telephone radio channels in the United States today occupy 60 kHz of spectrum, 30 kHz for transmissions to the car and 30 kHz for transmissions from the car. This means that each megahertz of spectrum supports only about 17 calls. The FCC is only able to allocate about 20 MHz of paired spectrum to the car telephone service; this implies 700 simultaneous calls per major metropolitan area. How can hundreds of thousands of users be handled with so little spectrum? The answer is to reuse the spectrum within the metropolitan area. Rather than link into the telephone system from a single high site that covers the entire metropolitan area (for example, like television and radio broadcasters do), users are instead linked via many low sites. A single low site of course can only cover a limited area, termed a cell, but many low sites taken together can cover the entire metropolitan area. With multiple sites the same frequency can be used simultaneously at all sites that are sufficiently separated.

An example of how Chicago, Illinois might be covered with 100 sites and a 4-

Coverage Analysis and Simulation 171

cell reuse pattern (allowing a geographic reuse factor of $100/4 = 25$) is shown in Figure 10.19. To limit interference, each cell is actually divided into six sectors and directive antenna patterns are used. An attractive feature of cellular radio is the ability to vary the size of the cells in accordance with user density; hence, the increase in cell size away from the city center. To sustain the reuse pattern with mixed cell sizes, power levels must be taylored to produce comparable signal levels at all cell boundaries. Otherwise excessive interference will result in the larger cells near large-cell/small-cell boundaries.

10.10.2 Call Control and Handoff

Of course as cell size decreases, the likelihood of driving beyond a cell's coverage before the call is completed goes up. Hence, the call needs to be handed off from one cell to another. Calls in need of handoff can be recognized by monitoring call quality and comparing it to some minimum threshold. The precise control algorithm is best studied by simulation. One example of a control algorithm is shown in Figure 10.20.

Notice that the algorithm reacts to both talk-in and talk-out link quality. This is important because the interference sources for each transmission direction are different, both in type (mobile or portable versus base) and location. The composite link quality that is maximized refers to some measure like:

$$\text{Composite} = \left[\frac{1}{\text{new talk} - \text{in quality measure}} + \frac{1}{\text{new talk} - \text{out quality measure}} \right]^{-1}$$

Handoff control procedures for first-generation analog FM cellular systems were developed in the late 1970s. They have stood the test of time fairly well as cellular technology matured. Nonetheless, there is ample room for improvement. Three basic problems are: (1) the wrong quantity is monitored (the so-called *received signal strength indication* (RSSI) represents total received signal power, which is the desired power plus the interference and noise powers), (2) only talk-in performance is monitored (although to some degree the use of different supervision tones allows bad talk-out quality to be sensed), and (3) monitoring is updated at too slow a rate (typically only once per 7 sec). Field tests have shown that signal strengths decorrelate over periods of just tens of seconds [40] and shifts of several decibels are highly probably in just 1 sec. Second generation digital cellular systems, such as *United States Digital Cellular* (USDC) [41] and *Groupe Speciale Mobile* (GSM) [42, 43], have attempted to avoid these shortcomings by using more meaningful measurements of quality such as *bit error rate* (BER) and by having the subscriber units assist in the handoff process. A more recent second-generation proposal for cellular in the United

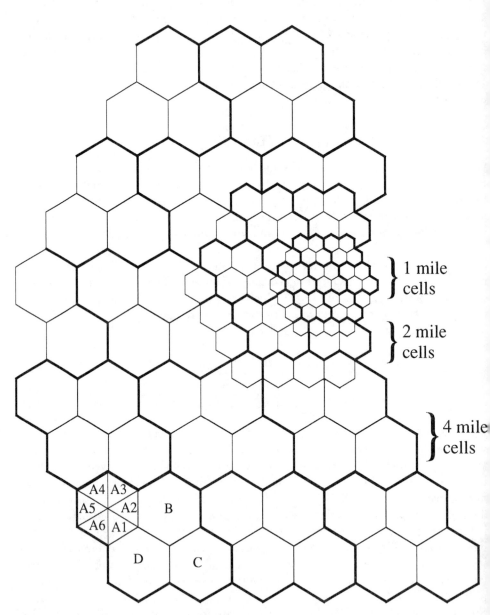

Figure 10.19 Possible 100-site Chicago, Illinois cellular system layout.

Coverage Analysis and Simulation

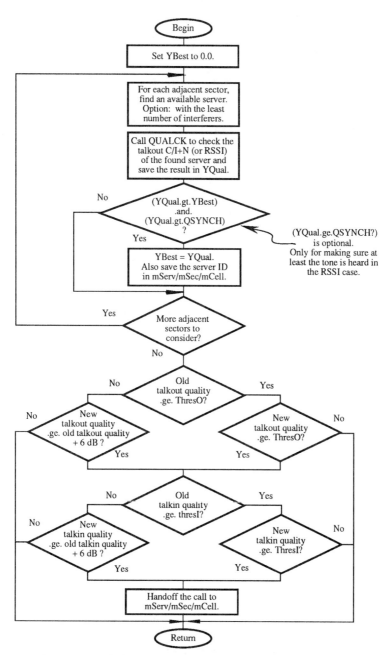

Figure 10.20 An improved channel handoff assignment method.

States, involving spread spectrum (code division) multiplexing, has even introduced the concept of soft handoff [44, 45].

Cellular literature generally implies that assignments and handoffs are best handled by selecting the server for which the carrier-to-interference plus noise ratio is maximum (for analog voice) or, equivalently, the BER is minimum (for digital voice or data). See, for example, United States patents 3,764,915; 3,819,872; 3,906,166; 4,144,496; 4,355,411; 4,485,486; 4,696,027; and 4,736,453. Several technical papers presented at the 1989 IEEE Vehicular Technology Conference also support the preceding statement. Lastly, we note that the mobile-assisted handoff specifications for the next-generation USDC also imply that minimum bit error rate assignments are optimal. This is not true in general. As bit rates are pushed higher, the importance of frequency selective fading rises. Although equalization can minimize the effect, such fading nonetheless results in an irreducible error floor [1]. Thus, BER is not a simple function of only carrier-to-interference plus noise $C/I + N$; it is a function of delay spread as well. The importance of this to cellular assignment and handoff is as follows.

Consider Figure 10.21 which is based on one form of quadrature amplitude digital modulation. At the minimum acceptable digital-voice BER (which typically runs 3%) an energy per bit per noise spectral density (E_b/N_0) ratio of 13 to 15 dB is required for flat Rayleigh fading (delay spread $S = 0$ μs) to very bad hilly fading

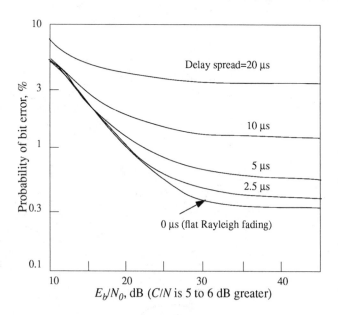

Figure 10.21 Sample digital QAM performance.

conditions ($S = 10$ μs), respectively. In terms of $C/I + N$, the ratios correspond to 18 to 20 dB, respectively. If one chooses to trigger handoffs when the probability of bit error exceeds 2%, the pertinent ratios become 16 dB E_b/N_0 and 21 dB $C/I + N$ for flat fading, and 19 dB E_b/N_0 and 24 dB $C/I + N$ for very bad hilly fading.

Now consider two servers available for assignment. One involves a flat faded link with 0.5% BER; the other involves a link with delay spread similar to very bad hilly terrain and has a BER of 1.3%. It is reasonable to assert that the best assignment is the link with 1.3% BER because it has an E_b/N_0 of about 35 dB, which represents a margin over handoff threshold of 16 dB. The link with 0.5% BER has an E_b/N_0 value of 26 dB, which is just 10 dB above the handoff threshold. The former link can tolerate a desired signal fade (or interference signal rise) 6 dB greater than the latter can tolerate and thus is the preferred choice.

It is true that speech quality improves monotonically as bit error probability decreases. Thus, the 0.5% link would indeed have a higher initial speech quality. However, the subjective quality of digital speech versus bit error probability tends to be rather binary in nature. That is, below a certain error probability the quality is nearly indistinguishable from zero errors, while above that error probability the speech almost immediately becomes unacceptable. Thus, the quality difference between the 0.5% and 1.3% links would not generally be significant.

Therefore, it is valid to assert that link margin over handoff threshold is the proper criterion on which to base assignments and handoffs, at least for digital speech geographic reuse systems. When delay spread is minimal this reduces to that which has been tauted as obvious for quite some time. However, as spectrum efficiency needs push one toward modulation schemes that are increasingly sensitive to delay spread, the more general criterion will become important.

10.10.3 Reuse Patterns and Performance

The vagaries of real-world propagation make actual cell coverage contours erratic. Nonetheless, assuming the contours match regular polygon patterns is mathematically useful. In particular, the hexagon shape has proven useful. This shape is a reasonable approximation to circular coverage (which really would exist if propagation was homogeneous) but avoids the overlap or underlap problems when circles are used to cover an area. With hexagonal geometry, the classical relationship between the number of cells per reuse cluster N and the first-tier cochannel reuse distance D is

$$\frac{D}{R} = \sqrt{3N} \qquad (10.43)$$

where R equals the cell center-to-vertex distance and N can take on only the values $(k+l)^2 - kl$, where k and l range over the positive integers (i.e., $N = 3, 4, 7, 9, 12, \ldots$). This would seem to imply that 2-cell, 5-cell, etc. reuse patterns do not exist. However, a hidden assumption is that all six first-tier interferers are at the same distance from the desired cell. Asymmetrical reuse arrangements, where interferers lie at various distances, do allow for 2-cell, 5-cell, etc. reuse. For example, Figure 10.22 indicates how five-cell reuse with 90-deg sectors can be achieved. In that arrangement four interference sources are at $D/R = 4.58$ and two interference sources are at $D/R = 3.0$.

The interference performance of various reuse patterns can be evaluated by Monte Carlo simulations like the CELLSIM program cited in Reference [46]. Such simulations randomly place users about the desired coverage cell or sector-cell and for each placement evaluate the ratio of desired talk-out signal power to the summation of first-tier interference talk-out powers. A common design guideline is to require the 90 percentile C/I provided by the reuse geometry to equal or exceed the C/I required for minimum acceptable voice quality by the actual modulation scheme to be used. For example, the current analog FM target is 17 dB. Table 10.10 lists the interference performance of several cellular reuse arrangements (assuming three- and six-sector antenna directivities, as given in Table 10.11).

As cell sizes decrease, the fixed channel assignment method previously discussed becomes impractical. Consequently, a great deal of attention has been focused on the development of adaptive (dynamic) channel assignment strategies for microcellular systems [47, 48, 49]. The basic idea is to allow the servers at each site to choose from most or all of the possible assignments in the system. With *frequency division multiplexing* (FDM) this of course implies the need for frequency-agile base stations. Thus, there is no *a priori* D/R value forced on cochannel assignments. For example, if user A is near site A and user B is near adjacent site B, they most likely can operate cochannel without interference. However, just because user B is in a distant cell does not guarantee it can operate cochannel with user A. For example, it could be that very favorable propagation conditions exist between cells A and B despite their physical separation. In fact, it is the increasing likelihood of such favorable propagation as cell sizes shrink that leads to the need for adaptive assignment and handoff.

10.10.4 Cochannel Isolation Characteristics

Sectored-antenna reuse patterns would seem to offer increased spectrum efficiency over omnidirectional reuse patterns. However, this has been questioned, particularly in the case of 60-deg sectorization. The main concerns are decreased trunking efficiency (traffic engineering is discussed in Chapter 13) and loss of directivity.

The three common reuse patterns currently used in analog cellular are the 12-cell omnidirectional pattern, the 7-cell/3-sector pattern, and the 4-cell/6-sector pattern. Such patterns require the available channels to be split into 12, 21, and 24 groups, respectively. Because the number of channels per site is highest with the 12-cell omnidirectional pattern, it will have the highest trunking efficiency. However, only if the number of channels is very small or the grade-of-service criterion is very high

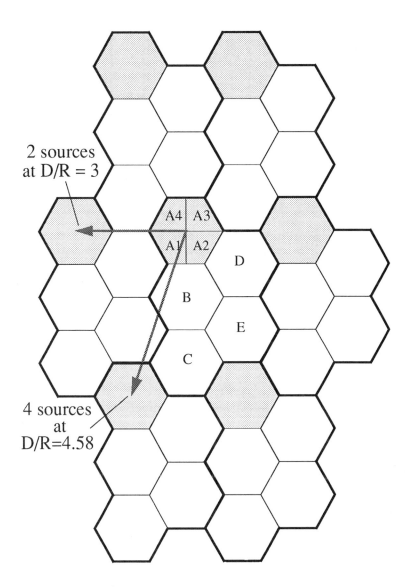

Figure 10.22 Five-cell, four-sector reuse pattern.

Table 10.10 90 Percentile C/I Performance of Various Cellular Reuse Arrangements

Cellular arrangement[a]	90% C/I (dB)	Cellular arrangement	90% C/I (dB)
12 cell omni, D/R=6.0	20.8	4 cell/6 sector, D/R=3.46	19.4
7 cell omni, D/R=4.58	14.2	3 cell omni, D/R=3.0	5.6
7 cell/3 sector, D/R=4.58	20.8	3 cell/3 sector, D/R=3.0	12.5
7 cell/6 sector, D/R=4.58	24.9	3 cell/6 sector, D/R=3.0	16.7
5 cell/4 sector, 2 D/R values	17.5	2 cell/6 sector, 2 D/R values	12.4
5 cell/3 sector, 2 D/R values	16.2	1 cell omni, D/R=1.73	-5.7
4 cell omni, D/R=3.46	8.3	1 cell/3 sector, D/R=1.73	2.7
4 cell/3 sector, D/R=3.46	15.0	1 cell/6 sector, D/R=1.73	7.8

a. Conditions: 3.8 mi cells, base antenna height=100 ft, lognormal shadowing standard deviation=6.5 dB

Table 10.11 Three- and Six-Sector Antenna Directivities

Degrees off-axis[a]	Type 1 (dBr)	Type 2 (dBr)	Degrees off-axis	Type 1 (dBr)	Type 2 (dBr)
0	0	0	90	18.4	10.0
10	0.3	0	100	19.6	12.0
20	1.1	0.1	110	20.0	15.0
30	3.0	0.3	120	20.0	15.0
40	5.0	1.0	130	20.0	15.0
50	7.5	2.0	140	20.0	15.0
60	10.5	3.0	150	20.0	15.0
70	13.7	6.0	160	20.0	15.0
80	16.3	8.0	170	20.0	15.0
90	18.4	10.0	180	20.0	15.0

a. Directional antenna type 1 corresponds to 60 deg sector, type 2 corresponds to 120 deg sector

Coverage Analysis and Simulation 179

will this efficiency advantage offset the substantial geographic efficiency disadvantage of 12/7 or 12/4. The difference between the 21 and 24 groups is small and with sector sharing a 60-deg sector approach can actually appear to have more servers than a 120-deg approach. Such sector sharing also minimizes trunking efficiency loss caused by unequal load across the sectors.

Both 120- and 60-deg sector approaches depend on antenna directivity and thus are at risk if the environment appreciably impacts antenna performance. Field tests have been conducted to quantify this effect [40]. Results for typical urban and suburban sites indicate satisfactory directivity. Substantial nearby buildings do produce reflections that reduce antenna protection, but the impact on C/I performance is minor. At urban sites surrounded by taller buildings, directivity suffers greatly and such sites should be avoided. Instead, sites that are above most of the surrounding clutter should be chosen. Antenna downtilt can boost C/I in such cases, though mainly in the center of the cell rather than at the edge.

Interestingly, omnidirectional reuse schemes would also fare poorly at cluttered sites with directivity shortfalls. This is because propagation is generally far from homogeneous; i.e., substantial variations in median signal strength occur at fixed range around the site. Thus, omnidirectional ERP does not produce received signal strengths independent of azimuth. Sectored approaches of course afford a means of combating such median signal level variations through custom adjustments of transmit power.

10.10.5 Spectrum Efficiency Measures

Various measures of spectrum efficiency are discussed by Hatfield in Reference [50]. He concludes that for land-mobile radio applications the preferred metric is erlangs per frequency bandwidth per area. This is in line with the more general viewpoint of Berry [51], who suggests communications achieved (or information delivered) per spectrum space use. By the term spectrum "space" Berry is referring to the dimensions of RF bandwidth; physical space such as an area, a building volume, an orbital arc, etc.; and time.

Although the Hatfield metric is generally accepted, it is not without its difficulties. For example, Mikulski [52] criticizes the use of erlangs because it is a poor indicator of how much information is really being conveyed per unit time. One can appreciate this complaint by contrasting mobile telephone channels with mobile dispatch channels. It is not unreasonable to argue that the six average dispatch calls required to equal just a single average interconnect call convey more information than that single call and therefore should be rated higher on the spectrum efficiency scale.

Of course this predicament does not occur when comparing the efficiency of different modulation, multiplexing, and coding schemes for the same traffic. A paper by Chang and Porter [53] addresses coding comparisons as follows. Define the code rate r as the ratio of information bit rate to the instantaneous channel bit rate. Channel bandwidth is proportional to the denominator. The number of channel sets N needed to cover some large geographic area with multiple sites is related to the total allocated spectrum A and the number of channels per set C through $(1/N) = (C/A)$. Efficiency is then proportional to (r/N).

Now N is proportional to (D/R), where D equals the base-to-base separation between cochannel assignments and R equals the individual site coverage radius (Section 10.10.3). Furthermore, assuming power-law propagation and interference-limited conditions, one has $(C/I) \propto [D/(D-R)]^\gamma$, where γ represents the propagation law. Thus, the ratio of spectrum efficiencies for two coding plans, i and j, is given by:

$$\frac{E_i}{E_j} = \frac{r_i}{r_j} [\frac{(\frac{C}{I})_j^{\frac{1}{\gamma}} + 1}{(\frac{C}{I})_i^{\frac{1}{\gamma}} + 1}]^2 \qquad (10.44)$$

and for equal efficiency one has:

$$\frac{r_i}{r_j} = [\frac{(\frac{C}{I})_j^{\frac{1}{\gamma}} + 1}{(\frac{C}{I})_i^{\frac{1}{\gamma}} + 1}]^2 \qquad (10.45)$$

Only if the (C/I) terms are large relative to unity is the simplified expression in Reference [53], and the related Figure A1, approximately correct.

The comparison of schemes that differ in more than just the coding used for accomplishing the same task is more complicated. For example, Reference [54] considers the spectrum efficiency of various ways of implementing cellular telephone service. In addition to effective channel spacing, protection ratio (C/I), and reuse distance, one must account for possible differences in trunking efficiency and system control overhead.

10.10.6 Generation of Correlated Signal Strength Draws for Multiple Site Reuse System Simulations

Introduction

Early multiple site reuse simulations were based on snapshots of activity (for example, CELLSIM [46]). While this provides useful comparisons of various geographic reuse patterns, it offers little insight regarding how often handoffs are required and even less on how different strategies for assignment and handoff compare. To answer such questions, a simulation that tracks calls from start to finish is required. Such call-tracking simulations are the subject of much research activity today (for example, see Reference [55]). In what follows, a method is presented for incorporating the temporal behavior of the average desired signal power, as well as that of each interfering signal, into a call-tracking simulation. In addition, cross-correlations between desired and interfering signals are taken into consideration.

Statement of the Problem

Consider the task of generating random draws for the interference power at some time t, given knowledge of the desired power at time t and the interference power for some previous time $t - \tau$. Let the present interference power be denoted by the random variable $I(t)$, the present desired power by the random variable $D(t)$, and the previous interference power by the random variable $I(t - \tau)$.

Land-mobile signals typically have two distinct behaviors. The small scale or instantaneous signal strength is Rayleigh faded about some mean or average power. This power in turn is lognormally shadowed over a larger distance scale. By taking the random variables to represent average power in decibels, all three will be Gaussian distributed. Furthermore, our concentration will be on the deviations of the three signals from their median values; i.e., all three random variables are zero-mean Gaussian distributed.

Analysis

The general expression for the joint PDF of three zero-mean Gaussian random variables is [56]:

$$f(x_1, x_2, x_3) = \frac{1}{\sqrt{(2\pi)^3 |\mu|}} \exp(-\frac{1}{2} X \mu^{-1} X^T) \qquad (10.46)$$

where $X = (x_1, x_2, x_3)$, μ is the covariance matrix with entries μ_{ij} equal to the expected values of the products $x_i x_j$, and $|\mu|$ is the determinant of the covariance matrix. Note that applied to the problem at hand one has $x_1 = d(t)$, $x_2 = i(t)$, and $x_3 = i(t - \tau)$. The correlation coefficient as a function of time has approximately the form $\exp(-|\tau|/T) = \gamma$, where T represents the decorrelation time constant [57]. Furthermore, the desired signal and interference signals are also in general correlated, with correlation coefficient α [58]. If the lognormal shadowing variance is equal for both desired and interference signals, then the covariance matrix can be written as:

$$\mu = \begin{bmatrix} \sigma^2 & \sigma^2\alpha & \sigma^2\beta \\ \sigma^2\alpha & \sigma^2 & \sigma^2\gamma \\ \sigma^2\beta & \sigma^2\gamma & \sigma^2 \end{bmatrix}$$

where σ^2 represents the variance of the shadowing effect and the correlation between variables x_1 and x_3 has been taken as β. β will actually be equal to the product $\alpha\gamma$.

The determinant of the matrix is:

$$|\mu| = \sigma^2(1 - \alpha^2 - \beta^2 - \gamma^2 + 2\alpha\beta\gamma) \qquad (10.47)$$

The matrix inverse can be found by dividing the transpose of the cofactor matrix by the determinant. The result is:

$$\mu^{-1} = |\mu|^{-1} \begin{bmatrix} \sigma^2(1-\gamma^2) & \sigma^2(\beta\gamma - \alpha) & \sigma^2(\alpha\gamma - \beta) \\ \sigma^2(\beta\gamma - \alpha) & \sigma^2(1 - \beta^2) & \sigma^2(\alpha\beta - \gamma) \\ \sigma^2(\alpha\gamma - \beta) & \sigma^2(\alpha\beta - \gamma) & \sigma^2(1 - \alpha^2) \end{bmatrix}$$

Carrying out the multiplications inside the argument of the exponent in Equation (10.46), the joint PDF is found to be:

$$f(x_1, x_2, x_3) = \frac{1}{\sqrt{(2\pi)^3 \sigma_e^2}} \exp{-\frac{1}{2\sigma_e^2}[x_2^2(1-\beta^2) + x_2 A + B]} \qquad (10.48)$$

where $\sigma_e^2 = \sigma^2(1 - \alpha^2 - \beta^2 - \gamma^2 + 2\alpha\beta\gamma)$, $A = 2x_1(\beta\gamma - \alpha) + 2x_3(\alpha\beta - \gamma)$, and $B = x_1^2(1-\gamma^2) + x_3^2(1-\alpha^2) + 2x_1 x_3(\alpha\gamma - \beta)$.

The PDF for variable $x_2 = i(t)$ is sought, conditioned upon knowledge of the values for variables $x_1 = d(t)$ and $x_3 = i(t-\tau)$. Again from Reference [56] one finds:

$$f(x_2 | x_1 = x_{10}, x_3 = x_{30}) = \frac{f(x_{10}, x_2, x_{30})}{f(x_{10}, x_{30})} \qquad (10.49)$$

where $f(x_{10}, x_{30})$ refers to the bivariate PDF for zero-mean Gaussian random variables X_1, X_3 compared to the explicit values x_{10}, x_{30}, respectively. This latter density function is:

$$f(x_{10}, x_{30}) = \frac{1}{2\pi\sigma^2 \beta} \exp[-\frac{1}{2\sigma^2 \beta}(x_{10}^2 - 2\sqrt{1-\beta^2} x_{10} x_{30} + x_{30}^2)] \qquad (10.50)$$

Substituting Equations (10.48) and (10.50) into Equation (10.49) one obtains:

$$f(x_2 | x_1 = x_{10}, x_3 = x_{30}) = \frac{1}{\sqrt{(2\pi)\sigma^2 \phi}} \exp(-\frac{1}{2\sigma^2 \phi}[x_2^2 + x_2(2x_{10}\frac{\beta\gamma - \alpha}{1-\beta^2}$$
$$+ 2x_{30}\frac{\alpha\beta - \gamma}{1-\beta^2}) + C(x_{10}, x_{30})]) \qquad (10.51)$$

where:

$$C(x_{10}, x_{30}) = x_{10}^2(\frac{1-\gamma^2}{1-\beta^2}) + x_{30}^2(\frac{1-\alpha^2}{1-\beta^2}) + 2x_{10}x_{30}(\frac{\alpha\gamma - \beta}{1-\beta^2}) - \frac{\phi}{1-\beta^2}(x_{10}^2 + x_{30}^2 - 2x_{10}x_{30}\beta)$$

and

$$\phi = \frac{1 - \alpha^2 - \beta^2 - \gamma^2 + 2\alpha\beta\gamma}{1 - \beta^2}$$

With the help of the computer mathematics program *Mathematica* ^{TM}, the exponential argument can be shown to be a perfect square. This of course is expected because it means that the conditional distribution is also Gaussian.

Solution

The mean and variance of the conditional distribution are, respectively:

$$E(x_2 | x_1 = x_{10}, x_3 = x_{30}) = \frac{1}{1-\beta^2}[x_{10}(\alpha - \beta\gamma) + x_{30}(\gamma - \alpha\beta)] \qquad (10.52)$$

Coverage Analysis and Simulation

$$Var(x_2|x_1 = x_{10}, x_3 = x_{30}) = \sigma^2 \phi \qquad (10.53)$$

Consequently, new interference power deviations from the median level can be obtained by noting (1) the present desired power deviation from its median value (this will be correlated with the previous desired power deviation via γ) and recalling (2) the previous interference power deviation. Setting x_{10} equal to the former and x_{30} equal to the latter in Equations (10.52) and (10.53) allows one to establish the mean and variance that characterize the Gaussian distribution of the new interference power. Making one draw from such a distribution yields the new interference power.

10.11 Analysis of Intermodulation Interference

A form of interference called *intermodulation* (IM) can arise in communication systems due to the presence of nonlinear transfer functions. For example, the transmitted signals from transmitters at frequencies $f_1 = f_0 + \delta f$ and $f_2 = f_0 + 2\delta f$ might be coupled back through the antenna system of some other nearby transmitter into its final amplifier. Third-order nonlinearity at that point will cause the undesired signals to mix and produce an output at frequency f_0 that will then be transmitted. This can completely overwhelm a nearby receiver tuned to frequency f_0 attempting to copy a signal at the edge of its coverage contour. A similar problem can occur when the receiver is subjected to two strong signals, perhaps from nearby control stations, on the upper (or lower) adjacent and the second upper (or lower) adjacent channels. Imperfections in the receiver mixer will result in an IF response related to the adjacent channel signals and the on-channel signal. If the former signals are sufficiently strong they will interfere with the on-channel signal.

The behavior of such undesired responses can be studied by modeling the transfer function of the circuit in question as a power series (for example, see Reference [59]). The third-order IM response is the result of cubing the input signal. Recall from basic trigonometry relations that multiplication of two sinusoidal signals produces an output with frequencies equal to the sum and difference of the input frequencies. Thus, squaring a signal produces a dc component and a double frequency component. Multiplying that result by another sinusoidal signal gives an output component at a frequency equal to twice the first signal's frequency minus the second signal's frequency. The amplitude of this response will increase 3 dB for just a 1-dB increase in the amplitude of both source signals. The fictitious level at which the desired output and IM signal levels are equal is termed the intercept point (specifically in this case the third-order intercept point).

10.11.1 Wide-Area Coverage Site Receiver Intermodulation Probability

An analysis of the probability of third-order IM interference at a land-mobile base receiver can be carried out as follows. First, we note that signal strength observations (in decibels) at the base site for wide-area coverage system subscribers are well described by a dual normal distribution with two means, two standard deviations, and a classification parameter equal to the probability of being in class 1 (equal to 1 minus the probability of being in class 2). It is conservative to assume that levels from users on other systems and channels (possibly not collocated) are similarly distributed.

Now consider the random variable $Z = 2X_1 + X_2$, which is the logarithmic analog of the third-order relation $z = x_1^2 x_2$. For x_1 and x_2 lognormally distributed, X_1 and X_2 will be Gaussian random variables and so will Z. Thus $X_i \sim N(\mu_i, \sigma_i^2), i = 1, 2$ and $Z \sim N(2\mu_1 + \mu_2, 4\sigma_1^2 + \sigma_2^2)$, where the N notation implies Gaussian (normal) and the arguments denote the mean and variance, respectively. The probability of third-order IM interference for any particular desired signal level X_d is given by:

$$\text{Prob}(\text{IM}|X_d) = \text{Prob}(Z - C \geq X_d - P) \quad (10.54)$$

where $C = 3(S + IMR) - (S - SNR)$ represents the third-order conversion factor of the receiver (in decibels), S equals the receive system sensitivity used in measuring the *intermodulation rejection ratio* (*IMR*) (in decibels; see the EIA test method in Reference [12]), SNR equals the signal-to-noise power ratio at sensitivity (in decibels), and P equals the required protection to avoid interference (in decibels).

Because Z is normally distributed:

$$\begin{aligned}\text{Prob}(\text{IM}|X_d) &= 1 - \int_{-\infty}^{X_d - P} \frac{\exp\{-\frac{[y-(2\mu_1+\mu_2-C)]^2}{2(4\sigma_1^2+\sigma_2^2)}\}}{\sqrt{2\pi(4\sigma_1^2+\sigma_2^2)}} dy \\ &= \frac{1}{2}\text{erfc}[\frac{X_d - P - 2\mu_1 - \mu_2 + C}{\sqrt{2(4\sigma_1^2+\sigma_2^2)}}]\end{aligned} \quad (10.55)$$

This conditional probability is next weighted over the likelihood of various values of X_d, which is also normally distributed (with mean μ_d and standard deviation σ_d), to obtain the unconditional probability of interference as:

$$\begin{aligned}\text{Prob}(\text{IM}) &= \int_{-\infty}^{\infty} \text{Prob}(\text{IM}|X_d) f_{X_d}(x_d) dx_d \\ &= \frac{1}{2}\text{erfc}[\frac{\mu_d - 2\mu_1 - \mu_2 - P + C}{\sqrt{2(4\sigma_1^2+\sigma_2^2+\sigma_d^2)}}]\end{aligned} \quad (10.56)$$

Because the received signal strength appears as a mixture of two Gaussian components, the preceding equation must actually by applied in eight different ways and

Coverage Analysis and Simulation

weighted appropriately to yield the desired interference probability. If one labels the mixture components A and B, and lets a represent the probability of component A being active, then the eight cases that must be evaluated and their weighting factors are: (1) $X_1 X_2 X_d = AAA$, weight a^3, (2) $X_1 X_2 X_d = AAB$, weight $a^2(1-a)$, (3) $X_1 X_2 X_d = ABA$, weight $a^2(1-a)$, (4) $X_1 X_2 X_d = ABB$, weight $a(1-a)^2$, (5) $X_1 X_2 X_d = BAA$, weight $a^2(1-a)$, (6) $X_1 X_2 X_d = BAB$, weight $a(1-a)^2$, (7) $X_1 X_2 X_d = BBA$, weight $a(1-a)^2$, and (8) $X_1 X_2 X_d = BBB$, weight $(1-a)^3$.

The preceding analysis applies to interference caused by a single pair of properly spaced signals. Generally, the receiver passband will admit a large number of signals and many IM products might fall on channel. Even if all these products are individually too weak to cause problems, their sum may not be too weak. Also, a special form of third-order intermodulation called triple beat might be important. This results from three signals mixing, rather than just two, and allows for more IM combinations at responses 6 dB greater than conventional IM responses [60]. Monte Carlo simulation is in order to address these matters.

10.11.2 Small-Area Coverage Site (Cellular) Receiver Intermodulation Probability

To assist the evaluation of base receiver IM probability for small cellular-like sites, subscriber signal levels have been measured simultaneously over 600 channels (806 to 821 MHz) from a representative Chicago, Illinois site. That site has a clear view of a busy suburban expressway, which approaches to within 500 ft. A CELWAVE PD10176 105-deg sectored antenna was used at about 150-ft height. Figure 10.23 shows graphs that are typical of the hourly cumulative signal strength distributions observed. When data for the worst-case hour (5 to 6 p.m.) are plotted on a logarithmic probability scale versus signal level (Figure 10.24), a linear relation is apparent that holds for at least 40 dB of dynamic range. Ultimately, of course, this relation must break down because there is a limit to the signal level possible.

Based on these signal level observations, one can conservatively model the received signal level distribution as exponential. That is, the natural logarithm of the probability P of observing a level in excess of x dBm has the form:

$$\begin{aligned}
\ln(P_>) &= A^* + Bx \\
P_> &= \exp(A^*)\exp(Bx) \\
&= A\exp(Bx)
\end{aligned} \quad (10.57)$$

where $A = \exp(A^*)$ and x must exceed some minimum value x_0 so that the PDF integrates correctly to unity. This relation is conservative because the likelihood of the very strongest signals falls off rapidly in probability from the indicated slope. Based on Figure 10.24, the coefficient values are $A = 5.357E - 6$, $B = -0.08608$, and $x_0 = -141$ dBm. The probability of observing a level less than x dBm is

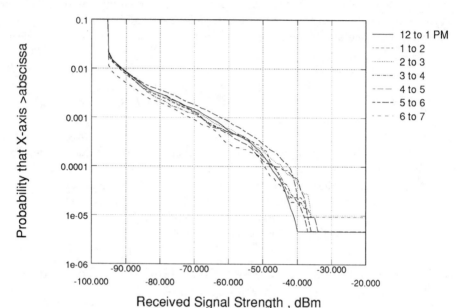

Figure 10.23 Hourly CDFs of received signal strength.

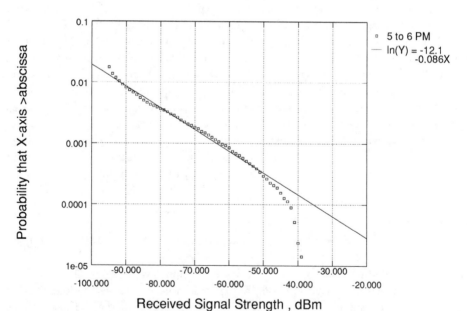

Figure 10.24 Worst-case hourly CDF (log base 10 probability) of received signal strength.

$P_< = 1 - A\exp(Bx)$; the related PDF is:

$$f(x) = \frac{dF(x)}{dx}$$
$$= \frac{dP_<}{dx}$$
$$= \alpha\exp(-\beta x) \quad (10.58)$$

where $\alpha = AB$, $\beta = -B$, and $x_0 \leq x \leq \infty$.

Now consider the third-order IM quantity in decibels: $X_3 = 2X_1 + X_2$, where X_1 and X_2 are independently, identically distributed as the $f(x)$ just noted. Rather than working with offset exponentials, it is convenient to work with the random variables $X = 2X_1 - 2x_0$ and $Y = X_2 - x_0$, which are defined over $(0, \infty)$. Via Papoulis (5-7) [56] one obtains:

$$f_X(x) = \frac{1}{2}\alpha\exp[-\frac{\beta}{2}(x + 2x_0)]$$
$$f_Y(y) = \alpha\exp[-\beta(y + x_0)] \quad (10.59)$$

Also, for $Z = X + Y$, using (7-8) of Papoulis leads to:

$$f_Z(z) = \int_0^z (\frac{\alpha}{2})\exp[-\frac{\beta}{2}(z - y + 2x_0)]\alpha\exp[-\beta(y + x_0)]dy$$
$$= -(\frac{\alpha^2}{\beta})\exp(-2\beta x_0)[\exp(-\beta z) - \exp(-\frac{\beta z}{2})] \quad (10.60)$$

The cumulative distribution function is:

$$F_Z(z) = \int_0^z f(s)ds$$
$$= (\frac{\alpha}{\beta})^2 \exp(-2\beta x_0)[\exp(-\frac{\beta z}{2}) - 1]^2$$
$$= [\exp(-\frac{\beta z}{2}) - 1]^2 \quad (10.61)$$

A sample IM probability prediction is as follows. Consider a receive system with 10-dBm third-order intercept point, referenced to the antenna output (effective receiver input). If the noise floor is -125 dBm, two equal inputs at -35 dBm will produce a comparable level of IM. The IM threshold is thus a $2x_1 + x_2$ level exceeding -105 dBm or 3[-35 - (-141)] = 318 dB above the offset value of $3x_0$. The probability of a single pair of channels exceeding this threshold is $1 - \{\exp[-(0.08608 \cdot 318)/2] - 1\}^2 = 2.275E - 6$. With 300 possible IM contributors in a 600-channel receiver passband, the overall IM probability is still only 0.07%. One can legitimately argue that certain metropolitan areas like Los Angeles will have a greater prevalence of high-powered subscriber units than Chicago and, at least for some sites, may have a greater traffic intensity. Both these circumstances imply a boost in curves like those shown in Figure 10.24. Assuming the boost to be a factor of 20 (13 dB), the new fit coefficients are $\alpha = 1.412E - 6$, $\beta = -0.08608$, and $x_0 = -128$ dBm. But even here the overall IM probability is only 0.4%. Hence, the use of a full-band receive preselector need not in general pose an IM interference risk.

Power-law propagation is commonly assumed for land-mobile links and empirical studies like Okumura's show this to be reasonable for a large spread of ranges. The median received power for a unit at range r can thus be taken as $p = (K_p/r^n)$, where n is the propagation-law exponent (typically about 3.5 for cellular sites) and K_p is a proportionality constant related to transmit power, antenna gain, and operating frequency. The power-law model is also useful for describing the distribution of subscriber range: $f(r) = K_r r^m$, where the range varies between some lower limit r_l and some upper limit r_u (implying that the proportionality constant equals $(m+1)/(r_u^{m+1} - r_l^{m+1})$, for $m \neq 1$). The probability of receiving a signal level greater than x dBm thus equals the probability of the subscriber unit keying up at a range less than $(K_p/x)^{\frac{1}{n}}$. This in turn equals:

$$\text{Prob}(P \geq x) = \int_{r_l}^{[(\frac{K_p}{x})^{\frac{1}{n}}]} K_r r^m dr$$

$$= \left(\frac{K_p^{\frac{m+1}{n}}}{r_u^{m+1} - r_l^{m+1}}\right) x^{-\frac{m+1}{n}} - \left(\frac{r_l^{m+1}}{r_u^{m+1} - r_l^{m+1}}\right)$$

$$\approx a x^{-\frac{m+1}{n}} \qquad (10.62)$$

for r_l small relative to r_u. The natural logarithm of this probability thus has the approximate form:

$$\ln[\text{Prob}(P \geq x)] \approx \ln(a) - \frac{m+1}{n} \ln(x)$$

$$\ln[\text{Prob}(P \geq x_{\text{dB}})] \approx \ln(a) - \left(\frac{m+1}{n}\right)\left[\frac{\ln(10)}{10}\right] x_{\text{dB}}$$

$$\approx A + B x_{\text{dB}} \qquad (10.63)$$

The observations cited in this section yield $B = -0.08608$, which implies $(m+1)/n = 0.374$. Using $n = 3.5$, this gives $m = 0.31$ and indicates some tendancy for users to be near the site ($m = 1$ corresponds to the situation of all locations equally likely). However, note that the slope B increases to about -0.25 for the strongest signals, for which m does indeed approximate unity when $n = 2$ (i.e., free-space propagation circumstances).

10.11.3 Transmitter IM

So far the discussion has concerned quantifying the probability of IM interference caused by base receiver limitations. Transmitter limitations can also cause IM interference problems, particularly at crowded sites. A common base transmitter lineup

Coverage Analysis and Simulation   189

involves the generation of 70 W of RF power, followed by 7 dB of antenna combiner and feedline loss, and then a 9-dBd omnidirectional antenna. The ERPd is thus approximately 100 W (50 dBm). At major sites, the port-to-port coupling between pairs of antennas on the antenna "farm" is typically on the order of -40 dB. This is also a fair representation of the usual isolation provided between transmitters combined onto the same antenna. Thus, if preventive steps are not taken, about 50 - 9 - 40 -7 = -6 dBm may reach the transmitter from other collocated transmitters. The efficiency of IM generation in the transmitter depends on the frequency band and type of amplifier. Tube amplifiers are often biased for rather linear operation and produce less IM than class-C transistor amplifiers. A typical value is -10 dB, which implies -23 dBm reaching the antenna for reradiation at -14-dBm ERPd. Now static receive sensitivies are often on the order of 0.35 μV (-116 dBm) for minimum acceptable voice quality. The associated noise power is typically 7 dB less for 5-kHz peak deviation FM. Thus, for IM to be relatively unnoticed it should not exceed -123 dBm. This implies an isolation of -23 - 40 + 123 = 60 dB is needed. That is why dense site base stations typically use a cascade of three circulators on the transmitter output ports. Each stage provides on the order of 20 dB protection, for a net protection of about 60 dB.

10.12 Homework Problems

10.12.1 Problem 1

Can the area coverage analysis of Section 10.1 be used to address portables on the street? How about portables in buildings?

Solution: The receive sensitivity of portables is generally comparable to that of mobiles. However, antenna efficiency is lower, especially when the unit is worn on the hip, a common practice while monitoring dispatch traffic. For talk-in purposes, not only is the antenna gain lower, but transmit power is much lower than that of mobile units. Calling the portable shortfall δ, Equation (10.6) can be used with $a^* = a + (\delta/\sqrt{2}\sigma)$, rather than just a, to address the portable on-the-street situation.

Portables operated in buildings suffer a building penetration loss in addition to the usual power-law loss to street level at the edge of the building. Measurements of this building loss show it is approximately normally distributed in decibels [61, 19]; hence, it can readily be incorporated into the lognormal shadowing part of the analysis by using:

$$\begin{aligned} a^* &= a(\frac{\sigma}{\sigma_T}) + \frac{(\delta + \mu_B)}{\sqrt{2}\sigma_T} \\ b^* &= b(\frac{\sigma}{\sigma_T}) \end{aligned} \qquad (10.64)$$

where $\sigma_T = \sqrt{\sigma^2 + \sigma_B^2}$ and building loss in decibels is distributed as $N(\mu_B, \sigma_B)$.

Figure 10.25 shows portable on-the-street coverage probability versus portable shortfall assuming fourth-law propagation with a lognormal-shadowing standard deviation of 9 dB. Figure 10.26 shows similar curves for the in-building situation (a building-penetration standard deviation of 9 dB is assumed).

10.12.2 Problem 2

Construct a 25-kHz channel link budget to yield the required median signal strength around the periphery of an area with 90% coverage.

Solution: Two such link budgets are described. The first pertains to current analog FM mobiles; the second pertains to a proposal for digital voice modulation with portable units (see the FCC Waiver request by Fleet Call, Inc., April 5 and June 7, 1990, Reference No. LMK-90036).

```
Analog Voice, Frequency Modulation,
Mobile Subscriber Units


Receive sensitivity (static 20-dB quieting)    -116.0 dBm
Receive antenna gain (effective gain in field)    0.0 dBd
Receive feedline loss (14 ft, RG-58, 850 MHz)    -2.8
Allowance for Rayleigh fading                   10.0 dB
Shadowing standard deviation                     5.5 dB
Multiplication factor (using 3.5-law             0.53
propagation, 90% area coverage
equates to 70% contour coverage)
Required shadowing allowance                     2.9 dB
-----------------------------------------------------------
Required median signal level on                -100.3 dBm
contour for 90% area coverage


Digital Voice, Multilevel Linear Modulation,
Portable Subscriber Units


Boltzmann's constant                           -228.6 dB-K
Reference temperature (290 K)                    24.6 dB-K
Noise equivalent bandwidth (16 kHz)              42.0 dB-Hz
Conversion to mW                                 30.0
Thermal noise floor                            -132.0 dBm
Receiver sensitivity (faded 4% BER)              12.5 dB
Bit rate to bandwidth ratio                       6.0 dB
Receive noise figure (spec)                      10.5 dB
Receive antenna gain (portable on hip)           -8.5 dBd
```

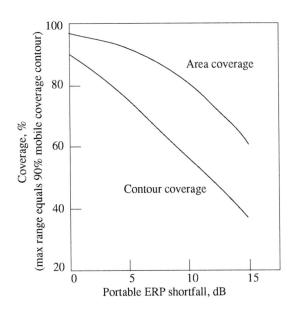

Figure 10.25 On-the-street portable coverage versus portable ERP shortfall.

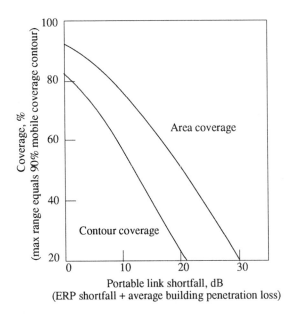

Figure 10.26 In-building portable coverage versus portable link shortfall.

```
Shadowing standard deviation                                5.5 dB
Multiplication factor (using 3.5-law                        0.53
propagation, 90% area coverage
equates to 70% contour coverage)
Required shadowing allowance                                2.9 dB
----------------------------------------------------------------
Required median signal level on                           -91.6 dBm
contour for 90% area coverage
```

10.12.3 Problem 3

What polygon shapes other than hexagons might be used to cover (tesselate) an area without overlap and underlap?

Solution: One can show by induction that the sum of angles interior to a polygon of n sides equals $180(n-2)$ deg. To fit perfectly, the equal interior angles about any vertex must sum to 360 deg. Thus, we require $k\{180[(n-2)/n]\} = 360$, where k is an integer. Starting with $n = 3$, the smallest value possible, we find the requirement met for $k = 6$, so triangles are one possibility. Trying $n = 4$ again yields a solution, $k = 4$, so squares work too. Next, $n = 5$ fails to yield integer k so pentagons do not work. Of course, we already know hexagons work ($k = 3$). But no higher order polynomial works because the only possibility left, $k = 2$, cannot be exactly matched to $2n/(n-2)$ by any choice of n. Of the three solutions, hexagons best approximate circular coverage.

10.12.4 Problem 4

Cellular-like operation is proposed adjacent to an SMR channel with high-powered (30 W), low-specification (75-dB splatter and sideband noise) subscriber units. Discuss the interference situation when an SMR unit attempts to transmit just 500 ft from the base site.

Solution: First, estimate the talk-in interference level as follows.

```
interference power (30 W)                                  44.8 dBm
cable loss (14 ft RG-58, 850 MHz)                          -2.8 dB
maximum antenna gain                                        2.0 dBd
free-space loss
(500 ft, 850 MHz, dipole to dipole)                       -70.4 dB
base antenna gain                                          10.0 dBd
cable loss (inc. connectors)                               -2.7 dB
direct interference to multicoupler port                  -19.1 dBm
adjacent channel splatter                                 -94.1 dBm
```

Coverage Analysis and Simulation 193

The splatter would appear to be quite detrimental since -95 dBm is a common desired signal design level. Note, however, that no elevation directivity benefit is included. The off-horizontal angle from a 100-ft tower to a 5-ft-high mobile antenna at 500-ft range is nearly 11 deg, so a significant benefit (10 dB) is possible. Even so, some weaker calls would be bothered and presumably be handed off.

A more subtle problem might be the direct interference power of -19.1 dBm. If linear modulation is used, the receiver must be able to tolerate such a level or *all* channels might experience interference, not just the adjacent channels with relatively weak desired signal levels. Strong-signal compression problems can of course be circumvented by switching in an attenuator ahead of the receiver, as necessary. Then only some weakest fraction of desired signal levels would be bothered.

In fact, one might turn the tables around and ask would an adjacent channel SMR unit really transmit from within 500 ft of the cellular base site. If that site radiated 100 W ERPd with the same 75-dB splatter, it would produce 50 dBmd − 70.4 dB + 2 dBd − 2.8 dB − 75 dB = -96.2 dBm to compete with the desired SMR base talk-out signal. This is well above the usual faded FM sensitivity and thus the SMR unit may not be able to successfully decode the outbound supervision information needed to stay on the channel.

From a probabilistic point of view, there are generally only about 100 SMR units per channel over the entire metropolitan area, each transmitting only a few minutes throughout the workday. The likelihood of one such unit keying up within 500 ft of a cell site should therefore be low, even though there are of course a multiplicity of such sites throughout the metropolitan area.

10.12.5 Problem 5

Verify that the level of third-order IM due to triple beat is 6 dB greater than that due to two-signal mixing.

Solution: Consider the general power-law series expansion of a voltage transfer function:
$$E_o = a_0 + a_1 E_i + a_2 E_i^2 + a_3 E_i^3 + \cdots \tag{10.65}$$
where E_o represents the output voltage, E_i represents the input voltage, and a_i represents the i-th order transfer function coefficient. Taking as the input a sum of three sinusoids of comparable frequency, we have: [15]
$$E_i = V_0 \cos(\omega_0 t) + V_1 \cos(\omega_1 t) + V_2 \cos(\omega_2 t) \tag{10.66}$$
which leads to:
$$E_o = a_0 + [a_1 V_0 \cos(\omega_0 t) + a_1 V_1 \cos(\omega_1 t) + a_1 V_2 \cos(\omega_2 t)]$$

---

[15]The phase relationships among the sinusoids do not impact the output spectrum levels, so for convenience the sinusoids are considered in phase.

$$
\begin{aligned}
&+[a_2 V_0^2 \cos^2(\omega_0 t) + a_2 V_1^2 \cos^2(\omega_1 t) + a_2 V_2^2 \cos^2(\omega_2 t) \\
&+ 2a_2 V_0 V_1 \cos(\omega_0 t)\cos(\omega_1 t) + 2a_2 V_0 V_2 \cos(\omega_0 t)\cos(\omega_2 t) \\
&+ 2a_2 V_1 V_2 \cos(\omega_1 t)\cos(\omega_2 t)] \\
&+ [a_3 V_0^3 \cos^3(\omega_0 t) + a_3 V_1^3 \cos^3(\omega_1 t) + a_3 V_2^3 \cos^3(\omega_2 t) \\
&+ 3a_3 V_0 V_1^2 \cos(\omega_0 t)\cos^2(\omega_1 t) + 3a_3 V_0 V_2^2 \cos(\omega_0 t)\cos^2(\omega_2 t) \\
&+ 3a_3 V_0^2 V_1 \cos^2(\omega_0 t)\cos(\omega_1 t) + 3a_3 V_0^2 V_2 \cos^2(\omega_0 t)\cos(\omega_2 t) \\
&+ 3a_3 V_1 V_2^2 \cos(\omega_1 t)\cos^2(\omega_2 t) + 3a_3 V_1^2 V_2 \cos^2(\omega_1 t)\cos(\omega_2 t) \\
&+ 6a_3 V_0 V_1 V_2 \cos(\omega_0 t)\cos(\omega_1 t)\cos(\omega_2 t)] + \cdots
\end{aligned}
\tag{10.67}
$$

Applying the trigonometric identity:

$$
\cos(x)\cos(y) = \frac{1}{2}[\cos(x+y) + \cos(x-y)]
\tag{10.68}
$$

and dropping dc terms and other terms whose frequencies differ significantly from $\omega_0 \approx \omega_1 \approx \omega_2$ (such terms can easily be filtered from the output), the output can be expressed as:

$$
\begin{aligned}
E_o \approx\ & a_1 [V_0 \cos(\omega_0 t) + V_1 \cos(\omega_1 t) + V_2 \cos(\omega_2 t)] \\
&+ \frac{3a_3}{4}[V_0^3 \cos(\omega_0 t) + V_1^3 \cos(\omega_1 t) + V_2^3 \cos(\omega_2 t)] \\
&+ \frac{3a_3}{2}[V_0 V_1^2 \cos(\omega_0 t) + V_0 V_2^2 \cos(\omega_0 t) + V_0^2 V_1 \cos(\omega_1 t) \\
&+ V_1 V_2^2 \cos(\omega_1 t) + V_0^2 V_2 \cos(\omega_2 t) + V_1^2 V_2 \cos(\omega_2 t)] \\
&+ \frac{3a_3}{4}\{V_0 V_1^2 \cos[(2\omega_1 - \omega_0)t] + V_0 V_2^2 \cos[(2\omega_2 - \omega_0)t] \\
&+ V_0^2 V_1 \cos[(2\omega_0 - \omega_1)t] + V_0^2 V_2 \cos[(2\omega_0 - \omega_2)t] \\
&+ V_1 V_2^2 \cos[(2\omega_2 - \omega_1)t] + V_1^2 V_2 \cos[(2\omega_1 - \omega_2)t]\} \\
&+ \frac{3a_3 V_0 V_1 V_2}{2}\{\cos[(\omega_0 + \omega_1 - \omega_2)t] \\
&+ \cos[(\omega_0 + \omega_2 - \omega_1)t]\}
\end{aligned}
\tag{10.69}
$$

In order of first to last, the respective output terms are: (1) the true linear ouput, (2) the distortion due to the third-order transfer coefficient alone (self mixing), (3) six two-signal mixing distortion terms due to the third-order transfer coefficient and dc from a signal mixing with itself, (4) six two-signal mixing distortion terms due to the third-order transfer coefficient and the sum frequency of one signal mixing with itself (these are the usual third-order IM products of interest), and (5) two three-signal mixing distortion terms (these are the so-called triple beat IM products). Note that with equal input amplitudes ($V_0 = V_1 = V_2$) the voltage amplitude of the triple beat products is twice that of the usual IM products. Hence, the power level of such products is 6 dB greater.

Coverage Analysis and Simulation

10.12.6 Problem 6

Compare the adjacent channel interference performance of frequency and *amplitude modulation* (AM) passing through a receiver with a third-order nonlinearity.

Solution: In the FM case let the undesired signal be represented as $x = A\cos[\omega t + \phi(t)]$. The receiver response before detection can be written as $y = a_1 x + a_3 x^3$. Substituting x into that equation and simplifying yields:

$$\begin{aligned}
y &= a_1 A \cos[\omega t + \phi(t)] + a_3 A^3 \cos[\omega t + \phi(t)]^3 \\
&= a_1 A \cos[\omega t + \phi(t)] + a_3 A^3 \cos[\omega t + \phi(t)](\frac{1}{2} + \frac{1}{2}\cos\{2[\omega t + \phi(t)]\}) \\
&= (a_1 A + \frac{3}{4} a_3 A^3) \cos[\omega t + \phi(t)] + \frac{1}{4} a_3 A^3 \cos\{3[\omega t + \phi(t)]\} \\
&\approx (a_1 A + \frac{3}{4} a_3 A^3) \cos[\omega t + \phi(t)]
\end{aligned} \quad (10.70)$$

assuming bandpass filtering around ω. Note that the preceding analysis indicates that no spectral spreading into adjacent channels occurs. Rather, adjacent channel interference will be the result of the sideband noise associated with strong signals. However, if a signal becomes too strong it will cause gain compression (i.e., the a_1 and a_3 coefficients will change) and interference will result regardless of local oscillator purity. In general, with FM one can expect the adjacent channel interference protection ratio to hold steady as the desired signal-level reference is increased substantially above sensitivity.

In the AM case let the undesired signal be represented as:

$$x = A[1 + m\cos(\omega_m t)]\cos(\omega t)$$

The receiver response before detection is then:

$$\begin{aligned}
y &= a_1 A[1 + m\cos(\omega_m t)]\cos(\omega t) + a_3 A^3[1 + m\cos(\omega_m t)]^3 \cos(\omega t)^3 \\
&= [a_1 A + \frac{3}{4} a_3 A^3 (1 + \frac{3m^2}{2})] \cos(\omega t) \\
&+ [\frac{1}{2} a_1 A m + \frac{9}{8} a_3 A^3 (1 + \frac{m^3}{4})]\{\cos[(\omega - \omega_m)t] + \cos[(\omega + \omega_m)t]\} \\
&+ [\frac{9}{16} a_3 A^3 m^2]\{\cos[(\omega - 2\omega_m)t] + \cos[(\omega + 2\omega_m)t]\} \\
&+ [\frac{3}{32} a_3 A^3 m^2]\{\cos[(\omega - 3\omega_m)t] + \cos[(\omega + 3\omega_m)t]\}
\end{aligned} \quad (10.71)$$

again assuming bandpass filtering around ω. Adjacent splatter here is, in reality, an IM phenomenon caused by the receiver front-end nonlinearity. Consider 25-kHz channels with digital modulation occupying 16 kHz. For this situation, take $f_m = (\omega_m/2\pi)$ equal to 8 kHz and $m = 1$. A strong adjacent channel signal will produce significant power 16 and 24 kHz away from its center. The former may fall just outside the desired signal receiver passband (from 17 to 33 kHz away), but the latter will fall inside it. This unwanted power is given by:

$$U = \frac{1}{2}(\frac{3}{32}a_3 A_u^3)^2 \tag{10.72}$$

compared to the desired power in the passband of:

$$D = \frac{1}{2}(\sqrt{\frac{3}{2}}a_1 A_d)^2 \tag{10.73}$$

Now suppose the receiver is such that a desired signal at static sensitivity plus 3 dB is degraded to sensitivity performance by an adjacent channel signal 60 dB above static sensitivity. This equates to an adjacent channel splatter or desense performance of 60 dB. With FM, if the desired signal level is increased 10 dB, the coupled power can also be increased about 10 dB before performance degrades to that at sensitivity. However, with AM the third-order nature assumed for the receiver will cause such performance degradation with just a 5-dB increase in the adjacent channel signal level. Hence, with AM the ACIPR is not constant, but rather decreases as desired signal levels increase. This extra sensitivity to splatter must be properly accounted for in interference simulations.

10.12.7 Problem 7

Several sections in this chapter discuss calculation of the interference outage probability caused by an undesired signal bothering a desired signal. Frequency division multiplexed operation is implicit in this work. However, the introduction of *time division multiplexed* (TDM) channels to both private and cellular 800-MHz spectrum seems imminent. Discuss how one might quantify the interference outage probability caused by a TDM neighbor channel to a conventional radio channel.

Solution: Assuming that the channels are collocated, we know from observations that inbound signal levels tend to be distributed in a dual-lognormal fashion. In fact, because the primary lognormal component is much larger than the secondary lognormal component, it is not too unreasonable to ignore the latter to simplify the analysis (this, of course, is unnecessary if a Monte Carlo simulation is conducted). For a TDM interferer with N timeslots, up to N different interference sources may be sequentially present and generally if any one of them exceeds the single-carrier tolerance level interference will be noted. Consequently, what is needed is the distribution function for the maximum value obtained when N independent draws from a normal (in decibels) population are made.

When N is large, extreme value theory can be applied (for example, see References [62, 63]). When N is small, say 2, exact analysis is feasible. Let the PDF for inbound interference power in decibels be given by:

$$f_{X_{i=1,2}}(x) = \frac{1}{\sqrt{2\pi\sigma^2}}\exp[-\frac{(x-\mu)^2}{2\sigma^2}] \tag{10.74}$$

Coverage Analysis and Simulation 197

where μ represents the median interference level and σ equals the effective standard deviation of the interference level for each of the two TDM slots. The corresponding CDF for a single interference draw (one slot) is thus:

$$F_{X_i}(x) = \frac{1}{2}\text{erfc}(-\frac{x-\mu}{\sigma\sqrt{2}}) \qquad (10.75)$$

The probability that the largest of two draws from the signal strength population falls at or below an arbitrary level y is given by:

$$\begin{aligned} F_Y(y) &= \text{Prob}(X_1 \leq y) \bullet \text{Prob}(X_2 \leq y) \\ &= F_{X_1}(y) \bullet F_{X_2}(y) \\ &= \frac{1}{4}[\text{erfc}(-\frac{y-\mu}{\sigma\sqrt{2}})]^2 \end{aligned} \qquad (10.76)$$

The related PDF is found by differentiating with respect to y:

$$f_Y(y) = \frac{1}{\sqrt{2\pi\sigma^2}}\exp[-\frac{(y-\mu)^2}{2\sigma^2}]\text{erfc}(-\frac{y-\mu}{\sigma\sqrt{2}}) \qquad (10.77)$$

The median value of effective interference distribution is found by solving:

$$\frac{1}{2} = \frac{1}{4}[\text{erfc}(-q)]^2 \qquad (10.78)$$

where $q = (y-\mu)/(\sigma\sqrt{2})$. The solution obtained by using error function tables in inverse fashion is $q \approx 0.386$. Because a typical value for σ is 10 dB, this implies a boost of about 5.46 dB over the median value of each of the two interferers. Solving for the average of Y is rather tedious, though straightforward, by integrating the density function times y from $-$infinity to $+$infinity. The result is $Y_{av} = \mu + (\sigma/\sqrt{\pi})$. Again, assuming a σ of 10 dB, we find a boost in level of 5.64 dB over the single-source median. Such boosts in level directly subtract from the interference protection afforded in non-TDM situations.

10.12.8 Problem 8

It is tempting to install a towertop LNA to boost the sensitivity of base station reception. Base station equipment generally has noise figures on the order of 10 dB and cable loss increases it directly in proportion to the loss. Yet noise figures of just a few decibels are achieved with LNAs and, if they have sufficient gain, the overall system noise figure will approximate that of the LNA (recall the Friss noise figure formula for amplifiers in cascade). Can such a boost in sensitivity really be achieved in practice?

Solution: The answer is maybe! Certainly impressive increases in sensitivity versus internally generated noise are possible. However, true sensitivity may be set by externally generated noise. This is particularly likely in major metropolitan areas at UHF and below. Even more likely in major metropolitan areas are dynamic range problems, generally caused by third-order IM products.

For example, consider a 25-kHz analog FM base station with the following performance values: (1) -116 dBm for 12-dB static SINAD and (2) 75-dB *IMR*. The SNR typically required of such equipment to produce 12-dB static SINAD is 5 dB and the noise equivalent bandwidth is generally about 12 kHz. Thus, the noise figure is about 12 dB (15.8 numeric). The IM specification indicates that two signals on adjacent and alternate channels, and 75 dB (EIA test method) above the sensitivity level of -116 dBm, produce a third-order on-channel product equal to the noise floor of -121 dBm. This implies a third-order intercept point of -1 dBm (0.79 numeric; from solution of $y = -116 + 75 + x = -116 - 5 + 3x$ one finds $x = 40$ and $y = -1$). The dynamic range of reception might be defined as the difference between the sensitivity signal level and the level that produces noticeable IM at sensitivity plus 3 dB (EIA test method). Its value is then simply the *IMR* value.

Next, consider the use of a 4-dB (2.51 numeric) noise figure LNA whose output third-order intercept point is 35 dBm. With 25-dB preamplifier gain this implies a third-order intercept point referenced to the input of 10 dBm (10 numeric). Let us assume that cable and other losses result in a net gain of just 10 dB (10 numeric). The system noise figure is thus $2.51 + (15.8 - 1)/10 = 4.0$ (6 dB), an improvement of 6 dB over the original base station receiver alone. That is the good news. The bad news is that the intercept point degrades to just -11.1 dBm (0.078 numeric, from solution of $1/IP = 1/IP1 + G1/IP2 = 1/10 + 10/0.79$; see Reference [59] for the intercept point formula of a cascade of two stages). This only slightly drops the dynamic range (to 72.3 dB) but substantially lowers the input signal level required to avoid IM desense (to -49.7 dBm).

The net LNA gain that maximizes dynamic range can be found by differentiating the ratio of total noise figure to intercept point (both in numeric form) with respect to the gain and setting the result to zero. Interestingly, this best cascade arrangement may not yield specifications better than those of the receiver alone, in either dynamic range or noise figure. However, using very little takeover gain, the base station can at least be made to appear as though it is located right at the antenna output.

References

[1] Jakes, Jr., W. C. (editor), *Microwave Mobile Communications*, New York, NY, John Wiley & Sons, 1974.

[2] Okumura, Y., et al., "Field Strength and Its Variability in VHF and UHF Land-Mobile Radio Service," *Rev. of Elec. Comm. Lab.*, Vol. 16, Nos. 9-10, September-October, 1968, pp. 825-873.

[3] Gradshteyn, I. S. and I. M. Ryzhik, *Table of Integrals, Series, and Products*, New York, NY, Academic Press, 1965.

[4] Ng, E. and M. Geller, "A Table of Integrals of the Error Functions," *J. of Res. of the NBS*, B. Mathematical Sciences, Vol. 73B, No. 1, January-March 1969, pp. 1-20.

[5] Beckmann, P., *Probability in Communication Engineering*, New York, NY, Harcourt, Brace & World, Inc., 1967.

[6] Conte, S. D., *Elementary Numerical Analysis*, New York, NY, McGraw-Hill Book Co., 1965.

[7] Rubinstein, R. Y., *Simulation and the Monte Carlo Method*, New York, NY, John Wiley & Sons, 1981.

[8] Marsan, M., et al., "Shadowing Variability in an Urban Land Mobile Environment at 900 MHz," *Electronics Letters*, Vol. 26, No. 10, May 1990, pp. 646-648.

[9] Bevington, P. R., *Data Reduction and Error Analysis for the Physical Sciences*, New York, NY, McGraw-Hill Book Co., 1969.

[10] Abramowitz, M. and I. Stegun (editors), *Handbook of Mathematical Functions*, New York, NY, Dover Publications, Inc., 1972.

[11] Hiben, B., et al., "Adjacent Channel Interference Considerations in Narrowband FM Radio Systems," *37th IEEE Vehicular Technology Conference*, Tampa, FL, June 1987.

[12] EIA/TIA Standard-204-D, Electronic Industries Association, Engineering Dept., Washington, D.C., April 1989.

[13] EIA/TIA Bulletin-TSB28, Electronic Industries Association, Engineering Dept., Washington, D.C., July 1989.

[14] Cerny, F., "Adjacent-Channel Interference Protection Between Closely-Spaced Single Sideband and FM Mobile Telephone Channels," Appendix B, Reply of American Telephone & Telegraph Company in the matter of Contemporary Communications Corporation, FCC File No. 21850 et al., June 1983.

[15] Brooner, M. and L. Kolsky, "Comments of Motorola, Inc., in the matter of General Docket No. 84-1233, RM-4829," Washington, D.C., pp. A-5.

[16] Hess, G., "Spectrum Efficiency Potential of 25 kHz Offset Channel Assignments in the 821-824/866-869 MHz Public Safety Bands," *38th IEEE Vehicular Technology Conference*, Philadelphia, PA, June 1988, pp. 358-363.

[17] Nasser, J. (chairman), National Public Safety Planning Advisory Committee, *Final Report to the Federal Communications Commission*, September 9, 1987.

[18] Box, F., "A Heuristic Technique for Assigning Frequencies to Mobile Radio Nets," *IEEE Tr. on Veh. Tech.*, Vol. VT-27, May 1978.

[19] Walker, E. H., "Penetration of Radio Signals into Buildings in the Cellular Radio Environment," *Bell System Technical Journal*, Vol. 62, No. 9, November 1983, pp. 2719-2734.

[20] Hess, G., "Further Sharing of UHF Television by Private Land Mobile Radio Services," *37th IEEE Vehicular Technology Conference*, Tampa, FL, June 1987, pp. 324-329.

[21] Wong, H. K., "A Computer Program for Calculating Effective Interference to TV Service," OST Technical Memorandum, FCC/OST TM 82-2, July 1982.

[22] Damelin, J., et al., "Development of VHF and UHF Propagation Curves for TV and FM Broadcasting," FCC Report R-6602, September 7, 1966 (third printing May 1974).

[23] Alder, H. L. and E. B. Roessler, *Introduction to Probability and Statistics*, New York, NY, W. H. Freeman and Co., 1972.

[24] Egli, J. J., "Radio Propagation Above 40 MC Over Irregular Terrain," *Proc. IRE*, Vol. 45, No. 10, October 1957, pp. 1383-1391.

[25] Fine, H., "UHF Propagation Within Line of Sight," FCC T.R.R. Report No. 2.4.12, June 1951.

[26] Canadian Task Force on UHF-TV Taboos, Project 6, "Assessment of Potential Land-Mobile Interference to/from UHF Television," 1976.

[27] Land Mobile/UHF Television Technical Advisory Committee, Final Report, May 7, 1986.

[28] Fredendall, G. L., "A Comparison of Monochrome and Color Television with Reference to Susceptibility to Various types of Interference," *RCA Review*, September 1953, pp. 341-359.

[29] Hata, M., "Empirical Formula for Propagation Loss in Land Mobile Radio Services," *IEEE Tr. on Veh. Tech.*, Vol. VT-29, No. 3, August 1980, pp. 317-325.

[30] FitzGerrel, R. G., et al., "Television Receiving Antenna System Component Measurements," NTIA Report 79-22, June 1979.

[31] Motorola, Inc. Comments on General Docket 85-172, July 11, 1986.

[32] Mass Media Bureau Technical Report MM 88-56, Federal Communications Commission, Washington, D.C., "Algorithm for Computing Field Strength for FM and TV Broadcast Stations," November 1987.

[33] Carey, R. B., "Technical Factors Affecting the Assignment of Facilities in the Domestic Public Land Mobile Radio Service," FCC Report R-6406, Washington, D.C., June 1964.

[34] Delisle, G. Y., et al., "Propagation Loss Prediction: A Comparative Study with Application to the Mobile Radio Channel," *IEEE Tr. on Veh. Tech.*, Vol. VT-34, No. 2, May 1985, pp. 86-95.

[35] Giovaneli, C. L., "An Analysis of Simplified Solutions for Multiple Knife-Edge Diffraction," *IEEE Tr. Ant. and Prop.*, Vol. AP-32, No. 3, March 1984, pp. 297-301.

[36] de Assis, M. S., "A Simplified Solution to the Problem of Multiple Diffraction over Rounded Obstacles", *IEEE Tr. Ant. and Prop.*, March 1971, pp. 292-295.

[37] Kalagian, G., "Field Strength Calculation for TV and FM Broadcasting (Computer Program TVFMFS)," FCC/OCE Report RS 76-01, January 1976.

[38] MacDonald, V. H., "The Cellular Concept," *Bell System Technical Journal*, Vol. 58, No. 1, January 1979, pp. 15-39.

[39] Lee, W.C.Y., *Mobile Cellular Telecommunications Systems*, New York, NY, McGraw-Hill Book Co., 1989.

[40] Marsan, M. and G. Hess, "Cochannel Isolation Characteristics in an Urban Land Mobile Environment at 900 MHz," *41st IEEE Vehicular Technology Conference*, St. Louis, MO, May 1991, pp. 600-605.

[41] EIA/TIA Project Number 2215, Cellular System, Dual-Mode Mobile Station - Base Station Compatibility Standard, Document IS-54, December 1989.

[42] Fuhrmann, W. and K. Spindler, "Standardization of an (sic) European Digital Mobile Radiocommunication System," *ICC '86*, pp. 3.6.1-3.6.7.

[43] Mende, W., "Evaluation of a Proposed Handover Algorithm for the GSM Cellular System," *1990 IEEE Vehicular Technology Conference*, Orlando, FL, May 6-9, pp. 264-269.

[44] QUALCOMM, "CDMA Digital Cellular Technology," CTIA Open Forum, San Diego, CA, June 6, 1989.

[45] CCIR Study Groups, Period 1986-1990, Interim Working Party 8/13, QUAL-COMM contribution: "Future Public Land Mobile Telecommunication Systems (FPLMTS) Employing Code Division Multiple Access (CDMA) and Advanced Modulation and Coding Techniques," January 30, 1990.

[46] Stern, M., "Analysis Shows Advantages of Four-Cell Site Repeat Patterns," *Mobile Radio Technology*, January 1984, pp. 42-45.

[47] Beck, R. and H. Panzer, "Strategies for Handover and Dynamic Channel Allocation in Micro-Cellular Mobile Radio Systems," *39th IEEE Vehicular Technology Conference*, San Francisco, CA, May 1989, pp. 178-185.

[48] Nettleton, R. and G. R. Schloemer, "A High Capacity Assignment Method for Cellular Mobile Telephone Systems", *39th IEEE Vehicular Technology Conference*, San Francisco, CA, May 1989, pp. 359-367.

[49] Eriksson, H. and R. Bownds, "Performance of Dynamic Channel Allocation in the DECT System," *1991 IEEE Vehicular Technology Conference*, St. Louis MO, May 19-22, pp. 693-698.

[50] Hatfield, D. N., "Measures of Spectral Efficiency in Land Mobile Radio," *IEEE Tr. on Electromag. Comp.*, Vol. EMC-19, No. 3, August 1977, pp. 266-268.

[51] Berry, L. A., "Spectrum Metrics and Spectrum Efficiency: Proposed Definitions," *IEEE Tr. on Electromag. Comp.*, Vol. EMC-19, No. 3, August 1977, pp. 254-260.

[52] Mikulski, J. J., "Technology and Spectrum Efficiency," *INTELCOM '79*, Dallas, TX, 1979, pp. 250-253.

[53] Chang, L. F. and P. T. Porter, "Performance Comparison of Antenna Diversity and Slow Frequency Hopping for the TDMA Portable Radio Channel," *39th IEEE Vehicular Technology Conference*, San Francisco, CA, May 1989, pp 464-469.

[54] Hammuda, H., et al., "Spectral Efficiency of Cellular Land Mobile Radio Systems," *38th IEEE Vehicular Technology Conference*, Philadelphia, PA, June 1988, pp. 616-622.

[55] *39th IEEE Vehicular Technology Conference*, San Francisco, CA, May 1989.

[56] Papoulis, A., *Probability, Random Variables, and Stochastic Processes*, New York, NY, McGraw-Hill Book Co., 1965.

[57] Hess, G., "Analysis of the Probability of Interference During a Telephone Interconnect Call," *34th IEEE Vehicular Technology Conference*, Pittsburgh PA, May 1984, pp. 74-79.

[58] Graziano, V., "Propagation Correlations at 900 MHz," *IEEE Tr. on Veh. Tech.*, Vol. VT-27, No. 4, November 1978, pp. 182-189.

[59] Sagers, R. C., "Intercept Point and Undesired Responses," *32nd IEEE Vehicular Technology Conference*, San Diego, CA, May 1982, pp. 219-230.

[60] Maxemchuk, N. F. and L. Schiff, "Third Order Intermodulation Interference - Bounds and Interference-Free Channel Assignment," *IEEE Tr. on Comm.*, Vol. COM-25, No. 9, September 1977, pp. 1041-1046.

[61] Durante, J., "Building Penetration Loss at 900 MHz," *IEEE Vehicular Technology Conference*, 1973, pp. 1-7.

[62] Gumbel, E., "Statistical Theory of Extreme Values and Some Practical Applications," NBS Applied Mathematics Series, 33.

[63] Bain, L. J., *Statistical Analysis of Reliability and Life-Testing Models*, New York, NY, Marcel Dekker, Inc., 1978.

Chapter 11

Diversity

This chapter examines various methods for combating the severe fading generally associated with land-mobile radio links. The methods fall under the generic heading of diversity and are discussed in terms of how multiple copies of the signal might be obtained and how those copies might then be combined to produce an output with improved performance. Two practical applications of diversity are discussed in detail: first, the case where separate selection of individual branches is not feasible and second, the case where cost and energy constraints preclude maximal-ratio combining but diversity gain is needed at both high and low speeds.

11.1 Classification of Techniques

The basic idea behind diversity techniques is that if a message is received via a number of transmission paths or branches that are subject to different fading statistics, then the various branches may be selected or combined in such a manner that improves the performance over that using any single branch. The key aspects of diversity systems are therefore the manner by which multiple copies of the signal are obtained and the manner by which those copies are selected or combined.

11.1.1 Space Diversity

One of the most common methods of obtaining multiple copies of the signal is to use two or more spatially separated antennas, hence the term space diversity. To provide sufficiently uncorrelated copies of the signal[1], base antenna separations on the order of tens of wavelengths are generally required. However, mobile and portable radios, because they tend to be surrounded by local scatterers, can achieve useful space diversity benefits with antenna separations on the order of just one-

---

[1]Most potential diversity gain can be realized provided the signal envelope cross-correlation coefficient is under 0.7; or, equivalently, the signal cross-correlation coefficient is under $\sqrt{0.7} = 0.84$.

quarter wavelength. For example, Figure 11.1 shows the correlation coefficient of the envelope of the vertical field strength in a Rayleigh field as a function of the spacing in wavelengths (or, equivalently, time delay at fixed speed) [1].

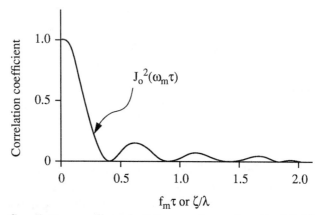

Figure 11.1 Correlation coefficient of the envelope of vertical field strength in a Rayleigh field.
After: Jakes, Jr., W. C. (editor), *Microwave Mobile Communications*, John Wiley & Sons, New York, NY, 1966, p. 472.

Lee [2] discusses the base antenna spacing question at some length in his Chapter 9. Design curves are presented there relating the antenna envelope correlation coefficient to the design parameter η (which is defined as the ratio of the antenna height to the antenna spacing) as a function of the angle of signal arrival compared to broadside. Adachi et al. [3] develop the following approximate expression for antenna envelope correlation at broadside:

$$\rho_{env} = \exp\{-[2\pi(d/\lambda)(r/L)]^2\} \qquad (11.1)$$

where (d/λ) equals the antenna separation in wavelengths, r equals the effective radius of local scatterers about the subscriber unit, and L equals the direct path length. Notice that antenna height is not a factor in the relationship since the base antennas are assumed to be free of local scatter. This situation usually holds for the very high antennas used by wide-area coverage systems like SMRs, but breaks down at the modest antenna heights associated with cellular systems; hence, Lee's height dependence. Reference [3] also examines the impact of horizontal separation versus vertical separation and find the former to be superior, except for cases nearly perpendicular to broadside.

Substituting the usual target of 0.7 correlation into Equation (11.1) and assuming a link frequency of 820 MHz leads to $(d/L) = (0.114/r)$. Representative effective

Diversity

scattering radii can be inferred from delay spread measurements (Section 12.6.2). Typical delay spread values are 0.1, 0.5, and 1.3 μs for rural, suburban, and urban environments, respectively, implying r values of 100, 500, and 1300 ft. These values in turn imply antenna separation-to-path length ratios of 6, 1.2, and 0.5 ft per mi for the three environment types.

11.1.2 Polarization Diversity

Collocated antennas that respond to orthogonal signal polarizations are diversity candidates that circumvent possible antenna spacing difficulties. Although three orthogonal polarizations exist, generally the number of branches is limited to just two: one for vertical polarization and one for horizontal polarization. When the transmitting antenna is oriented toward one polarization, say vertical, substantial scattering must take place along the propagation path for the average power intercepted by a horizontally polarized receive antenna to approximate the average power intercepted by a vertically polarized receive antenna. The amount of diversity gain possible decreases as the mismatch in average power between the branches increases. A way around this is to transmit equal power in both polarizations. This of course has a cost of at least 3 dB in sensitivity, which subtracts from the diversity gain realized. This technique is particularly attractive for portable radios because antenna spacing is at a premium and handset orientation is random [4].

11.1.3 Angle Diversity

A form of diversity that has been applied to tropospheric scatter links and microwave "hops" involves the use of multiple antenna feeds along with a single reflector-type antenna; for example, a parabolic dish. This produces multiple antenna beams that point to slightly different directions, hence the term angle diversity. Such diversity is useful in combating tropospheric bending of radio signals and the temporal variation of angle of arrival that such bending causes.

11.1.4 Frequency Diversity

If a message can be transmitted simultaneously over multiple channels sufficiently separated in frequency, then fading will be decorrelated and diversity gain can be achieved. The coherence bandwidth of mobile radio channels is generally on the order of 500 kHz, so to be useful the multiple channels should be separated at least several megahertz (Section 1.5 of Reference [1]). This technique requires only a single antenna; however, it pays a high price in spectrum efficiency. A way of getting around that price may be to use spread-spectrum modulation, either direct-sequence or fast frequency hopping. Here, the signal is purposely spread over sufficient bandwidth to combat fading. The use of orthogonal codes or hopping patterns, precise power control, and powerful forward error-correction codes holds some promise for achieving good spectrum efficiency [5, 6, 7].

11.1.5 Time Diversity

The equivalence of time and space indicated by Figure 11.1 means that repeat transmissions, sufficiently separated in time, can also be used to combat fading[2]. This idea is applied on a microscopic scale by digital encoding procedures that spread the information bits of each byte of information over the entire time taken to transmit an information packet. Errors in a string of bits due to a fade can then generally be tolerated with quite simple and efficient forward error-correction coding because individual bytes are unlikely to contain more than a single error. Time diversity is also behind the general idea of *automatic repeat request* (ARQ) schemes where packets that are not correctly received are automatically retransmitted at a later time.

11.1.6 Other Techniques

An interesting form of diversity reception suggested by J. R. Piece involves a special antenna that responds to the energy density of the signal. This requires antennas that respond to the three orthogonal field components; for example, a vertical dipole responding to the electric field and a pair of loop antennas responding to the perpendicular magnetic fields. Subjecting their outputs to square-law detection and summing the results yields an output proportional to the energy density [8]. This technique does produce shorter fade durations, but only for very deep fades [9]. Also, its nonlinear nature precludes its use as a simple "drop-in" item between the antenna and conventional receiver.

A special form of frequency diversity called sideband diversity has been described by Gosling et al. [10]. This technique involves simulcasting double sideband signals from transmitters whose audio phases differ by 90 deg. Such conditions produce negatively correlated sideband signal amplitudes and thus preclude signal cancellation.

11.2 Selection Diversity Combining

Perhaps the most common method of diversity combining does not really combine signals at all. Instead, selection diversity simply chooses the best individual branch signal from all the branch signals available. If this selection is updated instan-

---

[2] A problem may of course arise if the receiving unit is not moving at all. Then if one is located in a multipath fade and the environment itself does not change significantly for the repeat transmission, no diversity benefit will accrue. This issue is addressed in homework problem 2.

Diversity

taneously and involves two uncorrelated Rayleigh-faded branches of equal average power, the statistics of the output signal are already developed (Section 9.3.3). Since average thermal noise power per branch is simply some constant, those results also pertain to selection of the branch with the highest SNR. The meaning of α simply changes to "average SNR per branch." The result readily extends to an arbitrary number of branches, say M, by using:

$$F_Z(z) = [1 - \exp(-\frac{z}{\alpha})]^M \tag{11.2}$$

which equals the distribution function for a single branch raised to the power M.

In the case of unequal branch average powers (or average SNRs), one has:

$$\begin{aligned} F_Z(z) &= \text{Prob}(x_1 < z, x_2 < z, \ldots, x_M < z) \\ &= \prod_{k=1}^{M} [1 - \exp(-\frac{z}{\alpha_k})] \end{aligned} \tag{11.3}$$

where the average power (or average SNR) in the k-th branch is α_k. If $z \ll \alpha_k$ for all k (i.e., for fades that are deep even relative to the weakest branch) then:

$$\begin{aligned} F_Z(z) &\approx \prod_{k=1}^{M} \{1 - [1 - (\frac{z}{\alpha_k})]\} \\ &= (\frac{z}{\Gamma_g})^M \end{aligned} \tag{11.4}$$

where Γ_g is the geometric mean of the branch average powers (or average SNRs).

The impact of branch correlation substantially complicates matters. Reference [11] presents a general solution for the two-branch case where the branches have arbitrary average powers (or average SNRs) and arbitrary correlation. Figure 11.2 shows sample results. Notice that correlation coefficients as high as 0.7 cause little loss of diversity gain.

Instantaneous switching among branches is not necessarily practical. For example, it can produce damaging transients due to phase differences between the old selected branch and the new selected branch. Two other switching strategies called "switch and stay" and "switch and examine" are considered in reference [12].

An interesting form of selection diversity is possible on TDM systems [13]. The idea is to monitor signal quality on the available branches in the time slot immediately preceding the time slot assigned for reception and switch to the "winner" for actual reception. This procedure works well provided the time from measurement to the end of reception is modest relative to the time rate of signal decorrelation.

11.3 Maximal-Ratio Combining

This method really does combine the signals from the various branch inputs. It does so in a manner that cophases the desired signal components before combining them

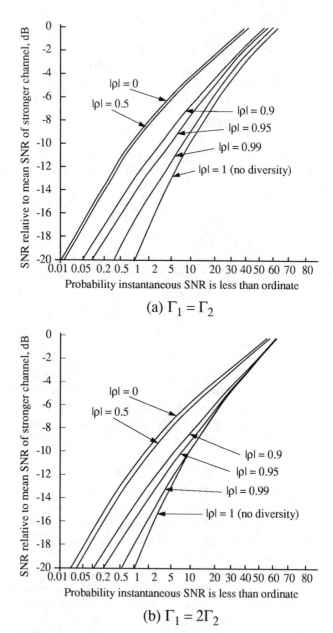

Figure 11.2 Cumulative distribution function of two-branch selection diversity. After: Schwartz, M., et al., *Communication Systems and Techniques*, McGraw-Hill Book Co., New York, NY, 1966, p. 472.

Diversity

so that they add coherently. On the other hand, noise from the various branches is generally uncorrelated and thus adds incoherently. Thus, the result maximizes the output SNR. One realization for accomplishing cophasing was described by Granlund in 1956 [14]. It involves both feedforward and feedback operations and has been implemented in commercial hardware; for example, the base station combiners in some cellular systems. Recent digital implementations of the concept have greatly increased performance reliability and reduced cost.

After cophasing, the envelope of the combined output signal can be written as:

$$r_o = \sum_{i=1}^{M} a_i r_i \quad (11.5)$$

where a_i represents a weighting factor for the i-th branch input. The output noise power equals the sum of the noise powers of each branch. Assuming all these powers equal σ_N^2, one can write the output noise power as:

$$N_o = \sigma_N^2 \sum_{i=1}^{M} a_i^2 \quad (11.6)$$

The output SNR is $\text{SNR}_o = (r_o^2/2N_o)$. It can be shown that SNR_o is maximized when the weights are chosen as $a_i = r_i/\sigma_N^2$ [15] [3]. The resulting output SNR is then:

$$\begin{aligned}
\text{SNR}_{o,\max} &= \frac{(\sum_{i=1}^{M} \frac{r_i^2}{\sigma_N^2})^2}{2\sigma_N^2 \sum_{i=1}^{M} (\frac{r_i}{\sigma_N^2})^2} \\
&= \sum_{i=1}^{M} \frac{r_i^2}{2\sigma_N^2} \\
&= \sum_{i=1}^{M} \text{SNR}_i \quad (11.7)
\end{aligned}$$

that is, the maximum output SNR equals the sum of the SNRs of the input branches.

Now the square of the envelope of any branch equals the sum of the squares of two independent, zero-mean Gaussian random variables. Thus, one can write:

$$\begin{aligned}
\text{SNR}_i &= \frac{1}{2\sigma_N^2} r_i^2 \\
&= \frac{1}{2\sigma_N^2}(x_i^2 + y_i^2) \quad (11.8)
\end{aligned}$$

The maximum output SNR can be recognized as a chi-square distribution of $2M$ zero-mean, independent Gaussian random variables with variance equal to the one-half the average branch signal power divided by twice the average branch noise

---

[3] Such weighting is generally not trivial to accomplish. A simplified form of combining (equal gain combining) is often used instead with only modest loss in performance.

power. The PDF for the output SNR of the maximal-ratio combiner is given by:

$$f_{\text{SNR}_{o,\max}}(y) = \frac{y^{M-1}\exp(-\frac{y}{\gamma})}{\gamma^M(M-1)!} \tag{11.9}$$

where γ represents the average input SNR of each branch (for example, see Section 8-4 of Reference [16]) and y must equal or exceed zero. Integration of this equation up to an arbitrary value, say y_r, gives the CDF:

$$F_{\text{SNR}_{o,\max}}(y_r) = 1 - \exp(-\frac{y_r}{\gamma})\sum_{i=1}^{M}\frac{(\frac{y_r}{\gamma})^{i-1}}{(i-1)!} \tag{11.10}$$

This function is plotted for several sizes of combiner in Figure 11.3. Two-branch performance in the presence of varying amounts of branch correlation is shown in Figure 11.4.

11.4 Analysis of a Special Form of Selection Diversity

To offer selection diversity reception with a single-block duplexer, an arrangement has been proposed where the two branches consist of antenna 1 and the voltage sum of antennas 1 and 2, respectively. This section will quantify the performance of such an arrangement.

Begin by considering the signals from each antenna to be represented in quadrature form with X_1 and X_3 the in-phase components of antennas 1 and 2, respectively, and X_2 and X_4 the quadrature components of antennas 1 and 2, respectively. In the multipath field typical of land-mobile operation, all four components are zero-mean Gaussian-distributed random variables. Assuming the antennas are similar and their signals are uncorrelated, the covariance matrix for the selection diversity problem is:

$$U = \begin{bmatrix} u_{11} & u_{12} & u_{15} & u_{16} \\ u_{21} & u_{22} & u_{25} & u_{26} \\ u_{51} & u_{52} & u_{55} & u_{56} \\ u_{61} & u_{62} & u_{65} & u_{66} \end{bmatrix}$$

$$U = \begin{bmatrix} \sigma^2 & 0 & \sigma^2 & 0 \\ 0 & \sigma^2 & 0 & \sigma^2 \\ \sigma^2 & 0 & 2\sigma^2 & 0 \\ 0 & \sigma^2 & 0 & 2\sigma^2 \end{bmatrix}$$

where $X_5 = X_1 + X_3$, $X_6 = X_2 + X_4$, and signals from the two antennas are assumed to be of equal power (σ^2 per component) and uncorrelated. The determinant of the

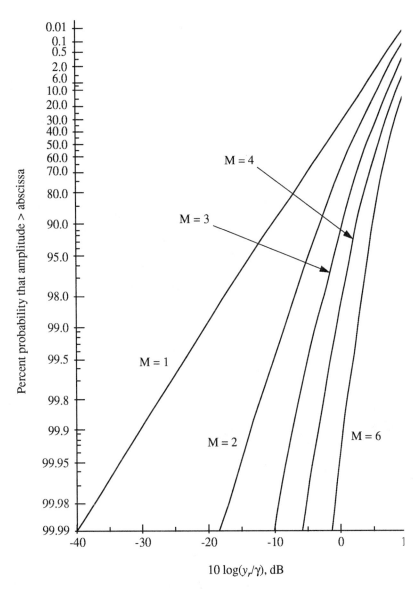

Figure 11.3 Cumulative distribution function for M-branch maximal-ratio diversity combiner, all branches equal average SNR and uncorrelated. After: Jakes, Jr., W. C. (editor), *Microwave Mobile Communications*, John Wiley & Sons, New York, NY, 1974, p. 320.

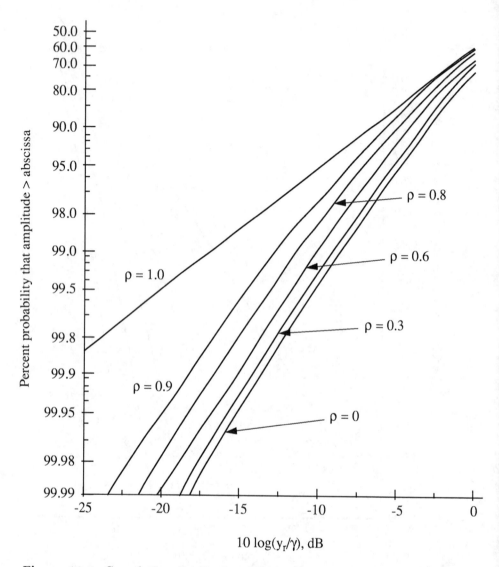

Figure 11.4 Cumulative distribution function for two-branch maximal-ratio diversity combining as a function of branch correlation.
After: Jakes, Jr., W. C. (editor), *Microwave Mobile Communications*, John Wiley & Sons, New York, NY, 1974, p. 327.

Diversity

covariance matrix is σ^8 and the inverse is:

$$U^{-1} = \begin{bmatrix} 2\sigma^2 & 0 & -\sigma^2 & 0 \\ 0 & 2\sigma^2 & 0 & -\sigma^2 \\ -\sigma^2 & 0 & \sigma^2 & 0 \\ 0 & -\sigma^2 & 0 & \sigma^2 \end{bmatrix}$$

Hence, the joint PDF for the random variables X_1, X_2, X_5, X_6 can be written as (p. 255 of Reference [16]):

$$f(x_1, x_2, x_5, x_6) = \frac{1}{4\pi^2\sigma^4} \exp[-\frac{1}{2\sigma^2}(2x_1^2 + 2x_2^2 + x_5^2 + x_6^2 - 2x_1x_5 - 2x_2x_6)] \quad (11.11)$$

Applying the transformation $x_1 = r_1 \cos\theta_1$, $x_2 = r_1 \sin\theta_1$, $x_5 = r_2 \cos\theta_2$, and $x_6 = r_2 \cos\theta_2$ leads to the PDF:

$$f(r_1, r_2, \theta_1, \theta_2) = \frac{r_1 r_2}{4\pi^2\sigma^4} \exp[-\frac{1}{2\sigma^2}(2r_1^2 + 2r_2^2 - 2r_1 r_2(\cos\theta_1 \cos\theta_2 + \sin\theta_1 \sin\theta_2)] \quad (11.12)$$

where r_1 and r_2 are greater than or equal to zero and θ_1 and θ_2 vary between 0 and 2π. This density function can be integrated over variables θ_1 and θ_2 with the help of relation 9.6.18 in Reference [17] to yield:

$$f(r_1, r_2) = \frac{r_1 r_2}{\sigma^4} \exp(-\frac{2r_1^2 + r_2^2}{2\sigma^2}) I_0(\frac{r_1 r_2}{\sigma^2}) \quad (11.13)$$

Integration over r_2 gives a PDF for r_1 that is Rayleigh distributed with average power σ^2. Likewise, integration over r_1 gives a PDF for r_2 that is Rayleigh distributed with average power $2\sigma^2$. These results are obtained via the substitution $s = r^2$ and using relations 6.614 #3, 9.220 #2, and 9.215 #1 in the tables of Reference [18].

The probability that the selected signal is less than or equal to some value z is then given by integrating the density function of r_1 and r_2 over 0 to z:

$$F_Z(z) = \int_0^z \int_0^z \frac{r_1 r_2}{\sigma^4} \exp(-\frac{2r_1^2 + r_2^2}{2\sigma^2}) I_0(\frac{r_1 r_2}{\sigma^2}) dr_1 dr_2 \quad (11.14)$$

This integration is not expressible in simple closed form. However, the main interest is in z values that are small relative to the average level σ; i.e. in the deep fades. Under these circumstances the exponential term is approximately unity, as is the Bessel function. Thus, for deep fades the CDF is approximately the square of the fade depth normalized to the average power. For example, a fade exceeding 10 dB would have a probability of just 1 %. A fade exceeding 20 dB would have a probability of just 0.01 %. Interestingly, this is precisely the behavior of conventional two-branch selection diversity. When the power from antenna 1 is low, the power in the second branch is essentially all the result of antenna 2. When the branch 2 power is low, it is most likely because the antenna signals are out of phase, rather than both antennas have low signal levels.

The full behavior of $F_Z(z)$ can readily be obtained through Monte Carlo simulation. It is easy in such simulations to include the possibility of nonzero correlation between the two antennas. Even when such is the case, the scheme behaves well in comparison to the conventional approach. In fact, the performance with three branches (the third branch representing the sum of the voltages from three antennas) can surpass that of the conventional three-branch approach.

11.5 Homework Problems

11.5.1 Problem 1

The maximal-ratio combiner maximizes signal-to-noise ratio. Is it also the optimal combining technique for radio systems subject to interference as well as noise (for example, cellular telephone systems)?

Solution: The optimal combiner, when both noise and interference are considerations, would maximize the signal-to-noise plus interference power ratio. A maximal-ratio combiner does not necessarily do this. For example, cophasing the desired signal contributions may also to some degree cophase the interference, since unlike noise, interference can be expected to be correlated to some degree among the various antennas being used. The structure and performance of the optimal combiner is discussed by Winters [19, 20].

11.5.2 Problem 2

With the increase of communication systems that cater to both mobile and portable radios, techniques that maintain good link quality at both fast and slow speeds have become important. Link enhancement at slower speeds is particularly important because it is not feasible to correct error strings caused by the associated long fades. Slow-speed problems occur not just for portables, but also for mobiles as they encounter traffic signals, stop signs, etc. Diversity techniques that are effective at all speeds of interest and are also cost and energy efficient are unknown to us. For example, maximal-ratio combining gives maximum diversity gain and does function at all speeds of interest. Unfortunately, its use in subscriber units is tantamount to requiring multiple receivers and therefore is neither cost nor energy efficient. Describe a possible solution for single-site wide-area coverage applications that avoids such negatives.

Solution 1: Consider the adaptive use of two different diversity techniques in the same unit: one geared toward slow speed signal enhancement and one geared toward high-speed signal enhancement. The slow-speed scheme is activated when the unit is moving at less than some threshold speed; the high-speed scheme is used at higher speeds.

In a TDM system, the preferred slow-speed technique is time-switched selection diversity [13]. This diversity technique is attractive because it only requires two antennas and a switch, rather than two whole front end receivers. The receiver uses an early time slot to average the signal power present via each antenna. It then switches to the antenna with greatest power just before the allocated time slot. This technique works only for slow speeds because the correlation between time of measurement and time of use degrades as the unit's speed increases. Simulation shows that for 15-ms time slots and 0.5-mph speeds, a 5-dB gain at 4% probability of bit error can be achieved. This gain decreases to 4 dB at 5 mph and no gain can be expected at 45 mph.

At high speeds in an internal noise-limited environment, simply combining the two antenna ports can provide a boost of up to 3 dB. This is because while the signal distribution is still Rayleigh, the average power reference is 3 dB higher for uncorrelated inputs of equal average power. In interference-limited circumstances, a technique that switches between antennas quickly relative to symbol duration can provide high-speed diversity gain, at least for certain modulation schemes [21].

The speed of mobile radios can be obtained by monitoring a physical connection to the vehicle odometer. A wireless technique is preferred, however. For example, one can use an algorithm that alternately samples the in-phase and quadrature-phase signals at some known rate [22] (Section 12.4). The rate of change of phase in time yields an estimate of the Doppler frequency, which in turn is related to vehicle speed.

Solution 2: The essence of the problem is that, at speeds below some value, multipath fades are sufficiently long that forward error-correction coding cannot cope. Consequently, the problem is solved by techniques that guarantee the received field moves above a certain rate, regardless of the movement or lack of movement of the subscriber unit. One way of stirring the field is to use a second, parasitic antenna element whose phase is modulated at a rate corresponding to normal mobile speeds. The directivity of the effective two-element antenna is thus caused to move about in time, which is equivalent to movement of the received field. This technique is particularly effective with digital voice modulation, since there the BER need only be maintained below some level (typically a few percent) to yield an output quality near the no-noise capability of the voice encryption scheme. With analog voice, continuous stirring of the field will limit the maximum achievable output SNR. Consequently, it might be preferable to simply shift the phase on the parasitic element when the received signal strength falls below some level and the subscriber unit speed is low. Such a mode of operation is comparable to switched antenna diversity in performance but avoids the need for a physical RF switch.

References

[1] Jakes, Jr., W. C. (editor), *Microwave Mobile Communications*, New York, NY, John Wiley & Sons, 1974.

[2] Lee, W.C.Y., *Mobile Communications Engineering*, New York, NY, McGraw-Hill Book Co., 1982.

[3] Adachi, F., et al., "Crosscorrelation between the envelopes of 900 MHz signals received at a mobile radio base stations (sic) site," *IEE Proceedings*, Vol. 133, Pt. F, No. 6, October 1986, pp. 506-512.

[4] Bergmann, S. A. and H. W. Arnold, "Polarization Diversity in the Portable Communications Environment," *Electronic Letters*, Vol. 22, May 22, 1986, pp. 609-610.

[5] Cooper, G. R. and R. W. Nettleton, "Spectral Efficiency of Spread-Spectrum Land-Mobile Communication Systems," *Intelcom '79*, Dallas, TX, 1979, pp. 267-270.

[6] Mazo, J. E., "Some Theoretical Observations on Spread-Spectrum Communications," *Bell System Technical Journal*, Vol. 58, No. 9, November 1979, pp. 2013-2023.

[7] Viterbi, A. J., et al., "CDMA in Satellite and Terrestrial Communications Networks," *Globecom '90 Conference* (workshop), San Diego, CA, December 2, 1990.

[8] Itoh, K. and D. K. Cheng, "A Slot-Unipole Energy-Density Mobile Antenna," *IEEE Tr. on Veh. Tech.*, Vol. VT-21, No. 2, May 1972, pp. 59-62.

[9] Lee, W.C.Y., "Statistical Analysis of the Level Crossings and Duration of Fades of the Signal from an Energy Density Mobile Radio Antenna," *Bell System Technical Journal*, Vol. 46, February 1967, pp. 417-448.

[10] Gosling, W., et al., "Sideband Diversity: A New Application of Diversity Particularly Suited to Land Mobile Radio," *The Radio and Electronic Engineer*, Vol. 48, No. 3, March 1978, pp. 000-007.

[11] Schwartz, M., et al., *Communication Systems and Techniques*, New York, NY, McGraw-Hill Book Co., 1966.

[12] Blanco, M. A. and K. J. Zdunek, "Performance and Optimization of Switched Diversity Systems for the Detection of Signals with Rayleigh Fading," *IEEE Tr. on Comm.*, Vol. COM-27, No. 12, December 1979, pp. 1887-1895.

[13] Akaiwa, Y., "Antenna Selection Diversity for Framed Digital Signal Transmission in Mobile Radio Channel," *39th IEEE Vehicular Technology Conference*, San Francisco, CA, May 1989, pp. 470-473.

[14] Granlund, J., "Topics in the Design of Antennas for Scatter," Technical Report 135, Lincoln Laboratory, MIT, November 1956.

[15] Brennan, D. G., "Linear Diversity Combining Techniques," *Proc. IRE*, Vol. 47, June 1959, pp. 1075-1102.

[16] Papoulis, A., *Probability, Random Variables, and Stochastic Processes*, New York, NY, McGraw-Hill Book Co., 1965.

[17] Abramowitz, M. and I. Stegun (editors), *Handbook of Mathematical Functions*, New York, NY, Dover Publications, Inc., 1972.

[18] Gradshteyn, I. S. and I. M. Ryzhik, *Table of Integrals, Series, and Products*, New York, NY, Academic Press, 1965.

[19] Winters, J. H., "Optimum Combining in Digital Mobile Radio with Cochannel Interference," *IEEE J. on Sel. Areas in Comm.*, Vol. SAC-2, No. 4, July 1984, pp. 528-539.

[20] Winters, J. H., "Upper Bounds for the Bit Error Rate of Optimum Combining in Digital Mobile Radio," *ICC 1984*, 1984, pp. 173-178.

[21] Adachi, F., et al., "A Periodic Switching Diversity Technique for Digital FM Land Mobile Radio," *IEEE Tr. on Veh. Tech.*, Vol. VT-27, No. 4, November 1978, pp. 211-219.

[22] Hess, G. and M. Geller, "The Urbana Meteor-Radar System: Design, Development, and First Observations," Library of Congress ISSN 0568-0581.

Chapter 12

Simulcast

This chapter deals with the use of multiple transmitters broadcasting the same signal for the purpose of providing improved area coverage. Of course since the path lengths to the various sources are generally not the same, some amount of simulcast distortion is inevitable. The distortion can be kept acceptable provided one of the signals is sufficiently strong to dominate, or capture the receiver. Capture probability equations are developed under a variety of assumptions.

Computer simulations are then described that take into account various imperfections of real equipment (for example, modulation sensitivity mismatch and carrier frequency drift) and real world problems (for example, differential propagation delay and Rayleigh fading). The purpose of such simulations is to guide the system design methodology to produce some required level of delivered audio quality. To prove in the field that such quality has indeed been provided, various means of objectively measuring signal quality are discussed.

Finally, the special case of digital simulcast is addressed. Various rules are considered for handling multiple sources when only two-source capture curves are available. Simulations indicate that a multipath spread model is best, particularly when thermal noise is also an issue. The use of such a model in coverage predictions is described.

12.1 Description

The term simulcast refers to situations where the same audio information is broadcast simultaneously from multiple transmitters operating on the same nominal carrier frequency. The purpose of the multiple sites is to boost the coverage probability over some area of interest, rather than to allow frequency reuse as in cellular systems. Coverage enhancement is of higher importance than spectral efficiency for certain public safety operations, such as radio coverage to fire and ambulance equipment. Simulcasting enhances coverage relative to that practical from a single site through a macroscopic diversity effect. With simulcast, one need only have sufficient signal from at least one source. Thus, 99% coverage is possible in areas where the coverage from two overlapping sites is only 90% for each considered individually.

Simulcasting generally requires highly stable transmitter frequencies along with matched audio delays and modulation sensitivities to avoid distortion in the receiver. Gray [1] provides a very readable tutorial on the practical aspects of simulcast systems. Descriptions of how the distortion results are discussed in References [2, 3]. A brief analysis of two-signal simulcast interference is presented in Section 12.2. Interestingly, Reference [4] points out that purposely mismatching the transmitter frequency can be helpful in certain paging applications. This ensures that a paging unit cannot be continuously located in a null because the received field moves even when the unit does not. Even if the system parameters are perfectly matched, different physical distances to the various sites means that distortion cannot be entirely avoided. However, if one of the signals is sufficiently large relative to the others then it captures the receiver and parameter mismatches become unimportant. Where possible, simulcast coverage designers attempt to place the noncapture areas in regions that are not important; for example, over a body of water or around a mountaintop with no roads.

12.2 Two-Signal Frequency Modulation Simulcast Analysis

Consider two received FM signals, differing perhaps in modulation index, carrier frequency, time of arrival, and audio phase. The time-amplitude behavior of the signals can be written as:

$$\begin{aligned} e_1(t) &= E_1(t)\sin[\omega_c t + \beta_1 \sin(\omega_m t)] \\ e_2(t) &= E_2(t)\sin\{(\omega_c + \delta\omega_c)(t - t_2) + \beta_2 \sin[\omega_m(t - \tau - t_2)] + \phi\} \end{aligned} \quad (12.1)$$

where β represents the modulation index, $\delta\omega_c$ represents the carrier frequency difference, τ represents the audio phase delay, t_2 represents the propagation delay difference, and ϕ represents the RF carrier-phase difference (signal 1 is taken as the reference). From the phasor diagram shown in Figure 12.1, one finds the resultant received signal can be written in amplitude-phase form as:

$$\begin{aligned} r(t) &= R(t)\sin[\angle r(t)] \\ R(t) &= \sqrt{E_1(t)^2 + E_2(t)^2 + 2E_1(t)E_2(t)\cos[\theta(t)]} \\ \angle r(t) &= \angle e_1(t) - \alpha(t) \end{aligned} \quad (12.2)$$

where:

$$\angle e_1(t) = \beta_1 \sin(\omega_m t)$$

Simulcast

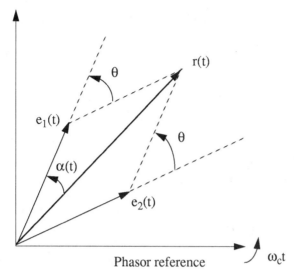

Figure 12.1 Phasor diagram of two-signal FM simulcast.

$$\tan[\alpha(t)] = \frac{E_2(t)\sin[\theta(t)]}{E_1(t) + E_2(t)\cos[\theta(t)]}$$

$$= \frac{x(t)\sin[\theta(t)]}{1 + x(t)\cos[\theta(t)]}$$

$$x(t) = \frac{E_2(t)}{E_1(t)}$$

$$\theta(t) = \angle e_1(t) - \angle e_2(t)$$

$$= \beta_1 \sin(\omega_m t) - \{\delta\omega_c t + \phi_{\text{net}} + \beta_2 \sin[\omega_m(t - \tau - t_2)]\}$$

$$\phi_{\text{net}} = \phi - (\omega_c + \delta\omega_c)t_2 \qquad (12.3)$$

It is advantageous to expand the β_2 term in the expression for $\theta(t)$ in such a manner that a $\sin(\omega_m t)$ term appears. This can be done using the identity:

$$\sin(a) \pm \sin(b) = 2\sin[\frac{1}{2}(a \pm b)]\cos[\frac{1}{2}(a \mp b)] \qquad (12.4)$$

Taking $a = \omega_m t$ and $b = \omega_m(t - \tau - t_2)$, one finds $\sin(b) = \sin(b) - \sin(a) + \sin(a) = [\sin(b) - \sin(a)] + \sin(a) = -2\sin[\frac{1}{2}(a-b)]\cos[\frac{1}{2}(a+b)] + \sin(a)$; that is, $\beta_2 \sin[\omega_m(t - \tau - t_2)] = \beta_2 \sin(\omega_m t) - 2\beta_2 \sin[\frac{\omega_m}{2}(\tau + t_2)]\cos[\omega_m(t - \frac{\tau + t_2}{2})]$. Consequently, the angle $\theta(t)$ can be written as:

$$\theta(t) = (\beta_1 - \beta_2)\sin(\omega_m t) - \delta\omega_c t - \phi_{\text{net}} + 2z_2 \cos[\omega_m(t - \frac{\tau + t_2}{2})]$$

$$z_2 = \beta_2 \sin[\frac{\omega_m}{2}(\tau + t_2)] \qquad (12.5)$$

Note that the terms of the angle $\theta(t)$ relate to mismatch in deviation, carrier frequency and phase mismatches, and audio and propagation delay differences, respectively.

Ideally, a limiter/discriminator subjected to the composite signal will produce a voltage proportional to the time rate of change of phase. This equals:

$$\begin{aligned} e_R(t) &= d/dt[\angle r(t)] \\ &= \beta_1 \omega_m \cos(\omega_m t) - d/dt \left(\tan^{-1}\{\frac{x(t)\sin[\theta(t)]}{1+x(t)\cos[\theta(t)]}\} \right) \end{aligned} \quad (12.6)$$

where the first terms equals the desired signal and the second term represents distortion. Note that if the signal envelopes are relatively constant over a period of interest (the so-called quasistatic approximation), then $x(t)$ equals a constant x and the distortion component equals $\{\dot{\theta}x[x+\cos(\theta)]\}/(1+2x\cos(\theta)+x^2)$. When the two signals arrive out of phase, $\theta = \pi$, the distortion term becomes $(\dot{\theta}x)/(x-1)$. This tends toward positive infinity for x values decreasing from above toward unity and toward negative infinity for x values increasing from below toward unity. In practice, imperfect limiter action and the finite bandwidth of the discriminator will limit the distortion excursions, but they nonetheless can be quite severe.

12.3 Capture Analyses

12.3.1 Independent Rayleigh Fading, Fixed Average Power. Envelope Analysis

The PDF of a Rayleigh-faded envelope can be expressed as:

$$p_X(x) = \frac{2x}{\sigma^2} \exp\left(-\frac{x^2}{\sigma^2}\right), x \geq 0 \quad (12.7)$$

where x equals the resultant amplitude and σ^2 equals the average power. The joint PDF for two independent, Rayleigh-faded signals is thus:

$$p_{X_1,X_2}(x_1,x_2) = \frac{4x_1 x_2}{\sigma_1^2 \sigma_2^2} \exp\left(-\frac{x_1^2}{\sigma_1^2} - \frac{x_2^2}{\sigma_2^2}\right), x_1, x_2 \geq 0 \quad (12.8)$$

Therefore, the probability that random variable X_1 exceeds random variable X_2 by at least a factor of k is found by solving the following integral:

$$\begin{aligned} P(x_1 \geq kx_2) &= \int_0^\infty \int_0^{x_1/k} p_{x_1,x_2}(x_1,x_2) dx_2 dx_1 \\ &= \int_0^\infty \frac{4x_1}{\sigma_1^2 \sigma_2^2} \exp\left(-\frac{x_1^2}{\sigma_1^2}\right) dx_1 I_1 \end{aligned} \quad (12.9)$$

where $I_1 = \int_0^{x_1/k} x_2 \exp(-x_2^2/\sigma_2^2) dx_2$. Via relation 5.(A4) in Reference [5] one finds:

$$I_1 = \frac{\sigma_2^2}{2}[1 - \exp(-\frac{x_1^2}{k^2\sigma_2^2})] \qquad (12.10)$$

$P(x_1 \geq kx_2)$ can therefore be expressed as the sum of two integrals:

$$I_2 = \int_0^\infty \frac{2x_1}{\sigma_1^2} \exp(-\frac{x_1^2}{\sigma_1^2}) dx_1 \qquad (12.11)$$

and:

$$I_3 = -\int_0^\infty \frac{2x_1}{\sigma_1^2} \exp[-\frac{x_1^2}{\sigma_1^2}(\frac{1+KA}{KA})] dx_1 \qquad (12.12)$$

where $K = k^2$ and $A = (\sigma_2^2/\sigma_1^2)$. Relation 5.(A4) in Reference [5] applies to both integrals yielding a final result of:

$$P(x_1 \geq kx_2) = \frac{1}{1+KA} \qquad (12.13)$$

This also means that $P(x_1 \leq kx_2) = KA/(1+KA)$, which represents the CDF for the random variable $Z = X_1/X_2$. The PDF for Z is therefore:

$$\begin{aligned} p_Z(k) &= \frac{d}{dk} P(z \leq k) \\ &= \frac{2kA}{(1+KA)^2} \end{aligned} \qquad (12.14)$$

Simulcast reliability R can now be determined from the likelihood that X_1 and X_2 differ by at least some factor $k = c$, the so-called capture requirement.

$$\begin{aligned} R &= 1 - \int_{1/c}^c p_z(z) dz \\ &= 1 - \int_{1/c}^c \frac{2Az}{(1+Az^2)^2} dz \\ &= \frac{1 + 2AC^{-1} + A^2}{1 + A(C + C^{-1}) + A^2} \end{aligned} \qquad (12.15)$$

where $C = c^2$ and Relation #47 in Reference [6] has been used.

12.3.2 Independent Rayleigh Fading, Fixed Average Power, Power Analysis

The results of the preceding section are easier to obtain if one considers the distribution of power in Rayleigh-faded random variables. The PDF is then:

$$p_Y(y) = \frac{1}{\sigma^2} \exp(-\frac{y}{\sigma^2}), y \geq 0 \qquad (12.16)$$

and the probability that the power in a random variable Y_1 exceeds K times the power in independent random variable Y_2 is:

$$P(y_1 \geq Ky_2) = \int_0^\infty \int_0^{y_1/K} p_{y_1,y_2}(y_1, y_2) dy_2 dy_1$$

$$= \int_0^\infty \frac{1}{\sigma_1^2 \sigma_2^2} \exp(-\frac{y_1}{\sigma_1^2}) dy_1 I_1 \quad (12.17)$$

where $I_1 = \int_0^{y_1/K} \exp(-y_2/\sigma_2^2) dy_2$. Using relation #519 in Reference [6] one finds:

$$I_1 = \sigma_2^2 [1 - \exp(-\frac{y_1^2}{K\sigma_2^2})] \quad (12.18)$$

$P(y_1 \geq Ky_2)$ can therefore be expressed as the sum of two integrals:

$$I_2 = \int_0^\infty \frac{1}{\sigma_1^2} \exp(-\frac{y_1}{\sigma_1^2}) dy_1 \quad (12.19)$$

$$I_3 = -\int_0^\infty \frac{1}{\sigma_1^2} \exp[-\frac{y_1}{\sigma_1^2}(\frac{1+KA}{KA})] dy_1 \quad (12.20)$$

where $A = (\sigma_2^2/\sigma_1^2)$ as before. Relation #519 in Reference [6] applies to both integrals yielding the expected result of:

$$P(y_1 \geq Ky_2) = \frac{1}{1+KA} \quad (12.21)$$

12.3.3 Independent Rayleigh Fading, Fixed Average Power, Envelope Analysis Including Threshold

To provide useful two-site simulcast communications it is necessary, but not sufficient, for one signal to capture over the other. The capturing signal must also be above system sensitivity. Such a requirement can be added by changing the lower limit of integration from zero to a threshold value x_t in Equation (12.9).

$$P(x_1 \geq kx_2, x_1 \geq x_t) = \int_{x_t}^\infty \int_0^{x_1/k} p_{x_1,x_2}(x_1, x_2) dx_2 dx_1$$

$$= \int_{x_t}^\infty \frac{4x_1}{\sigma_1^2 \sigma_2^2} \exp\left(-\frac{x_1^2}{\sigma_1^2}\right) dx_1 I_1 \quad (12.22)$$

where $I_1 = \int_0^{x_1/k} x_2 \exp(-x_2^2/\sigma_2^2) dx_2$. Via relation 5.(A4) in Reference [5]:

$$I_1 = \frac{\sigma_2^2}{2}[1 - \exp(-\frac{x_1^2}{k^2 \sigma_2^2})] \quad (12.23)$$

$P(x_1 \geq kx_2, x_1 \geq x_t)$ can therefore be expressed as the sum of two integrals:

$$I_2 = \int_{x_t}^{\infty} \frac{2x_1}{\sigma_1^2} \exp(-\frac{x_1^2}{\sigma_1^2}) dx_1 \qquad (12.24)$$

$$I_3 = -\int_{x_t}^{\infty} \frac{2x_1}{\sigma_1^2} \exp[-\frac{x_1^2}{\sigma_1^2}(\frac{1+KA}{KA})] dx_1 \qquad (12.25)$$

where $K = k^2$ and $A = (\sigma_2^2/\sigma_1^2)$. Relation 5.(A4) in Reference [5] applies to both integrals yielding a final result of:

$$P(x_1 \geq kx_2, x_1 \geq x_t) = \exp(-\frac{x_t^2}{\sigma_1^2})[1 - \frac{KA}{1+KA} \exp(-\frac{x_t^2}{K\sigma_2^2})] \qquad (12.26)$$

Because $P(x_1 \geq kx_2, x_1 \geq x_t)$ and $P(x_2 \geq kx_1, x_2 \geq x_t)$ are mutually exclusive events, the simplest way to express the simulcast reliability is to sum those two probabilities. Hence, one obtains:

$$\begin{aligned} R &= R_1 + R_2 \\ R_1 &= \exp(-\frac{x_t^2}{\sigma_1^2})[1 - \frac{KA}{1+KA} \exp(-\frac{x_t^2}{K\sigma_2^2})] \\ R_2 &= \exp(-\frac{x_t^2}{\sigma_2^2})[1 - \frac{K}{K+A} \exp(-\frac{x_t^2}{K\sigma_1^2})] \end{aligned} \qquad (12.27)$$

12.3.4 Independent Lognormal Fading, Power Analysis Including Threshold

The average power received in land-mobile transmissions is generally lognormally distributed. This means that the power expressed in dBm is Gaussian distributed and the PDF can be written as:

$$p_W(w) = \frac{1}{\sqrt{2\pi\sigma_{dB}^2}} \exp\left[-\frac{(w-\mu)^2}{2\sigma_{dB}^2}\right] \qquad (12.28)$$

where μ is the median average power in dBm and σ_{dB} is the standard deviation of the average power expressed in decibels. The PDF for the difference of two independent Gaussian random variables is also Gaussian, with variance equal to the sum of the individual variances and median equal to the difference of the individual medians. Given two such independent random variables, the probability of capture (i.e., that the stronger of the two variables exceeds the weaker variable by at least K_{dB}) is given by:

$$\begin{aligned} P(\text{capture}) &= 1 - \int_{-K_{dB}}^{+K_{dB}} \frac{1}{\sqrt{2\pi\sigma_T^2}} \exp\left[-\frac{(z-\mu_T)^2}{2\sigma_T^2}\right] dz \\ &= \frac{1}{2}[\text{erfc}(\frac{K_{dB}-\mu_T}{\sqrt{2}\sigma_T}) + \text{erfc}(\frac{K_{dB}+\mu_T}{\sqrt{2}\sigma_T})] \end{aligned} \qquad (12.29)$$

where $z = w_1 - w_2$, σ_T equals the square root of the sum of the variances in decibels of the two random variables, μ_T equals the difference between the medians in decibels of the two random variables, and relation 7.1.22 in Reference [7] has been used.

The probability that random variable W_1 exceeds independent random variable W_2 by K_{dB}, and also exceeds some threshold w_t, is given by:

$$P(w_1 \geq K_{\text{dB}} w_2, w_1 \geq w_t) = \int_{w_t}^{\infty} \int_{-\infty}^{w_1 - K_{\text{dB}}} p_{w_1, w_2}(w_1, w_2) dw_2 dw_1$$

$$= \frac{1}{2\pi \sigma_{\text{dB}}^2} \int_{w_t}^{\infty} \exp\left[-\frac{(w_1 - \mu_1)^2}{2\sigma_{\text{dB}}^2}\right] dw_1 I_1$$

(12.30)

where $I_1 = \int_{-\infty}^{w_1 - K_{\text{dB}}} \exp[-(w_2 - \mu_2)^2 / 2\sigma_{\text{dB}}^2] dw_2$. Via relation 7.1.22 in Reference [7]:

$$I_1 = \frac{\sqrt{2\pi \sigma_{\text{dB}}^2}}{2}\left[1 + \text{erf}\left(\frac{w_1 - K_{\text{dB}} - \mu_2}{\sqrt{2}\sigma_{\text{dB}}}\right)\right]$$

(12.31)

$P(w_1 \geq K_{\text{dB}} w_2, w_1 \geq w_t)$ can therefore be expressed as the sum of two integrals, the first of which has a closed form using the relation just cited and the second of which requires computer evaluation:

$$P(w_1 \geq K_{\text{dB}} w_2, w_1 \geq w_t) = \frac{1}{4}\text{erfc}\left(\frac{w_t - \mu_1}{\sqrt{2}\sigma_{\text{dB}}}\right)$$

$$+ \frac{1}{2\sqrt{2\pi \sigma_{\text{dB}}^2}} \int_{w_t}^{\infty} \exp\left[-\frac{(w_1 - \mu_1)^2}{2\sigma_{\text{dB}}^2}\right] \cdot$$

$$\text{erf}\left(\frac{w_1 - K_{\text{dB}} - \mu_2}{\sqrt{2}\sigma_{\text{dB}}}\right) dw_1$$

(12.32)

Simulcast reliability is given by the sum of two such equations where μ_1 and μ_2 are simply interchanged.

12.3.5 Independent Rayleigh Fading with Independent Lognormal Fading of Average Power, Power Analysis Including Threshold

Equation (12.27) expresses simulcast reliability for independent Rayleigh fading with fixed average powers. In land-mobile propagation the average powers themselves vary, typically in a lognormal fashion. Only the average power variation was considered in Section 12.3.4. To obtain a result that allows for both Rayleigh and lognormal fading one must integrate Equation (12.27), weighted by the product of the PDFs for both average powers. Letting $z_1 = \sigma_1^2$, $z_2 = \sigma_2^2$, $z_t = x_t^2$, and $z_{21} = (z_2/z_1)$ one has:

$$R(z_1, z_2) = \exp(-\frac{z_t}{z_1})\left[1 - \frac{Kz_{21}}{1+Kz_{21}}\exp(-\frac{z_t}{Kz_2})\right]$$
$$+ \exp(-\frac{z_t}{z_2})\left[1 - \frac{K}{K+z_{21}}\exp(-\frac{z_t}{Kz_1})\right] \quad (12.33)$$

where:

$$p_z(z) = \frac{1}{z\sqrt{2\pi\sigma^2}}\exp\left\{-\frac{[\ln(z)-m]^2}{2\sigma^2}\right\} \quad (12.34)$$

for $z = z_1$ and $z = z_2$. It is numerically preferable, however, to work in terms of $w = 10\log(z)$, rather than z directly. With such a substitution, one can write the reliability as:

$$R = \int_{-\infty}^{\infty}\int_{-\infty}^{\infty} R(w_1, w_2)p(w_1)p(w_2)dw_1 dw_2 \quad (12.35)$$

where

$$\begin{aligned}
R(w_1, w_2) &= R1 + R2 \\
R1 &= \exp[-\exp(\frac{w_t - w_1}{\alpha})]\left\{1 - R3\exp[-\frac{1}{K}\exp(\frac{w_t - w_2}{\alpha})]\right\} \\
R3 &= \frac{R4}{1 + R4} \\
R4 &= K\exp(\frac{w_2 - w_1}{\alpha}) \\
R2 &= \exp[-\exp(\frac{w_t - w_2}{\alpha})]\left\{1 - R5\exp[-\frac{1}{K}\exp(\frac{w_t - w_1}{\alpha})]\right\} \\
R5 &= \frac{K}{K + R6} \\
R6 &= \exp(\frac{w_2 - w_1}{\alpha}) \\
\alpha &= [10/\ln(10)] \quad (12.36)
\end{aligned}$$

and:

$$p_w(w) = \frac{1}{\sqrt{2\pi\sigma_{\text{dB}}^2}}\exp\left[-\frac{(w-\mu)^2}{2\sigma_{\text{dB}}^2}\right] \quad (12.37)$$

for $w = w_1$, $\mu = \mu_1$, and $\sigma_{\text{dB}} = \sigma_{1\text{dB}}$, and $w = w_2$, $\mu = \mu_2$, and $\sigma_{\text{dB}} = \sigma_{2\text{dB}}$.

12.4 Computer Simulation of Analog FM Simulcast Performance

This section describes a simple method of simulating the simulcast performance of conventional analog 25-kHz channel FM radios. The method is simple thanks to the use of a powerful high-level simulation language called TCSIM. Numerous other high-level simulation packages also exist. TCSIM source code for analog FM simulcast simulation is as follows.

```
#
# analog FM simulcast simulation
#
# begin by forming FM modulated signal of source 1 consisting of
# 4 kHz peak voice deviation (scaled for 6 dB clipping) and
# 1 kHz peak deviation by 75 Hz PL tone
#
input scratch:speech.ssp
scale 1.33336e-4
clip 0.8
filter scratch:splatter.cof
tone 0.00225 0.2 0
real_part
add 2
# from this point onward sources 1 and 2 differ
fork 2
# complete source 1 processing
scale 0.15
scale 6.283185308
filter scratch:integrate.cof
exchange exp
# modify source 1 signal by Rayleigh fading and
# save result in store 0
uniform_noise 1.219548 0
filter scratch:ray_33ks_450mhz_45mph.cof
scale 1.0000, 0.0
multiply
delay 35
store 0
# next complete source 2 processing;
# here the deviation is decreased 0.5 dB (11%)
scale 0.134
scale 6.283185308
filter scratch:integrate.cof
exchange
exp #add 1 Hz carrier offset here
tone 0.00003 1.0 0
multiply
```

```
# independently Rayleigh fade
uniform_noise 1.219548 0
filter scratch:ray_33ks_450mhz_45mph.cof
scale 1.0000, 0.0
multiply
# set up 70 us differential delay
delay 2
# the following all pass filter has a delay equal to
# 35 clock periods plus 10 us; hence the net differential
# delay is 2 clock periods plus 10 us = 70 us
filter scratch:delay_10us.cof
# recall source 1 signal and sum to obtain simlcast
# composite; also, add thermal noise at -25 dBr
recall 0
# complex Gaussian noise of unit variance
gaussian_noise 0.707106781 0
scale 0.056, 0.0
add 3
# run composite signal and noise through model of
# receiver selectivity
filter scratch:if_25.cof
# slow down the sampling rate and demodulate
decimate 2
log
mag
filter scratch:dif.cof
scale .15915493
offset 0.5
modulo 1.0
offset -0.5
scale 2.222222
# strip off the PL tone
filter scratch:pl_reject_75hz_16.cof
# run audio through de-emphasis filter
filter scratch:deem_16.cof
# save results
output scratch:output_test.sst
```

This code is set up to simulate two-source simulcast with 70 μs of differential propagation delay, 0.5 dB of deviation mismatch, and 1 Hz of carrier frequency offset. Both sources are independently faded at a rate corresponding to 45 mph at UHF and thermal noise 25 dB below the average signal power is also included. The modulation signal consists of speech from a female speaker that was sampled at 33.3 ksps, pre-emphasized, and saved in the file "speech.ssp." Six dB of clipping is applied to the speech, which is then run through a splatter filter described by the filter coefficient file "splatter.cof." A 75-Hz tone representative of *continuous tone-coded squelch system* (CTCSS) operation is also summed with the speech in such a manner that the peak deviation budget of 5 kHz is apportioned on a 4:1 basis.

Two versions of the modulating signal are then manipulated as follows. First, frequency modulation is carried out; then the amplitude and phase are varied in accordance with Rayleigh fading of 45 mph at UHF. In the second version, the modulation index is lowered by 0.5 dB and a 1-Hz carrier frequency offset is inserted. A delay difference of 70 μs is established by delaying two more clock periods and running the second version through allpass filter "delay_10us.cof." Finally, the two signals are summed along with Gaussian noise whose power is set 25 dB below the average power of a single 5-kHz peak deviation FM signal.

The simulated receiver selectivity is typical of 800-MHz FM radios and described in the filter coefficient file "if_25.cof." Prior to demodulation, the sample stream is decimated by a factor of two to minimize the size of the output file. The discrimination function is carried out by the filter "dif.cof." Before saving, the results are run through a 75-Hz CTCSS reject filter "pl_reject_75hz_16.cof" and a de-emphasis filter "deem_16.cof."

A program is then run to create a much more efficient file compatible with the Interactive Laboratory System of Signal Technology, Inc. With this system, one can easily scrutinize the output speech on a sample-by-sample basis if desired. One can also play back and record speech. SINAD performance can be measured by replacing the speech input with a 1-kHz tone through a distortion analyzer or through special software that emulates distortion analysis[1].

Analog FM simulcast simulations recorded to address design criteria are as follows:

- Recording 1: single UHF source, Rayleigh faded at 45 mph, average power 20 dB above static 12-dB SINAD.

- Recording 2: two-source UHF simulcast with equal average powers and 70-μ differential delay, independent Rayleigh fading at 45 mph, no thermal noise included.

---

[1] The reader is cautioned that SINAD is not a technically sound parameter for use in verifying coverage, though it has been claimed as such; for example, see [8]. Figure 2 in that reference purports to uniquely relate circuit merit and SINAD, but is without substantiation. Even two-source simulations readily show that circuit merit is a function of many conditions, such as differential delay, frequency netting error, deviation mismatch, Rayleigh-fading rate, etc. Consequently, to paraphrase George Orwell, all SINADS are not equal.

Simulcast

- Recording 3: two-source UHF simulcast with equal average powers and 70-μs differential delay, independent Rayleigh fading at 45 mph, and average powers 20 dB above static 12-dB SINAD.

- Recording 4: two-source UHF simulcast with equal average powers and 70-μs differential delay, 0.5-dB mismatch in deviation, 1-Hz mismatch in carrier frequency, independent Rayleigh fading at 45 mph, and average powers 20 dB above static 12-dB SINAD.

- Recording 5: two-source UHF simulcast with equal average powers and 150-μs differential delay, 0.5-dB mismatch in deviation, 1-Hz mismatch in carrier frequency, independent Rayleigh fading at 45 mph, and average powers 20 dB above static 12-dB SINAD.

All recordings involve approximately 10 sec of pre-emphasized speech from a female speaker, sampled at 33.3 ksps with 6-dB clipping applied and 4-kHz nominal peak deviation (the CTCSS tone is set to 1-kHz peak deviation so overall a 5-kHz peak deviation radio is simulated). Simulations are worst case in that both sources nominally have equal average power. Recording 1 illustrates the usual degradation caused by Rayleigh fading; i.e., occasional noise pops and sputtering sounds. Recording 2 is distinctly different in the abruptness of the pops and lacks the sputtering character. However, when thermal noise is included, as in recording 3, the sound is very much like recording 1. In fact, even with the deviation mismatch and carrier frequency offset of recording 4, the output differs little from recording 1. However, with 150-μs differential delay (as in recording 5) the quality is noticeably degraded compared to recording 1, though still intelligible. The background noise during speaker pauses is noticeably higher and additional noise associated with the presence of speech is also apparent.

SINAD has been computed and measured for the following situations:

Delay (μs)	Deviation error (dB)	Frequency error (Hz)	Computed (dB)	Measured (dB)
150	0.5	1	5.8	11.2
70	0.5	1	10.5	15.6
0	0.5	1	19.5	22.4
150	0.0	0	5.9	11.6
70	0.0	0	10.9	15.7
0	0.0	0	27.1	30.3
Single source only			22.9	28.0

The measured results were obtained by playing the output files into an HP339A distortion analyzer several times to note the maximum and minimum readings. The average of these extremes is taken as the measured result. Notice that the calculated SINAD is consistently less than the measured SINAD. A regression fit shows that this error is approximately 5 dB. Although the distortion analyzer nominally reads true rms, this operation breaks down with crest factors in excess of 3. When the time variation of SINAD is impulsive, as it is with Rayleigh fading, the nonlinear low-pass filtering action of the analyzer results in an overestimation of the true average SINAD. Interestingly, the subjective quality of the recordings indicates that the human ear acts in such a manner also.

12.5 Rapid, Robust Method for Establishing Signal Quality

This section examines one means of mitigating problems with the use of conventionally measured SINAD as the objective measure of signal quality.

12.5.1 Algorithm for Estimating Amplitude and Phase of a Sinusoid

Consider a sinusoid $A\cos(\omega t + \theta)$, where A represents the amplitude, t represents the time, ω represents the radian frequency, and θ represents an arbitrary phase angle. Three samples of this waveform can be mathematically described via the equations

$$M_1 = A\cos(\omega t_1 + \theta)$$
$$M_2 = A\cos(\omega t_2 + \theta)$$
$$M_3 = A\cos(\omega t_3 + \theta)$$

Assuming uniform sampling and taking t_2 as the time reference, these equations simplify to:

$$M_1 = A\cos(-\omega\Delta t + \theta)$$
$$M_2 = A\cos(\theta)$$
$$M_3 = A\cos(+\omega\Delta t + \theta)$$

Or, defining $\delta = \omega\Delta t$, one can write:

$$M_1 = A\cos(\theta - \delta)$$
$$M_2 = A\cos(\theta)$$
$$M_3 = A\cos(\theta + \delta)$$

When ω is known, it is advantageous to choose Δt such that $\delta = (\pi/2)$. Under such circumstances the equations reduce to:

$$M_1 = A\sin(\theta)$$

$$M_2 = A\cos(\theta)$$
$$M_3 = -A\sin(\theta)$$

One is now in a position to solve for A and θ. Considering first the amplitude, note through the Pythagorean relation $1 = \sin^2(x) + \cos^2(x)$ that:

$$A = \sqrt{M_1^2 + M_2^2} = \sqrt{M_2^2 + M_3^2}$$

In real applications the samples will be contaminated by noise and distortion, so these relations are in fact estimates of the true A value. For N samples spaced $\delta = \pi/2$ in phase (time), $N - 1$ estimates of A are possible by using:

$$A_i = \sqrt{M_i^2 + M_{i+1}^2}, i = 1, ..., N - 1; N \geq 2$$

Individually these estimates are not particularly reliable, nor is their average, which can be strongly impacted by even a single "wild" estimate. A robust estimate, however, is readily obtainable by ordering the $N - 1$ individual estimates and selecting the median value. In fact, to reduce the computational complexity, individual estimates of A^2 rather than A can be used by taking the square root of the median value afterwards to obtain the estimate of A.

Next, estimation of the phase angle θ is addressed. This can be obtained through:

$$\tan(\theta) = M_1/M_2 = -M_3/M_2$$

The quadrant of θ is unambiguous because M_1 and $M_2 > 0$ implies the first quadrant, $M_1 < 0$ and $M_2 > 0$ implies the second quadrant, etc. However, quadrant problems can arise if the divisor is too small. Then noise can cause 360-deg shifts in the answer. To avoid this, consider samples in groups of four, with $M_4 = -A\cos(\theta)$. If the absolute magnitude of sample M_2 exceeds that of sample M_3, then solve $\tan(\theta) = [(M_1 - M_3)/2M_2]$. Otherwise, solve $\tan(\phi) = [(M_4 - M_2)/2M_3]$, with $\theta = \pi/2 - \phi$.

12.5.2 Simulated Performance

A Fortran-77 program has been used to quantify the performance of the algorithm in the presence of thermal noise. The program generates samples at a rate of 4000 per second of a sinusoid with user-specified amplitude and phase that are summed with thermal noise at a user-specified SNR. The algorithm is applied to 100 groups of data, each consisting of 40 samples. Each group involves 10 determinations of amplitude and phase. The median values from these determinations allow subtraction of an estimated sinusoid from the actual samples. The residual power is calculated and used to estimate SNR every 10 ms.

Table 12.1 shows program outputs as the phase angle varies from 0 to 315 deg in steps of 45 deg. The amplitude is fixed at unity. Normally all 100 estimates of

Table 12.1 Algorithm Performance as a Function of Phase Angle

Test tone phase (degrees)[a]	True post detection S/N (dB)	Est. amp. mean (std dev)	Est. phase (rad.) mean (std dev)	Est. S/N (dB)
0	10	1.06 (0.05)	0.013 (0.06)	9.89
	20	1.01 (0.02)	0.007 (0.02)	20.00
45	10	1.06 (0.07)	0.81 (0.07)	9.89
90	10	1.06 (0.06)	1.59 (0.06)	9.81
135	10	1.06 (0.07)	2.37 (0.08)	10.00
180	10	1.05 (0.06)	0.02 (2.94)	8.41
225	10	1.06 (0.06)	-2.34 (0.08)	9.63
270	10	1.05 (0.06)	-1.54 (0.07)	9.96
315	10	1.07 (0.06)	-0.77 (0.07)	9.89

a. Conditions: peak amplitude of 1 kHz test tone equals unity

amplitude, phase, and SNR are output; however, here results are heavily edited to limit the table size. Overall averages and standard deviations of the three aforementioned quantities are retained for all runs. In all cases the mean signal-to-noise estimate is near the true input value of either 10 or 20 dB. There is a small bias toward underestimating signal-to-noise due to the low-pass filtering nature of the median sampling process.

To examine the performance of the algorithm in the presence of Rayleigh fading and nonlinear detection, the following has been done. First, a TCSIM simulation of a basic FM radio system is conducted. The input modulation consists of two tones, one at 1.042 kHz representing the test tone and the other at 75 Hz representing a CTCSS tone. A 1.042 kHz tone is used rather than 1 kHz because various filters needed by the simulation were already on hand, but based on a sampling rate of 33,333 samples per second. Decimating that rate a factor of 8 to 4167 samples per second produces the desired $\pi/2$ phase shift between samples for a tone of 1.042 kHz. Peak deviation is 4 kHz for the test tone and 1 kHz for the CTCSS tone. The simulated transmit signal is Rayleigh faded at a rate equal to 45 mph at 450 MHz and summed with thermal noise to produce a true average output SINAD of 10.9 dB, as obtained by computer analysis (approximately 12.5 dB on an HP339A distortion meter, the difference being due to the non-rms nature of the distortion meter and the ballistics of its meter display). The result passes through a filter representing typical receiver IF selectivity and is then demodulated. The CTCSS tone is suppressed by a notch filter.

The output file produced by these steps contains a startup transient that must be edited away. After this has been done, the file is used as input to a Fortran-77 program that implements the algorithm. Figure 12.2 shows a coarse view of the temporal behavior of the received waveform over a period of 0.8 seconds. Figure 12.3 shows a finer view where two severely perturbed periods are evident. These periods correspond to 10-ms analysis groups 7,11, and 12. The algorithm shows substantial decreases in SNR for these groups, but nonetheless accurately extracts the amplitude and phase of the test tone. For example, group 1, which involves no deep fades, has an estimated amplitude of 0.5095, an estimated phase of 0.425 rad, and an SNR of 25.82 dB. Group 7, which has a substantial fade in the middle of its 40 samples, estimates the amplitude as 0.5108 and the phase as 0.4172 rad, which is well in line with group 1 results. The SNR of course must show the effect of the deep fade and it does because it is only 5.468 dB.

A final test of the algorithm involves its performance in the presence of simulcast interference. For this the TCSIM program is modified to simulate the performance of a two-source simulcast system where the average powers from each source are equal, but a differential delay of 150 μs is present. In addition, degradations due to 0.5-dB mismatch in modulation index, 1-Hz RF carrier offset, thermal noise, and Rayleigh fading are included. The end result is a waveform whose true SINAD is just 5.8 dB.

Figure 12.4 shows the temporal behavior over 0.8 sec. Periods of capture by source 1 and capture by source 2 are readily apparent because of the 0.5-dB mismatch in modulation index (source 1 has the greater amplitude). Figure 12.5 shows closeup detail of one capture transition event. From the simulation output it is noted that the test tone phase is less than -2.0 rad for the first three analysis groups. The angle changes to about -1.2 rad for the fourth group and this shift is consistent with the 150-μs difference between sources 1 and 2 (about 1 rad in phase). Because both sources have equal average power, the time spent captured by each is comparable. Therefore, it is not surprising that the 100-group average phase (-1.644 rad) is roughly midway between the phase of each source individually. These phases are also, not surprisingly, roughly equal to the average value plus and minus one times the standard deviation of the angle estimates.

12.5.3 Application: Automated Measurements of Delivered Audio Quality

An ever increasing field problem involves proving to customers that the promised radio coverage has in fact been supplied. To date such acceptance tests have generally involved driving to a large number of locations within the intended coverage area and subjectively grading the delivered audio quality at each location. If a sufficiently reliable means of automatically rating the coverage at each test point existed, much time and labor could be saved.

800-MHz analog FM field data has allowed mean-opinion scores of subjectively

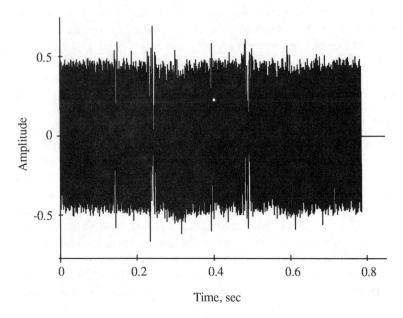

Figure 12.2 Coarse view of receiver output for single-source 10.9-dB SINAD Rayleigh-faded condition.

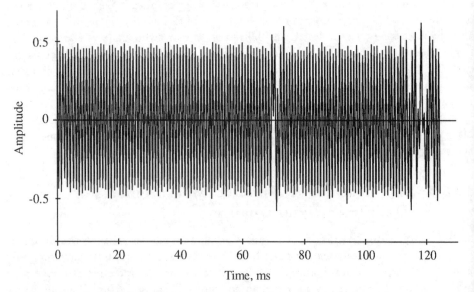

Figure 12.3 Fine view of receiver output for single-source 10.9-dB SINAD Rayleigh-faded condition.

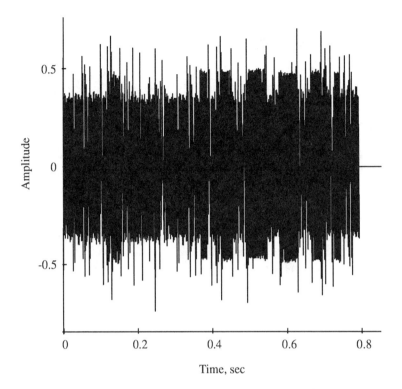

Figure 12.4 Receiver output for worst-case two-source simulcast.

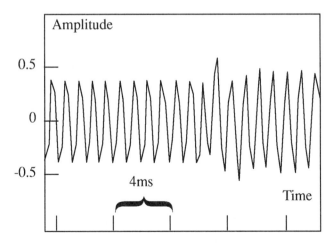

Figure 12.5 Expanded timescale view of receiver output for worst-case two-source simulcast.

rated speech to be compared with regression models based on various statistics related to signal strength. The result is a model whose standard deviation of error is on the order of 0.5 units (using a 1 through 5 rating scale) and whose maximum error magnitude is about 1 unit. This accuracy is marginally acceptable, but an important problem underlies the model. If it is applied to simulcast situations or modulations other than the standard 25-kHz FM used to generate it, it is unlikely that the same model coefficients would apply. Yet if the model must be customized to very specific circumstances its utility is reduced.

An alternative would be to measure end-to-end performance and thus respond to shortcomings wherever they appear (transmitter, receiver, microwave relays, etc.). The alternative should not be sensitive to the particular modulation method involved, nor should it matter whether transmissions are simulcast or from a single source. A candidate is the idea just discussed: start with a known sinusoidal modulation; because it is known, the above algorithm can be used on samples taken at the loudspeaker. The end result is rapid, robust estimates of the received amplitude and phase. This information can be used to create an error signal for the actual received waveform (i.e., take the difference between the actual received signal with its imperfections, caused by noise and distortion, and the estimate of the waveform that in fact should have been received).

For example, it is convenient to consider using a 1-kHz tone. The sampling rate of this tone for 90 degree phase advance between samples is 4 kHz; a value consistent with reasonable decorrelation between samples for the usual audio passband of 3 kHz. Speech is quasistationary over periods on the order of 30 ms, and during such an interval one can make about 120 estimates of amplitude and phase. Using the median choices out of such a large number of independent trials should provide very accurate estimates. Hence, the power of the error signal in the 30-ms interval should be a good reflection of the end-to-end transmission quality. Dispatch transmission lengths average on the order of 5 sec, suggesting that an automated testing procedure should tabulate error signal powers over such an interval. This would yield 167 values. To map these data to a mean-opinion score rating, a model using two parameters based on those values (for example, the median error signal and the maximum number of successive error signals whose value exceeds a specified threshold) might be helpful.

12.6 Computer Simulation of Digital Simulcast Coverage

12.6.1 Approximate Methods for Reducing Multisite Problems to Two-Site Problems

Coverage calculations for digital (data and/or voice) simulcast systems have traditionally relied on two-source capture models. Such models specify the required signal ratio between two simulcasting sources necessary to achieve some level of performance (for example, a BER of 1 %), as a function of differential signal delay. Figures 12.6 and 12.7 show capture curves for moving and stationary circumstances for a form of QAM at 1% and 3% BER, respectively.

Developing such a capture model for three or more simulcast sources is not very practical because of the large number of different signal ratios and delay combinations that must be considered. Consequently, multisource simulcast has usually been treated by reduction to a two-source case. For example, one could require that the strongest transmitter had to capture the next strongest source, which in turn had capture the third strongest source, etc. for adequate performance to be achieved. Such requirements would of course be overly conservative. A more reasonable approach is to require the strongest source to capture over all other sources considered on a one-to-one basis (we will call this the N signal greater than rule). This strategy tends to be optimistic because it fails to account for the effect of combined interference power. A third approach combines the interfering signals into a single effective signal and then applies two-source capture information (we will call this the $\sum$ powers rule). Because the interfering sources are at various time delays, they cannot simply be power summed. Instead, a vector sum can be made by assigning a magnitude and phase to each signal, with the phase based on delay relative to the symbol time. A relative delay of one symbol time implies a phase difference of 180 deg. This means of adding signals is arbitrary at best. For example, two signals of equal power and relative delay of one symbol time would cancel completely. Yet, in fact, such a circumstance could produce substantial simulcast interference. Hence, this method can strongly underestimate the interference.

12.6.2 Multipath Spread Model

A more desirable model would be one truly based on multiple interferers, rather than on two-source approximations. One way to achieve this is to view the received signal as a single transmission undergoing multipath delay spread. The relative signal strengths and delays would then correspond to the so-called power delay profile of the aggregate signal. Such effects have been studied extensively for very high-speed data systems, where the delay spread of the RF channel is a serious problem [9, 10, 11, 12]. One interesting result of these studies is that for delays limited to a fraction of the symbol time, the amount of signal degradation depends not on the actual delay profile, but on the rms value of the delays, weighted by their respective power levels. For this reason, multipath delay spread is often approximated simply by a two-ray model in the evaluation of data systems [13]. This offers a particularly attractive way of handling multisource simulcast because it reduces the multiple delays and signal powers to a single parameter called the multipath spread (equal to twice the rms of the delay profile). The multipath spread for N simulcasting signals

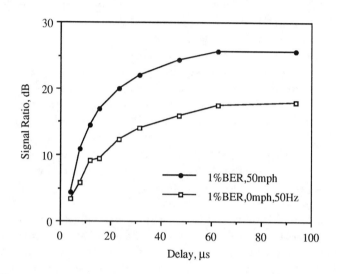

Figure 12.6 1% QAM capture curves.

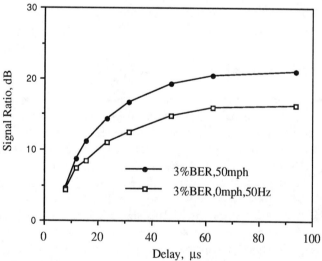

Figure 12.7 3% QAM capture curves.

is given by:

$$T_m = 2\sqrt{\frac{\sum_{i=1}^{N} P_i d_i^2}{\sum_{i=1}^{N} P_i} - \frac{(\sum_{i=1}^{N} P_i d_i)^2}{(\sum_{i=1}^{N} P_i)^2}} \qquad (12.38)$$

where P_i and d_i are the power and delay of the i-th signal, respectively.

Figure 12.8 shows BER data from two-source simulations using a form of QAM, as a function of multipath spread T_m, for different values of relative delays. One sees that for delays of 31 μs (one-half the symbol time, corresponding to about 6 mi) or less, the curves more or less lie on top of each other. For 1% BER, the corresponding value of T_m is around 4 μs. This is reasonable because the minimum relative delay in the QAM two-source capture model for a 0-dB signal difference is 3.4 μs. Therefore, by specifying a minimum T_m required for a given BER, a useful simplified criterion for performance with multisource simulcast interference is obtained.

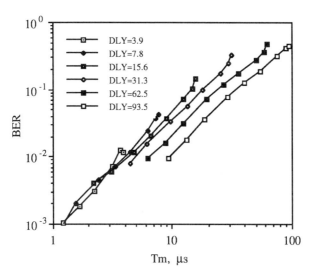

Figure 12.8 Two-source BER versus multipath spread.

Figures 12.9 and 12.10 show BER as a function of T_m for four-source and seven-source cases, respectively. Because of the extremely large number of possible combinations of power and delay with more than a few sources, the powers and delays for each case were picked on a random basis and the BER for each case was then computed via simulation. In each case the relative delay was limited to 31.3 μs. Even though the powers and delays varied widely, the BERs are closely dependent on the parameter T_m.

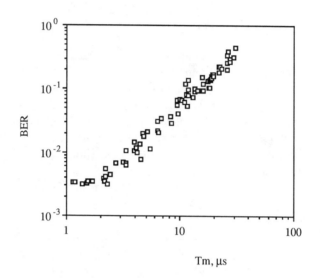

Figure 12.9 Four-source BER versus multipath spread.

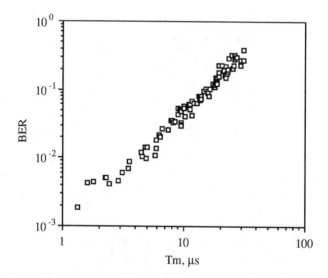

Figure 12.10 Seven-source BER versus multipath spread.

12.6.3 Comparisons of Model Performance

To evaluate the accuracy of this model against previous simulcast models, BER simulations were made for three simulcast sources as the power of two of the sources was varied in 6-dB steps from 0 to -30 dB and delay was varied in octaves from 3.9 to 62.5 μs. This resulted in 666 possible combinations to be simulated. The results were tabulated in terms of the numbers of times each model correctly predicted adequate performance (1% BER). Also recorded were the number of times each model was too liberal or too conservative and by what amounts (i.e., the largest BER that actually occurred when the model predicted BER less than 1% and the smallest BER when the model predicted greater than 1%). The models tested were the N signal greater than rule, the $\sum$ powers rule, and the minimum multipath spread rule, using a minimum T_m of 3.4 μs. The results are tabulated in Table 12.2. The multipath spread rule tends to be slightly more accurate than the N signal greater than rule and more on the conservative side. This is true even though delays greater than half a symbol time occurred. Also, when the T_m model is too liberal, the amount by which it is in error is smaller. The least accurate model is the $\sum$ powers rule, which tends to be too liberal more often and by much larger amounts.

Table 12.2 Comparison of Multisource Models

Rule	% right	% too liberal	Max BER, %	% too conservative	Min BER, %
N sig >	90.7	9.0	3.15	0.3	0.5250
$\sum$ powers	84.7	11.9	45.3	3.4	0.0136
T_m < 3.4	93.8	2.4	1.78	3.8	0.4650

An additional issue arises when one considers the impact of thermal noise. In coverage simulations employing capture models, adequate performance was assumed if the capture criterion was met and if the capturing signal was above the threshold level set by noise. However, this threshold value is determined in the absence of multipath interference. In reality, the threshold level rises as multipath interference increases. This is shown clearly for the QAM system in Figure 10.21, which shows BER as a function of E_b/N_0 for different values of multipath spread. To account for this, the threshold must be a function of the multipath interference. Multiple capture requirements would therefore have to include multiple threshold requirements. However, this concern is easily addressed in the multipath spread model by requiring that the aggregate signal power be greater than a certain threshold that is a function of the single parameter T_m.

The multipath spread model was again compared to the other multisource models for three simulcasting sources, as before, with 666 different combinations of relative delay and signal level. This time, however, noise was included by setting the E_b/N_0

of the desired signal to 23 dB (about 3 dB above the static threshold). The T_m model had a threshold that varied from -97.5 dBm for $T_m = 0$ to -87.5 dBm for $T_m = 3.4$ μs. The results are shown in Table 12.3. The T_m model with a varying threshold is much more accurate than the other models tested. The N signal greater than and Σ powers models, because they rely on a static threshold level, are seen to be excessively liberal in their prediction of performance with noise.

Table 12.3 Comparison of Multisource Models with Thermal Noise

Rule	% right	% too liberal	Max BER, %	% too conservative	Min BER, %
N sig >	81.4	18.3	3.80	0.3	0.761
Σ powers	75.7	21.0	41.5	3.3	0.156
$T_m < 3.4$	97.3	0.90	1.24	1.8	0.739

12.6.4 Use of the Multipath Spread Model in Coverage Predictions

A two-source coverage simulation program can be modified to treat multiple sources by applying the multipath spread model as follows. Because the value of T_m depends on the sum of powers, the small-sector probability of coverage, based on lognormal statistics, cannot be computed analytically. Instead, individual powers from each source are generated via Monte-Carlo random draws, according to the lognormal statistics and the multipath spread model is then applied. This process is repeated many times for each small-sector location in the area of interest to yield maps of the probability of coverage areawide.

12.7 Homework Problem

Can directive antennas be of use in shrinking the noncapture region in simulcast systems?

Solution: With omnidirectional simulcast, differential delay usually equals zero at the equal power points halfway between transmitters (equal ERP and homogeneous propagation situations). Consider a modulation that requires either less than 10-μs delay spread or at least 10-dB capture. With sites spaced just 10 mi, these criteria mean that moving ±0.93 mi from the center point must result in the near source being at least 10 dB stronger than the far source. But with fourth-law propagation the capture value is only:

$$\left(\frac{P_{\text{near}}}{P_{\text{far}}}\right) = \left(\frac{r_0 + r_d}{r_0 - r_d}\right)^4$$

$$= \left(\frac{1+x}{1-x}\right)^4$$
$$= 4.51 \ (6.54 \ \text{dB}) \tag{12.39}$$

where $x = (r_d/r_0) = 0.93/5 = 0.186$. Therefore the second criteria cannot be met.

Now consider the use of directional antennas aimed front to back such that forward beam 1 covers most of the region between sites 1 and 2, rather than just half. If the front-to-back ratio is 30 dB, then the equal power point is 8.49 mi from site 1 and 1.51 mi from site 2. Additional delay can be added electronically at site 2 so that zero differential delay occurs at the equal power point. Then 10 μs of differential delay will occur 0.58 and 2.44 mi from site 2. For the former distance, site 2 captures by $40 \log[(10 - 0.58)/0.58] - 30 = 18.4$ dB; for the latter distance, site 1 captures by 10.4 dB. Thus, directional antennas can indeed be helpful in minimizing simulcast distortion.

References

[1] Gray, G. D., "The Simulcasting Technique: An Approach to Total-Area Radio Coverage," *IEEE Tr. on Veh. Tech.*, Vol. VT-28, No. 2, May 1979, pp. 117-125.

[2] Corrington, M. S., "Frequency-Modulation Distortion Caused by Multipath Transmission," *Proc. of IRE*, December 1945, pp. 878-891.

[3] Ade, J., "Some Aspects of the Theory of Simulcast," *32nd IEEE Vehicular Technology Conference*, San Diego, CA, May 1982, pp. 133-163.

[4] Arredondo, G. A., "Analysis of Radio Paging Errors in Multitransmitter Mobile Systems," *IEEE Tr. on Veh. Tech.*, Vol. VT-22, No. 4, November 1973, pp. 226-234.

[5] Ng, E. and M. Geller, "A Table of Integrals of the Error Functions," *Journal of Research of the National Bureau of Standards*, B. Mathematical Sciences Vol. 73B, No. 1, January-March 1969.

[6] Selby, S. M. (editor in chief of mathematics), *CRC Standard Mathematical Tables*, 18th edition, 1970.

[7] Abramowitz, M. and I. Stegun (editors), *Handbook of Mathematical Functions*, New York, NY, Dover Publications, Inc., 1972.

[8] Flath, Jr., E. H., "Proof-of-Performance Testing," *40th IEEE Vehicular Technology Conference*, Orlando, FL, May 1990, pp. 720-725.

[9] Andersen, J. B., et al., "Statistics of Phase Derivatives in Mobile Communications," *36th IEEE Vehicular Technology Conference*, Dallas, TX, May 1986, pp. 228-231.

[10] Chuang. J. C-I, "The Effects of Time Delay Spread on Portable Radio Communications Channels with Digital Modulation," *IEEE Journal on Selected Areas in Communications*, Vol. SAC-5, No. 5, June 1987.

[11] Chuang, J. C-I, "Modeling and Analysis of a Digital Portable Communications Channel with Time Delay Spread," *36th IEEE Vehicular Technology Conference*, Dallas, TX, May 1986, pp. 246-251.

[12] Chuang, J. C-I, "The Effects of Delay Spread on 2-PSK, 4-PSK, 8-PSK, and 16-QAM in a Portable Radio Environment," *GLOBECOM'87 Conference Record*, 1987.

[13] Afrashteh, A., et al., "Measured Performance of 4-QAM with 2-Branch Selection Diversity in Frequency-Selective Fading Using Only One Receiver," *39th IEEE Vehicular Technology Conference*, San Francisco, CA, May 1989.

Chapter 13
Traffic Engineering

This chapter addresses engineering aspects of a radio system's capacity to carry messages; i.e., traffic. First, the character of such messages is examined in terms of length, for the whole of the message as well as its pieces: repeater access durations, transmission lengths, and gap durations. Next, message length and frequency of occurrence are lumped together to form the key traffic parameter system load. Means are given for estimating peak system load in terms of the number of subscribers and the concept of load smoothing for large systems with many separate user groups is introduced. System-to-system and day-to-day variations of peak load are considered.

Grade of service is then related to traffic load and the number of trunked servers available via queueing theory. The Erlang-B (blocked calls cleared (lost)) and Erlang-C (blocked calls delayed) viewpoints are discussed, along with the Poisson middle-ground viewpoint that allows for call abandonment in proportion to system overload.

The control of trunked systems is discussed next. Simulation results for one type of control channel arrangement are described in detail, as is an experimental arrangement for measuring the actual performance of a commercial trunked system using that control arrangement.

The final sections deal with (1) quantifying the impact on grade of service if some load can be rerouted to another system during peak traffic periods and (2) procedures for analyzing the access delay of group calls on multisite systems such as cellular radiotelephone systems.

13.1 Message Length Characteristics

A detailed study of land-mobile radio dispatch message length characteristics for both conventional shared, community repeaters at UHF and SMR trunked repeater systems at 800 MHz is reported in Reference [1]. This study involved commercial business radio users and the parameter values reported throughout this chapter pertain to such users. Other user groups, such as public safety and utilities, can be expected to be characterized by somewhat different parameter values. The study consisted of two phases, each involving simultaneous over-the-air tape recording of three conventional repeaters and six trunked user groups. The first phase involved very busy repeaters and user groups, the second phase involved low-activity operation. Message data were obtained by actual listening, a tedious but necessary activity as the definition of what constitutes a message is not simple.

Six classes of messages were recognized: (1) a *basic* message is a straightforward conversation with clear-cut beginning and end, (2) a *multiple* subject message involves two or more basic messages separated by clear changes of subject matter during the overall exchange (each component is counted as a message by itself for tabulation purposes), (3) a *silent* key is when the transmitter is keyed but no audio is offered[1], (4) a *single transmission message* (*STM*) is one for which either no reply is expected or none is received (for example, "Mobile 5 signing off for lunch"), (5) a *repeat STM* is a message where contact is attempted two or more times (for example, "Mobile 5 to base", ..., "Come in base, this is mobile 5"), and (6) *other* (for example, a system identification message which might be sent on the half hour).

Examination of 1200 messages per phase, half on conventional repeaters and half on trunked repeaters, yielded homogeneous results between phases and confirmed that average trunked message lengths consistently exceeded the average for conventional repeaters. Table 13.1 summarizes the findings. Figure 13.1, which compares cumulative probability density functions of message length on conventional and trunked repeaters, shows that messages exceeding any particular length are on average less likely on conventional repeaters. The straight-line nature of the curves in Figure 13.1 indicates that message lengths are approximately exponentially distributed.

Studies of telephone interconnect message lengths also indicate an exponential character to the distribution. Hence, average message length fully characterizes the behavior. Interconnect messages are on average much longer than the dispatch messages just discussed. FCC reports of calls on *improved mobile telephone service* (IMTS) systems, the precellular radiotelephone system, indicate an average of 119 sec [2]. Cellular systems were originally designed assuming 3-min average message lengths, but in practice the average is generally about half that value. A limited amount of interconnect traffic is permitted on some trunked dispatch systems [3]. There calls only averaged about 50 seconds in a 1984 study, perhaps due to newness of the service and the dispatch-oriented nature of the subscribers.

Although dispatch messages have been shown to be approximately exponentially distributed, knowing the average message length is not necessarily sufficient

---

[1]This often occurs as users check to make sure they are within repeater coverage range and that their equipment and the system are functional. Momentarily depressing the push-to-talk button of a subscriber radio normally activates a repeater for at least some "hang time" interval; this in turn unsquelches the subscriber receiver and allows a weak noise background to be noticed, confirming proper operation.

Traffic Engineering

Table 13.1 Summary of Message Statistics

After: Hess, G. and J. Cohn, "Communication Load and Delay in Mobile Trunked Systems," *31st IEEE Vehicular Technology Conference*, April 1981, Washington, D.C., p. 272.

I. Average message length (sec)	All	Basic	Mult.	Silent	STM	Rept.	Other
Motorola conventional repeaters[a]	13.7	22.0	17.9	2.1	4.2	4.1	25.1
Motorola trunked repeaters[b]	21.9	35.4	36.4	2.7	4.7	5.1	13.1
"Scanning" type trunked repeaters[c]	33.5	53.9	46.4	11.6	16.3	14.9	17.5
II. Message category (% total)							
Motorola conventional repeaters		51.0	2.0	11.2	22.2	12.2	1.5
Motorola trunked repeaters		48.6	5.8	14.1	14.3	16.1	1.1
"Scanning" type trunked repeaters		41.7	7.7	11.0	15.3	13.9	8.7
III. Message time (% total)							
Motorola conventional repeaters		81.4	2.9	1.8	7.4	4.0	2.8
Motorola trunked repeaters		81.5	10.0	1.8	3.2	3.0	0.6
"Scanning" type trunked repeaters		67.1	10.7	3.8	7.4	6.2	4.6

a. Measured channels equally weighted
b. Measured fleets equally weighted
c. 10-sec hang time; measured messages equally weighted

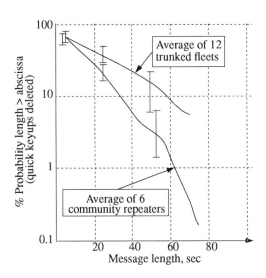

Figure 13.1 Cumulative distribution of message length. After: Hess, G. and J. Cohn, *ibid.*, p. 272.

to predict the performance of trunked systems. In general, a message will consist of some number of repeater accesses (only guaranteed to equal one with *pure message-trunking* (PMT) control)[2] and some number of transmissions. The latter can exceed the former due to repeater hang time. With a 1-sec hang time [3], the average trunked repeater access duration typically runs about 6 seconds. Interestingly, the number of accesses per message is reasonably well described by a Gaussian distribution with a median value of 1.0 and a standard deviation of 3.4, truncated below 1. The average number of accesses per message is about three. The gaps between the repeater accesses are exponentially distributed. The straight-line nature of the curves in Figure 13.2 confirms this. First gaps average longer than other gaps because it generally takes time for the called party to pick up their microphone and respond to a calling party.

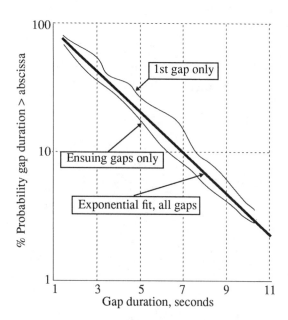

Figure 13.2 Cumulative distribution of intramessage gap durations.

---

[2]This is the mode of operation of the original General Electric trunking control scheme [4]. It is accomplished simply by making the repeater hang time long relative to the typical transmission gap duration. The scanning delay to start the call and the hang time are small enough that the distribution of repeater access lengths is still approximately exponential, but with a parameter that equals the sum of those fixed overheads plus the average message length.

[3]This is the normal operational mode for the Motorola trunked control scheme [5]. It results in what can be called *quasitransmission trunking* (QTT). Priority queueing of recent users is provided to offer the appearance of message-trunked operation.

Traffic Engineering

In a *pure transmission-trunking* (PTT) mode[4], the average access duration equals the average transmission length. This typically runs about 4 seconds and the average number of transmissions per message is generally about four.

A final comment about messages is that their interarrival times also tend to follow an exponential distribution (Figure 13.3). This is equivalent to message requests arriving randomly in time; i.e. Poisson statistics hold. A caution here is that when messages from only one or a few particular groups are considered, rather than all the groups in the system, interarrival times may be correlated due to one message stimulating followup messages by other group members.

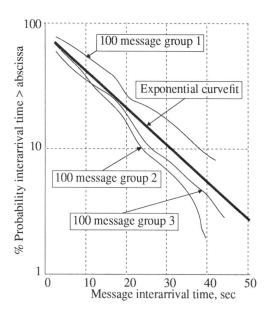

Figure 13.3 Cumulative distribution of message interarrival times.

13.2 Traffic Load Estimation

13.2.1 Nominal System Peak Load

Traffic load represents the product of the average repeater time required per message and the average message arrival rate. One erlang (the standard unit of traffic)

---

[4]This is the mode of operation with the E. F. Johnson trunking control scheme [6]. Motorola systems can be operated in this fashion too, if desired, and their priority queueing and callback features help maximize message continuity.

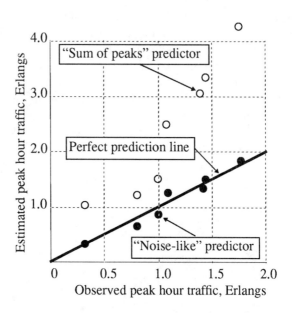

Figure 13.4 Comparison of sum-of-peaks and noise-like peak traffic load estimators.
After: Hess,G. and J. Cohn, "Communication Load and Delay in Mobile Trunked Systems," *31st IEEE Vehicular Technology Conference*, April 1981, Washington, D.C., p. 273.

implies 3600 sec of calls to be handled per hour, or 60 sec of calls to be handled per minute, etc. Extensive measurements of traffic on land-mobile radio repeaters show that load can vary substantially over time periods as small as 5 min. Also, peak load is generally poorly repeatable on a day-to-day basis in both magnitude and time of occurrence. Because traffic offered by any single user group is generally uncorrelated with traffic offered by other user groups, summing individual user peak loads is a poor way of estimating system peak load. The traffic sources tend to act like uncorrelated Gaussian noise with their peaks summing in an rms fashion. Figure 13.4 illustrates this behavior. Hence, an improved estimate of the system peak load for N user groups is through:

$$\hat{\rho}_{\text{sys,pk}} = \sum_{i=1}^{N} \rho_i + k \sqrt{\sum_{i=1}^{N} \sigma_i^2} \qquad (13.1)$$

where ρ_i equals the average traffic load from source i, and σ_i equals the standard deviation of traffic load from source i. The scaling factor k varies somewhat from system to system and with the averaging time. For estimation of the busy-hour load, a value near 1.5 is common. This is consistent with the workday typically spanning about 10 hr as one would then expect the busy hour and the 90 percentile of the load distribution (for which the k factor would be 1.28) to be comparable. Estimates for an arbitrary time interval T, in hours, should use $1.5/\sqrt{T}$. Thus, for example, the effective k value for estimating the busy 15-min traffic load is about 3.0.

A caution in the use of the system peak load equation concerns the difference between message load and airtime load. QTT systems with 1-sec hang times only require a repeater to be keyed up on average for about 80% of the message seconds. With PTT that factor drops even further, to perhaps 65%. Thus, if a typical QTT user averages 1.5 messages per hour of 22-sec average length, the average message load per user is 0.00917 Erl but the average airtime load is just 0.0073 Erl. If message trunking is accomplished with a scan delay of 2 sec and hang time of 6 sec, then the average airtime load will be about 0.0125 Erl. In grade-of-service matters it is airtime load that really counts. This will be addressed shortly by applying queueing theory.

13.2.2 System-to-System Variation of Peak Load

Extension of the load-smoothing concept to handle system-to-system variation is discussed in Reference [7]. This variation refers to the fact that comparably sized systems do not necessarily produce equal peak loads because traffic requirements vary widely from subscriber to subscriber. Quantifying this variation is important to the generation of general system loading guidelines. Clearly, it would be unwise to suggest loading all systems of a certain size to the average peak load for that size that produces just acceptable grade of service during the peak. If this were done about half the systems would experience even higher peak loads and therefore substandard grades of service.

For any particular subscriber user group or fleet, the traffic average load and standard deviation of load can be approximated as $\rho_i = \rho_{s,i} N_{s,i}$ and $\sigma_i = \sigma_{s,i}\sqrt{N_{s,i}}$, respectively, where $\rho_{s,i}$ is the average traffic load on a per subscriber basis of fleet i, $\sigma_{s,i}$ is the per-subscriber standard deviation of traffic load of fleet i, and $N_{s,i}$ is the number of subscribers in fleet i. Figure 13.5 indicates that $N_{s,i}$ can be taken as approximately lognormally distributed. Observations of $\rho_{s,i}$ and $\sigma_{s,i}$ indicate that they are approximately lognormally distributed as well. Thus, ρ_i and σ_i must be approximately lognormally distributed too. The system peak load estimate is based on summations of those random variables. Monte Carlo generation of 500 subscriber systems built via random draws from a 70-fleet database indicates that is of sufficient size for the central limit theorem to take effect (these numbers are representative of a five-channel trunked SMR). This is confirmed by the reasonably straight-line character of the system busy-hour load distribution plotted on a normal probability scale, as in Figure 13.6.

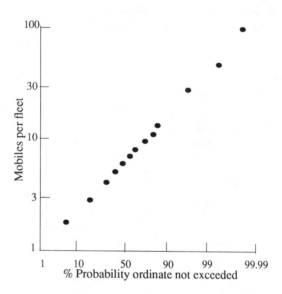

Figure 13.5 Cumulative distribution function for trunked SMR fleet sizes. After: Hess, G., "Estimation of Peak Load on Trunked Repeater Systems," *32nd IEEE Vehicular Technology Conference*, San Diego, CA, May 1982, p. 333.

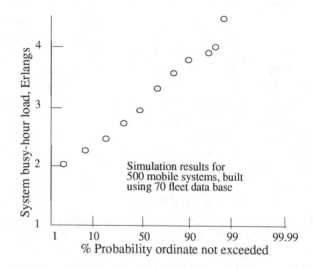

Figure 13.6 Cumulative distribution function of system busy-hour load. After: Hess, G., *ibid.*, p. 333.

Traffic Engineering

Because $\rho_{\hat{\text{sys}},\text{pk}}$ is approximately normal, a simple procedure can be used to estimate system-to-system load variations. In terms of subscriber units, the nominal peak traffic equation can be written:

$$\rho_{\hat{\text{sys}},\text{pk}} = N_s \rho_s + k\sqrt{N_s}\sigma_s \qquad (13.2)$$

where N_s is the number of subscriber units per system for the systems of interest, ρ_s is the nominal average traffic load per subscriber, and σ_s is the nominal traffic load standard deviation on a per-subscriber basis. Typical values observed for QTT systems are $\rho_s = 0.005$ Erl and $\sigma_s = 0.02$ Erl. Treating ρ_s and σ_s as correlated Gaussian random variables gives the following expression for the standard deviation of $\rho_{\hat{\text{sys}},\text{pk}}$:

$$\sigma_{\rho_{\hat{\text{sys}},\text{pk}}} = \sqrt{N_s^2 \sigma_{\rho_s}^2 + k^2 N_s \sigma_{\sigma_s}^2 + 2r N_s^{\frac{3}{2}} \sigma_{\rho_s}\sigma_{\sigma_s}} \qquad (13.3)$$

where σ_{ρ_s} is the standard deviation of ρ_s, σ_{σ_s} is the standard deviation of σ_s, and r is the cross-correlation coefficient between ρ_s and σ_s. Typical values observed for QTT systems are $\sigma_{\rho_s} = 0.001$ Erl and $\sigma_{\sigma_s} = 0.002$ Erl. The correlation coefficient is generally quite high so it is prudent to simply consider it unity. The 10% worst-case load (highest) expected over a large number of comparably sized systems is then $\rho_{\hat{\text{sys}},\text{pk}} + 1.28\sigma_{\rho_{rm\hat{s}ys,pk}}$. Other percentiles are readily found by adding or subtracting the appropriate multiple of $\sigma_{\rho_{\hat{\text{sys}},\text{pk}}}$, according to standardized Gaussian tables.

13.2.3 Day-to-Day Variation of Peak Load

Reference [7] also discusses variation of peak load on the same system from day to day. In general, neither the peak load amplitude, nor its time of occurrence is very stable. Figure 13.7 shows the day-to-day behavior of peak traffic load for 1-hr and 15-min periods using extreme value theory, as discussed by Gumbel [8]. The return period scale refers to the expected number of days between events of peak load as high as the ordinate. An improvement over the extreme value theory approach is reported by Barnes and O'Connor [9]. The idea is to describe the peak load distribution via a Gaussian raised to some power. The power 6, which they found useful for telephone exchanges, also fits land-mobile radio behavior fairly well.

13.2.4 Interconnect and Mixed Traffic Loads

Telephone traffic generally involves calls whose lengths average appreciably greater than that of dispatch traffic, and whose call rates average typically somewhat less. For example, a common design guide used to project the performance of the first cellular telephone systems assumed an average subscriber generated 0.03 Erl of busy-period traffic (about one 100-sec call per hour). FCC measurements of IMTS behavior averaged about 0.018 Erl/subscriber (the average message length was 119 sec, implying less frequent calls than on cellular systems). After FCC rules changed to

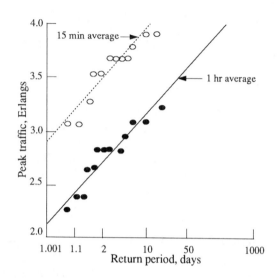

Figure 13.7 Trunked system daily peak load distribution.
After: Hess, G., *ibid.*, p. 333.

allow interconnect traffic on trunked SMR systems, the traffic behavior on a number of such systems was experimentally studied. The data shed light on how one might estimate not only interconnect load, but also total system load composed of both interconnect and dispatch traffic.

The six systems studied in detail involved 82 channels with 8627 dispatch units and 1705 interconnect-equipped units. Measurements were made over the period of 7 a.m. to 5 p.m. local time for a total of 54 weekdays.

Interconnect Traffic

Key findings are: (1) the load in erlangs over the workday is approximately normally distributed, (2) the median average message length equals just 43 sec, (3) the median call rate per hour per unit equals 0.56, (4) the median average interconnect load per unit equals just 0.0062 Erl, (5) and the median busy-hour interconnect load per unit equals 0.01 Erl.

Rather than estimate the busy-hour interconnect load as 0.01 times the number of units, the following load-smoothing type relation is preferable:

$$\rho_{\text{busyperiod}} = N_i(0.0062) + \sqrt{N_i}(0.0415)/\sqrt{T} \qquad (13.4)$$

where N_i is the number of interconnect-equipped units and T is the averaging period in hours ($T = 1$ for the peak hour load).

Traffic Engineering

Dispatch Traffic

Key findings are: (1) the load in erlangs over the workday is approximately normally distributed, (2) the median average dispatch load per unit equals just 0.004 Erl, (3) and the median busy-hour dispatch load per unit equals 0.0059 Erl.

The relevant load-smoothing relation for dispatch on these systems is:

$$\rho_{\text{busy-period}} = N_d(0.004) + \sqrt{N_d(0.0407)}/\sqrt{T} \quad (13.5)$$

Total System Load

Key findings are: (1) the load in erlangs over the workday is approximately normally distributed (2) and the zero-lag cross-correlation coefficient between dispatch and interconnect traffic is normally distributed with a median value of 0.46 and a standard deviation of 0.3. The implication is that the distribution of total system load throughout the workday can be estimated through $\rho \sim N(\mu, \sigma)$, where $\mu = N_i(0.0062) + N_d(0.004)$, $\sigma^2 = \sigma_i^2 + \sigma_d^2 + 2(0.46)\sigma_i\sigma_d$, $\sigma_i = 0.0223\sqrt{N_i/T}$, and $\sigma_d = 0.0395\sqrt{N_d/T}$. The average busy hour occurs consistently at about the 90 percentile of this distribution.

To include day-to-day variation Figure 13.8 can be used. This figure plots the

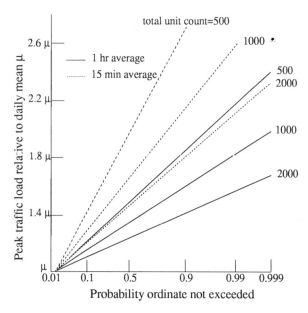

Figure 13.8 Distribution of peak system load using sixth-power Gaussian model.

distribution of peak system load in terms of average system load for three system sizes (500, 1000, and 2000 units; in all cases 20% of the units are interconnect equipped). As an example of how these curves were generated, consider a system with 1000 units. The average load on such a system would equal 1000(0.004) + 200(0.0062) = 5.24 Erl. The interconnect variance would be $0.0223\sqrt{200} = 0.3154$ for 1-hr periods; the dispatch variance would be $0.0395\sqrt{1000} = 1.2491$. The overall variance thus equals $0.3154 + 1.2491 + 2(0.46)\sqrt{0.3154(1.2491)} = 2.142$. The 90 percentile load is therefore $5.24 + 1.28\sqrt{2.142} = 7.113$ Erl, which is 1.36 times the average load level. Hence, the 90 percentile normal scale ($0.9^6 = 0.53$ normal-to-sixth-power probability scale) and the 1.36 μ peak load scale fall on the solid 1000-unit line (the dashed line pertains to 15-min averaging, which has twice the variability).

13.3 Grade-of-Service Evaluation

13.3.1 Blocked Calls Lost, Erlang-B Viewpoint

In the landline telephone world, grade of service is a phrase often used interchangeably with the probability that an offered call is blocked and thus lost. The queueing theory generally used to calculate such probabilities is called Erlang-B, after the Danish scientist that first solved the problem. The Erlang-B formula can be written as:

$$\begin{aligned}\text{Prob(blocked)} &= \text{Prob(lost)} \\ &= P_B \\ &= \frac{\rho^c}{c!}[\sum_{i=0}^{c}\frac{\rho^i}{i!}]^{-1}\end{aligned} \quad (13.6)$$

where c equals the number of servers, channels, or trunks available and ρ is the traffic load offered to the system. Of course, since calls can be lost in this arrangement, the actual traffic load handled is less than that offered by the factor $(1 - P_B)$. Development of this formula is standard in queueing theory texts, a premier example being Reference [10]. A very readable publication geared specifically toward telephone traffic is Reference [11]. Tables of blocking probability are commonly found in such texts, as are graphical portrayals of the formula (see Figure 13.9 and Reference [12]). Reference [13] is an elaborate publication devoted entirely to such matters. In this age of personal computers, one can readily calculate probabilities directly from the formula. Such calculations are prone to numerical difficulties when the number of servers is large, but Reference [14] indicates a way to circumvent them.

Several important assumptions underlie the above theory:

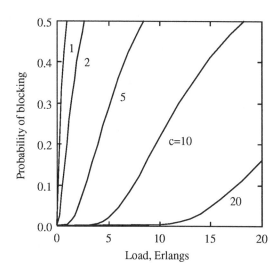

Figure 13.9 Erlang loss probability as a function of offered load ρ and number of servers c.

- First, calls are assumed to arrive at random; i.e., the arrival process is Poisson and therefore the time between call arrivals is exponentially distributed. For this to be strictly true, users whose calls have been blocked cannot immediately retry.

- Second, call service time, and hence message length, is exponentially distributed.

- Third, the number of traffic sources is essentially infinite and sources are homogeneous in message character; i.e., they all are drawn from the same exponential population. The requirement of an infinite user population can readily be removed; the result is the so-called Engset formula [15]. In small trunked systems removing this assumption can be quite important [16].

- Fourth, the system is in statistical equilibrium.

- Fifth, and last, the offered traffic load is known.

Table 13.2 Transition Table for c-Server Loss System

After: Beckmann, P., *Introduction to Elementary Queuing Theory and Telephone Traffic*, Golem Press, Boulder, CO, p. 63.

Number of customers in system at time t+dt	at time t	Event during dt causing transition	Transition probability
0	0	no arrival	(1-vdt)P(t;0)
	1	1 service completion	μdtP(t;1)
0<k<c	k-1	1 customer arrival	vdtP(t;k-1)
	k	no arrival, no completion	(1-vdt)(1-kμdt)P(t;k)
	k+1	1 service completion	(k+1)μdtP(t;k+1)
c	c-1	1 customer arrival	vdtP(t;c-1)
	c	no service completion	(1-cμdt)P(t;c)

Based on assumptions 1,2, and 3, Table 13.2 [11] shows the transition possibilities for the number of calls in a c-server Erlang-B trunked system at an arbitrary time. From the table, the equation for the probability of being in state $k = 0$ at time $t + dt$ is seen to be:

$$P(t + dt; 0) = (1 - \nu dt)P(t; 0) + \mu dt P(t; 1) \tag{13.7}$$

This can be rearranged into the form:

$$[P(t + dt; 0) - P(t; 0)]/dt = \mu P(t; 1) - \nu P(t; 0) \tag{13.8}$$

where the left-hand side equals $P'(t; 0)$. Thus, under steady-state conditions one obtains $P(1) = \rho P(0)$, where $\rho = (\nu/\mu)$. Similarly, disregarding second-order terms, one also obtains:

$$\rho P(k-1) - (\rho + k)P(k) + (k+1)P(k+1) = 0 \tag{13.9}$$

for $0 < k < c$, and:

$$\rho P(c-1) - cP(c) = 0 \tag{13.10}$$

These equations can actually be written down immediately with the help of Figure 13.10, [10] which shows the steady-state (assumption 4) state-transition behavior.

The probability of being in any state other than zero can be expressed solely in terms of the probability of the zero state. For example, with $k = 2$ one finds $\rho P(0) - (\rho + 1)P(1) + 2P(2) = 0$. But substituting $P(1) = \rho P(0)$ gives $P(2) = [\rho^2 P(0)]/2$. Proceeding to higher k values allows the generalization $P(k) = [\rho^k P(0)]/k!$. To

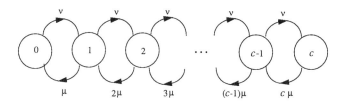

Figure 13.10 State-transition rate diagram for c-server loss system.
After: Kleinrock, L., *Queueing Systems, Volume 1: Theory*, John Wiley & Sons, New York, NY, 1975, p. 105.

obtain $P(0)$ note that the probabilities of each state must sum to unity. Thus:

$$1 = \sum_{i=0}^{c} \frac{\rho^i P(0)}{i!}$$

$$P(0) = \frac{1}{\sum_{i=0}^{c} \frac{\rho^i}{i!}}$$

$$P(k) = \frac{\frac{\rho^k}{k!}}{\sum_{i=0}^{c} \frac{\rho^i}{i!}} \quad (13.11)$$

The probability of blocking is of course equal to the probability that an arriving call will find all c servers already in use ($P(c)$).

13.3.2 Blocked Calls Delayed, Erlang-C Viewpoint

Users offer calls expecting them to be serviced, so if they are blocked what usually happens in real life is that they are repeatedly reoffered until a server is successfully obtained. Such reoffering is unnecessary if queueing is provided; i.e., calls that are blocked are not lost but rather are delayed until servers become available. Erlang addressed this type of delay system with his Erlang-C formula:

$$\text{Prob(delayed)} = P_D$$
$$= \frac{\rho^c}{c!}[\frac{\rho^c}{c!} + (1 - \frac{\rho}{c})\sum_{i=0}^{c-1} \frac{\rho^i}{i!}]^{-1} \quad (13.12)$$

Figure 13.11 shows this equation for a family of c values.

The distribution of delay depends on the queueing discipline, but the average delay does not and is a reasonable indication of grade of service. It is given by:

$$<T> = P_D <T_D>$$
$$= P_D \frac{h}{c - \rho} \quad (13.13)$$

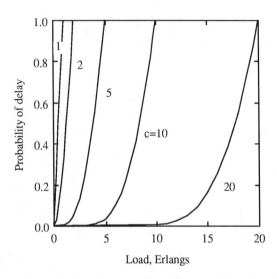

Figure 13.11 Erlang delay probability as a function of offered load ρ and number of servers c.

where $<T>$ is the average delay of all calls (both those delayed and those not delayed), $<T_D>$ is the average delay of only delayed calls, h is the average service time, and ρ is the total offered load (which of course in a delay system will also equal the handled load). For first-come first-served queueing, the distribution function of delay for delayed calls is exponential:

$$P(T_D > t) = \exp[-(c-\rho)\frac{t}{h}] \qquad (13.14)$$

and for all calls the distribution function is:

$$P(T > t) = P_D \exp[-(c-\rho)\frac{t}{h}] \qquad (13.15)$$

Riordan [17] has derived curves approximating the distribution of delay with random-order queueing(Figure 13.12).

13.3.3 Poisson Viewpoint

The Poisson viewpoint combines the character of both Erlang-B and Erlang-C viewpoints. It is a delay system that allows call abandonment at a rate proportional to the overload. Its behavior can be described by a particular choice of the weighting function in the model of Nesenbergs [18].

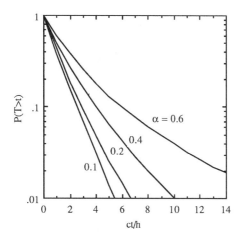

Figure 13.12 Random order of service delay distribution as a function of (1) server count times delay divided by average message length and (2) the parameter offered load per server.
After: Riordan, J., "Delay Curves for Calls Served at Random," *Bell System Technical Journal*, Vol. 32, January 1953, p. 102.

13.3.4 Application to Trunked Dispatch Grade of Service

Erlang-C theory applies directly to message-trunked systems. Practical implementations of message trunking may involve hang and scan times that are fixed values, so even if message lengths are exponentially distributed the service time is not exactly exponential. Nonetheless, trunked message lengths on average are sufficiently long to dominate the distribution; thus, Erlang-C results agree well with results from detailed simulations.

PTT systems also fall nicely under the umbrella of Erlang C theory. However, multiplying the average delay per transmission by the average number of transmissions per message to obtain the average message delay can be unwise. In PTT systems without queueing, transmissions that are delayed must be sustained by random retries on the part of subscribers. Even with priority queueing and callback, the subjective impact of interruptions to message continuity is generally severe. Hence, one is forced to limit the load to a value that gives less delay than that acceptable with PMT. This tends to offset the advantage of PTT not requiring server assignments during pauses between transmissions.

QTT with priority queueing and callback is an attempt to get the best of both worlds. It appears to operate as a message-trunked system, but takes advantage of most of the gap time between transmissions. In the event of delay, calls already in progress are considered first for server assignment. This naturally biases the access delays that occur such that initial access delays are far more lengthy than intramessage access delays. Unfortunately, dual-service queues are not part of Erlang-C traffic theory. However, it has been observed that by somewhat extending the average repeater access duration (from about 6 to 7.6 sec), Erlang-C predictions can be useful [1] (Figure 13.13). Of course, detailed Monte Carlo simulation results are preferable.

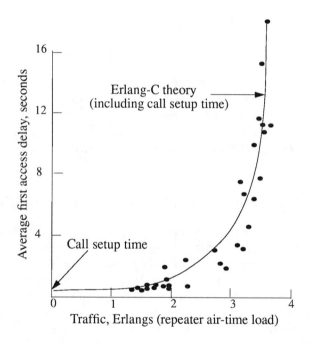

Figure 13.13 Scatter plot of average first access delay versus average 15-min system load and solid-curve prediction using Erlang-C theory.
After: Hess, G. and J. Cohn, "Communication Load and Delay in Mobile Trunked Systems," *31st IEEE Vehicular Technology Conference*, April 1981, Washington, D.C., p. 273.

13.4 Trunked System Control Channel Performance

13.4.1 Introduction

So far the traffic discussion has concerned voice messages (dispatch and interconnect) and the repeaters that serve them. Of course, all trunked systems need some means of control to indicate call requests and assign particular servers to particular calls. A common means of control involves setting aside one of the channels for signaling purposes. This seems a substantial penalty for small systems. For example, a five-channel trunked system could lose 20% of its capacity. Bear in mind, however, that QTT offers roughly a 20% advantage over ideal message trunking (the advantage in practice is greater because message trunking is accomplished by long hang times which represent control overhead). Also, small trunked systems can be operated in a voice-on-control method, where the signaling channel is used as a voice channel during traffic overloads.

Signaling under voice is an interesting way to avoid an explicit control channel penalty. The typical voice passband starts at 300 Hz, so subcarrier modulation around 150 Hz allows low-rate information (about 300 bps) to be transmitted along with voice. This is sufficient to handle plain old trunking service, adapting a phrase from the telephone industry. But the addition through the years of many complicated features and the additional signaling they require poses a problem since to maintain sensitivity error coding bits must be sent, making the number of information bits per second even less than 300 per channel.

The control channel is not really a penalty at all for large systems because it allows the servers to be utilized quickly as needed. A scanning-type system like LMTS, where mobiles wishing to place calls scan all the channels until an idle one is found, could take several seconds just to start delayed calls. Those seconds represent unused air time and hence inefficiency.

13.4.2 Example Control Channel Operation

The Motorola arrangement for accomplishing control signaling is described by Thro [5]. It consists of *outbound signaling words* (OSWs) of approximately 23 ms duration sent in time division multiplexed form, and inbound signaling words of approximately equal duration sent by subscriber radios in a slotted ALOHA protocol [19, 20, 21][5]. OSW signaling is contention free and thus the capacity is about 42 packets per

---

[5]Pure ALOHA refers to an automatic repeat request scheme where signaling packets are sent at random without regard to channel activity information and, if unacknowledged (due to collisions with other transmissions; errors caused by noise are not considered in the theory), are repeated at random intervals until an acknowledgment is received. The slotted version of ALOHA ties all transmissions to some system clock so that they can only begin at certain particular times.

second; however, contention on the inbound channel limits the ISW capacity to a lesser value and thus it is this capacity that is most generally of interest.

To allow for synthesized radios, the inbound signaling transmission occupies two basic 23-ms time slots. The first slot allows the synthesizer to settle onto frequency and the second slot contains the coded signaling bits. The result is that inbound signaling capacity appears to be 21 ISWs per second. In fact, because ISW requests are generated randomly in time, collisions occur and further reduce the signaling capacity. Assuming such collisions require retransmission (after random waiting periods) by all the parties involved, the capacity is reduced by the factor $(4/3)(1/2e)$ to 5.25 successful requests per second. The reduction factor falls between the capacity with nonslotted random signaling $(1/2e)$ and the capacity with pure slotted random signaling $(1/e)$. This is because the signaling packet is two slots in duration. Hence three overlap situations exist: (1) complete overlap, (2) 50% overlap, and (3) no overlap. In the usual slotted ALOHA development only situations (1) and (3) are considered. Certain features require additional signaling information for implementation and thus three slot packets are required. The theoretical capacity for such transmissions is $(42)(0.33)(6/5)(1/2e) = 3$ requests per second.

These capacities may still be higher than practical because of excessive delay between request initiation and successful receipt of requests. Equation (54) of Reference [20] can be used to evaluate the average delay in terms of throughput and request rate. Using parameter values relevant to the Motorola retry algorithm, one finds that at throughput capacity the delay averages about 0.6 sec for two-slot ISWs. This may be too large to maintain acceptable message continuity in a QTT mode of operation, a mode of operation very important to boosting the spectral efficiency possible with trunking. However, there are two crucial items not accounted for by theory that impact the practical signaling capacity: (1) retries necessitated by noise contamination of contention-free packets and (2) successful decodes of packets in the presence of contention via the capture phenomenon. Item 1 reduces capacity while item 2 raises it.

The latter aspect has been addressed by Namislo [22]. He finds that capture can eliminate the foldback behavior of offered load versus throughput and substantially reduce transmission delay. Unanswered is to what degree the benefits of capture are lost due to decode errors resulting from noise (item 1). To answer this question, a detailed simulation program was developed that includes both capture and decode errors. In addition, experiments were conducted on a commercial trunked system to measure actual delay versus throughput performance. This allowed verification of the accuracy of the simulation.

13.4.3 Simulation Description

The simulation program begins by requesting system information from the user. The data to be specified include: (1) the offered load in terms of new ISWs per second, (2) the probability that a new ISW is of the regular, two-slot format, (3) the probability that a new ISW originates from a mobile (as opposed to a fixed control station), and (4) the number of antenna sectors (this is to allow the impact of frequency reuse on the inbound control channel to be evaluated; however, in what follows, omnidirectional reception is presumed).

Two ways to specify the range from the base station of each requesting unit were considered. Initially, the range was obtained by draws from a probability distribution with the form of a power law, truncated at some maximum range. Using a linear law means that each x,y coordinate is equally likely; using a zeroth law means that all ranges from zero to the maximum are equally likely. The second, preferred approach, treats the range as a truncated Gaussian random draw. That is, the range was normally distributed between some minimum and maximum value with user-specified median and standard deviation.

Having chosen a range, the initial received power in dBm is specified via the relation $P_r = A + B \log(R)$, where R is the range in miles. The parameters A and B are set by solving Hata's empirical propagation loss equation [23] with some desired base antenna height; mobile antenna height, gain, and transmit power; operating frequency; and environmental conditions. The median path losses for 1- and 30-mi ranges yield two equations in the unknowns A and B. For the choices, 100 m, 2 m, 3 dBd, 15 W, 821 MHz, and suburban, respectively, the solution is $A = -56, B = -32.5$. Having specified a reference received signal power, one then adjusts it for lognormal shadowing by adding a decibel value drawn from a zero-mean Gaussian population with the appropriate standard deviation. If the source is a control station, this received power is used on each signaling attempt until success or timeout. If the source is a mobile, the received power is further adjusted on each attempt by including the effect of Rayleigh fading. This is done by using the lognormal-shadowed power as the average power parameter in an exponential distribution draw, the so-called quasistatic approximation.

The credibility of the simulation results depends strongly on how realistically the received power distribution is modeled. This actually depends on modeling two distributions: one for source location and one for signal propagation to the central site. Empirical studies give a degree of confidence in modeling the propagation aspect, but no such support for the source location aspect existed at the time of the simulation. Consequently, the received power distribution of a commercial trunked system was measured. These measured results compare well with the distribution produced by the simulation (where the propagation parameters were set to $A = -50.2$ and $B = -28.0$ to reflect the actual base antenna height of 425 m). The character of the data is approximately lognormal, as anticipated, with a standard deviation of about 15 dB.

One adjustment required concerns control stations. Realistically, signaling from control stations will only have problems due to contention, not noise. Hence, a check is made to ensure that the lognormal-adjusted received power is sufficiently above the central receiver noise floor so that the contention-free signaling error rate from control stations is at most 10%. In all cases signaling error rates have been estimated by curvefits to performance measured on the bench.

Some parameters set by the program, but of course easily changeable, are: (1) overall central receiver sensitivity equals -113 dBm (static 20-dB quieting number) (2) maximum coverage range equals 30 mi, and (3) lognormal-shadowing standard deviation equals 8 dB.

The simulation uses a call activity matrix to keep track of each call's progress. Call states consist of idle, synthesizer settle, first ISW, and, where appropriate, second ISW. Call attempts that fail due to contention or noise are retried in accordance to the specific Motorola mobile retry algorithm. If there is no successful packet in a 4-sec period, the attempt times out.

13.4.4 Simulation Results

Figure 13.14 shows simulation results for nominal conditions consisting of: (1) all ISWs are the regular single-word type, (2) half the ISW requests are originated by mobile units, and (3) the range distribution of requesting units is truncated normal from 0.25 to 30 mi, with a median range of 15 mi and a standard deviation of 11.5 mi (this limits the likelihood of exceeding the 30-mi limit and having to redraw the range to about 10%). Two lines are shown in the figure, one related to theoretical performance and one fitted to the simulated performance. The simulation indicates that a new request rate of about 6.5 per second can be supported at an average delay of 200 ms.

The delay cited in Figure 13.14 is from the start of front-porch transmission [6] to the completion of successful ISW transmission. The delay experienced by a system user will also include a debounce period between push to talk and the start of the front-porch transmission, plus the time required to receive a channel assignment, decode it, move to the assigned channel, successfully handshake on it, and unmute the radio. A 200-ms period was chosen because it is felt that larger delays would not provide sufficient message continuity for QTT operation. It is encouraging to note that the benefits of capture outweigh the penalties due to noise-induced errors.

13.4.5 Experiment Description

Simulation results are of little value unless supported by actual behavior in the field. Consequently, an experiment was carried out to measure the relation between access delay and ISW throughput on a commercial trunked system. The natural load on this system peaked at about 2.5 ISWs per second. To confirm that levels on the order

---

[6] To allow time for frequency synthesizers to settle in frequency when switching from the receive channel to the transmit channel (45-MHz offset in the 800-MHz band), an idle pattern is sent briefly before the real data are transmitted. This idle period is referred to as the "front porch."

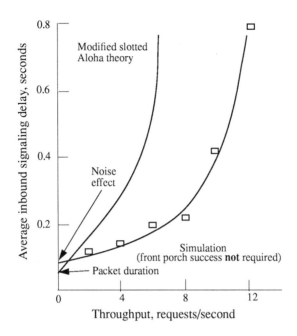

Figure 13.14 Simulated inbound signaling performance.

of 5 ISWs per second could be handled without detrimental effects, it was therefore necessary to add artificial signaling load. To this end, a radio was modified to output ISWs at switch-selectable rates between 0.5 and 4.9 ISWs per second. A special ISW type was chosen so that no airtime load penalty on voice channels would accrue from the experiment. To match the artificial load to the character of real load as closely as possible, equipment was developed to add both lognormal and Rayleigh fading to the transmitted signals (Figure 13.15). System gain was chosen so that the median average power received at the central site matched that observed for actual system users.

A second radio was used to probe the system, as shown in Figure 13.16. As before, Rayleigh fading is added, but in this case the lognormal variation was accomplished manually by keying up the radio the proper number of times for each particular switched-attenuator setting. For each keyup the logic analyzer timed the push-to-talk to receipt-of-grant (or busy) interval. The analyzer also generated a histogram of these times for each test run. The test runs consisted of 50 keyups under some particular artificial load setting. The true ISW throughput for each such run was measured via a counter on the decode line of the inbound recovery board at the central site.

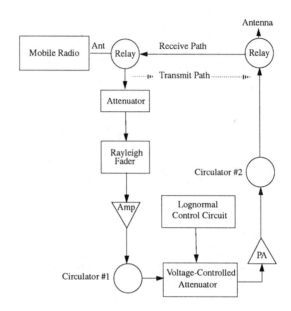

Figure 13.15 Artificial signaling load generation equipment.

13.4.6 Experiment Results

Figure 13.17 is a scatter plot of the average delay versus throughput for 24 trials of 50 keyups each. The artificial load was purposely set so that the trials fit into cases of relatively high or relatively low throughput. Furthermore, the high and low cases are each readily separable into two groups, yielding four basic data groups overall. The "A" symbols denote the average behaviors of these four groups. Those symbols are in reasonable accord with the solid line predicted via simulation.

In summary, a simulation technique has been described for evaluating the inbound signaling capacity of a Motorola-type trunked system. Simulations indicate that the benefit of capture outweighs the detriment of noise because the throughput of 6.5 signaling requests per second exceeds the 5.25 value predicted via simple theory. Experiments conducted on a commercial trunked system support key simulation assumptions and confirm that the practical throughput is in excess of 5 requests per second. This is sufficient for fully loaded 20-channel systems even with heavy use of new features. Even higher throughputs can be achieved via the use of sectored receivers.

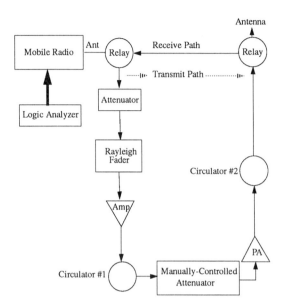

Figure 13.16 Probe radio equipment.

13.5 Analysis of Trunked System Performance with Load Shedding

This section concerns the potential gain in traffic-handling ability of a trunked system with load shedding. The stimulus for this analysis is the idea of handling overload traffic from a low-orbit land mobile communications satellite system via terrestrial SMRs.

Normally the peak workday traffic load determines the number of users supportable by a trunked system. Since the average load is generally substantially less than the peak load, the grade of service (on average) offered by the system substantially exceeds that deemed minimum acceptable. Now consider the impact of load shedding during busy periods. Load shedding means redirecting a fraction of the heavy traffic to some other service facility. This offers the possibility of equalizing the grade of service throughout the day, and hence handling more users per system.

Studies of trunked SMR systems have noted that traffic load over the workday is approximately Gaussian (normally) distributed. Similar behavior appears on large cellular systems and presumably would also appear on the low-orbit satellite system. Letting random variable ρ_1 denote the traffic load for a typical trunked system, one can therefore write $\rho_1 \sim N(\mu, \sigma)$, where μ equals the median traffic load throughout

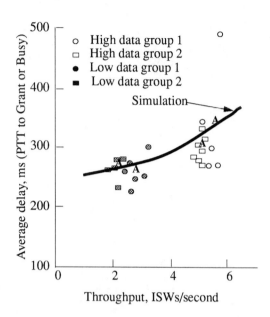

Figure 13.17 Measured inbound signaling performance.

the workday and σ equals the standard deviation. The peak workday load equals the median load plus some multiple of the standard deviation, $\rho_{1_{pk}} = \mu + k\sigma$. For the busy hour this multiplier is typically on the order of 1.3; a rationale being that the workday is about 10 hr and therefore the busy hour should approximate the 90 percentile of the distribution.

Now consider the impact of handling increased load with load shedding. It is convenient to let the load be scaled by a factor α, in which case the total boosted load is $\rho_2 \sim N(\alpha\mu, \alpha\sigma)$. Because of load shedding, however, only a truncated version of the boosted load is actually handled by the trunked system. To ensure that the grade of service on the trunked system never degrades below the minimum acceptable level, it must shed all loads in excess of the normal peak workday load for which it was originally designed. The actual load distribution is thus $\rho_3 \sim \rho_2| \leq \rho_{1_{pk}}$. The PDF for ρ_3 is of standard Gaussian form, but with a nonunity multiplier so the distribution integrates to unity over its range:

$$\int_{-\infty}^{\rho_{1_{pk}}} K \frac{1}{\sqrt{2\pi\alpha^2\sigma^2}} \exp[-\frac{(\rho_3 - \alpha\mu)^2}{2\alpha^2\sigma^2}] d\rho_3 = 1 \qquad (13.16)$$

With the help of relation 7.1.22 in Reference [24], this equation can be solved for K, yielding:

$$K = \frac{2}{1 + \operatorname{erf}(\frac{\rho_{1_{pk}} - \alpha\mu}{\sqrt{2}\alpha\sigma})} \qquad (13.17)$$

The average load under load shedding is given by:

$$<\rho_3> = \int_{-\infty}^{\rho_{1_{pk}}} K\rho_3 \frac{1}{\sqrt{2\pi\alpha^2\sigma^2}} \exp[-\frac{(\rho_3 - \alpha\mu)^2}{2\alpha^2\sigma^2}] d\rho_3 \qquad (13.18)$$

Using relation 5.(A4) in Reference [25] one obtains:

$$<\rho_3> = \alpha\mu - \sqrt{\frac{2}{\pi}}\alpha\sigma[\frac{\exp(-x^2)}{1 + \operatorname{erf}(x)}] \qquad (13.19)$$

where $x = (\rho_{1_{pk}} - \alpha\mu)/\sqrt{2}\alpha\sigma$. The boost in average load handled by the system with load shedding as compared to without load shedding is thus:

$$\begin{aligned}\delta &= \frac{<\rho_3>}{<\rho_1>} \\ &= \alpha\left\{1 - \sqrt{\frac{2}{\pi}}(\frac{\sigma}{\mu})[\frac{\exp(-x^2)}{1 + \operatorname{erf}(x)}]\right\} \end{aligned} \qquad (13.20)$$

Now x is a function of k, α, and (σ/μ). But for any particular system, k and (σ/μ) will be known, so really α is the sole variable. Likewise δ is only a function of α. For example, observations of a 19-channel trunked SMR atop the Sears building in Chicago indicate $k = 1.37$ and $(\sigma/\mu) = (2/2.3) = 0.163$. Similar values occur in cellular systems and a future low-orbit satellite system would therefore probably be similar. If one chooses $\alpha = 1.223$, so the average boosted load value ($<\rho_2>$) equals the normal system peak load, then $\delta = 1.064$. This implies a gain of just 6.4% in load-handling ability even though the peak load is boosted 22.3%. This is not a particularly attractive payback since it means most of the load increase must be handled by other facilities. Even if one doubles the offered load by setting $\alpha = 2$, the average load is boosted only 12%. In fact, the average load can only just approach the peak load without shedding, thus limiting system gain to 22.3%.

13.6 Analysis of Group Call Access Delay

13.6.1 Introduction

In first-come first-served ordered queues driven by random (Poisson) arrivals requiring exponential service times (i.e., the usual circumstances for which Erlang traffic theory applies), the waiting time distribution is also exponential [11]. The probability that the delay T exceeds any particular value t is given by:

$$P(T > t) = P_D \exp(-\alpha t) \qquad (13.21)$$

where P_D is the probability of nonzero delay and α is the ratio of the difference between the number of servers and the offered load to the average service time. An exponential character to initial access delay has also been observed for priority queued transmission trunked systems [1] and for message-trunked cellular systems where handoffs occur at rates comparable to or greater than the average message length [26]. Consequently, it is useful to consider a PDF of access delay of the form:

$$f(t) = \frac{1}{t_0}\exp(-\frac{t}{t_0}), t \geq 0 \qquad (13.22)$$

where t_0 represents the average access delay.

Group calls placed on a cellular system do not in general violate the Poisson arrival assumption of Erlang traffic theory. This is because, with a large number of cells, the probability of more than one unit requiring service from a particular sector-cell can be arbitrarily low; or, the control can simply lump requests from units in the same sector-cells into single server requests. However, the exponential service time assumption is violated because group calls normally do not really start until a server has been found for each member of the group. Consequently, one would substantially underestimate group call delay by simply equating it to the delay that an equivalent number of individual call units would experience. Instead, two adjustments are in order: (1) delays for each member of the group should be associated with the maximum delay seen for N draws of the underlying individual call delay distribution, where N represents the number of units in the group and (2) an accounting of the excess load due to tying up servers 1 through $N-1$ prior to the actual start of the call should be made.

13.6.2 Average Group Call Delay Relative to Individual Call Delay

Adjustment (1) can be handled through application of extreme value theory. The PDF for the i-th ordered value obtained via N draws of the exponential population noted above is given by [27]:

$$f_i(t) = \frac{N!}{(N-i)!(i-1)!}F(t)^{i-1}[1-F(t)]^{N-i}f(t) \qquad (13.23)$$

where $f(t)$ has the density form assumed above and $F(t)$ is the corresponding CDF, $[1-\exp(-t/t_0)]$. The expected delay for the i-th ordered delay out of N trials is thus given by:

$$<t_i> = \frac{N!(-1)^{i-1}t_0}{(N-i)!(i-1)!}\int_0^\infty x\exp[-(N-i+1)x][\exp(-x)-1]^{i-1}dx \qquad (13.24)$$

Traffic Engineering

where the substitution $x = (t/t_0)$ has been made. The integration can be carried out by expanding the bracketed exponential term via the binomial expansion:

$$(a+b)^n = \sum_{k=0}^{n} \binom{n}{k} a^{n-k} b^k \qquad (13.25)$$

with $a = \exp(-x)$ and $b = -1$. Interchanging the order of integration and summation leads to:

$$<t_i> = \frac{N!(-1)^{i-1} t_0}{(N-i)!(i-1)!} \sum_{k=0}^{i-1} \binom{i-1}{k} (-1)^k \int_0^\infty x \exp[-(N-k)x] dx \qquad (13.26)$$

The integration remaining can be recognized as proportional to the gamma function with an argument of 2 [24] (which nicely equates to unity), so that the final result is:

$$<t_i> = \frac{N!(-1)^{i-1} t_0}{(N-i)!(i-1)!} \sum_{k=0}^{i-1} \binom{i-1}{k} (-1)^k \frac{1}{(N-k)^2} \qquad (13.27)$$

Solutions of Equation (13.27) indicate that group call average delays exceed individual call average delays by factors of 1.83 for 3-unit groups (the originator plus two other units) and 3.02 for 11-unit groups (the originator plus ten other units). When uniformly weighted for N varying between 3 and 11, the delay factor is 2.53. This compares to a factor of 2.59 for $N = 7$, the average number of units in a group call. Such a close comparison indicates the equation is not particularly nonlinear and the average call delay of the average group size can be used; avoiding the more complicated exact calculation. Hence, one can expect group call delays to average about 2.5 sec under loading circumstances that give just 1-sec average individual call delay.

13.6.3 Group Call Excess Traffic Load

Having developed an expression for the average delay of the i-th ordered unit in a group call of N units, one can proceed to evaluate the amount of time servers are tied up waiting for the group call to actually begin. For the first unit assigned, on average a time of $<t_N> - <t_1>$ is wasted; for the second unit, the time wasted is $<t_N> - <t_2>$, etc., through unit $N-1$, which wastes $<t_N> - <t_{N-1}>$ call seconds. Thus, the average excess call seconds per group call involving N units equals the sum of the preceding wasted times divided by N (remember the final unit assigned contributes no waste).

For example, with an average of 7 units per group call, 375 group call units would normally be expected to generate a traffic load equal to that of 2625 individual call units. However, due to wasted server time awaiting assignment of all group members, one would predict a load boost of 2.02 times the individual call average access delay divided by the average message length. Simulations show the former to be 1.52 sec with 22 sec for the latter; hence, the predicted load boost is about 14%.

13.6.4 Application

Detailed call-tracking simulation runs have been conducted for group and individual calls on a 28-cell system (4-cell/6-sector reuse pattern). The resulting load/delay relationships are shown in Figure 13.18, along with approximate logarithmic curve-fits. The ratios between group call average delay and individual call average delay at equal handled load are as follows: 45 Erl yields 2.06, 50 Erl yields 2.42, 55 Erl yields 2.65, and 60 Erl yields 2.97. These values are reasonably in accord with previous analysis that predicted a ratio of 2.53. One reason for the simulation tendency toward greater ratios as the load increases is that its current logic requires multiple servers to be assigned in the same cell when multiple group members happen to reside there. In terms of excess load, note that 2625 individual units generated an average load of 47.7 Erl (cells 1, 14, and 17), whereas 375 group units averaging 7 units per group generated 55.5 Erl. Analysis indicates that the latter should really be viewed as 375 x 7 x 1.14 = 2992 individual units. The fact that 3000 individual units averaged 54.0 Erl indicates the excess load analysis is accurate.

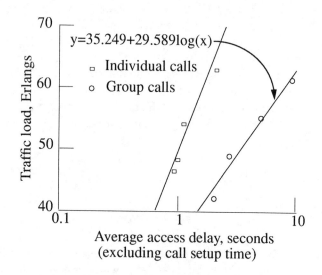

Figure 13.18 Load versus delay relationships for group and individual traffics.

In cases where individual and group traffic are present simultaneously, the following is suggested. First, estimate the average individual call delay assuming that all units are individual type. For example, if the system consists of 1500 individual units and 214 group units averaging 7 units per group, the effective unit count is $1500 + 214 \times 7 \approx 3000$. Simulation shows that the delay associated with such an individual call load averages 1.78 sec. Hence, the estimated group call average delay

Traffic Engineering

is 2.53 x 1.78 = 4.5 sec. Since both traffic types are equally likely, the composite average delay is 3.14 sec, which is nicely in agreement with the 3.04 value noted through simulation. The group call load boost factor is (2.02 x 1.78/22) times one-half, since only half the traffic is group call type. The result is about 8.1%, implying a traffic load of 58.4 Erl. Simulation gave a 54.0 Erl average with 4.2 Erl standard deviation.

13.7 Homework Problems

13.7.1 Problem 1

Consider the performance of a message-trunked system with 22-sec average length calls and 1.5 busy-hour calls per unit. Compare the traffic loads supportable on 5-channel and 20-channel trunked systems using the criteria (1) Erlang-B loss probability equals 5% and (2) average message delay equals 10% of the average message length.

Solution: With $c = 5$ and Prob(lost) = 5%, the single unknown in the Erlang-B formula is the traffic load ρ. Via trial and error one can show that $\rho = 2.22$ Erl is consistent with the specified parameters. Similarly, the answer when $c = 20$ is 15.3 Erl.

The normalized average delay (average delay divided by average message length) using the Erlang-C formula is a function only of traffic load when c is given. Setting this quantity to 0.1 and solving for ρ gives 2.89 Erl with $c = 5$ and 16.6 Erl with $c = 20$. The probabilities of delay are 21% and 9.7%, respectively.

Trunking efficiency can be defined as the ratio of traffic to server count. The 10% normalized delay loading criterion offers higher efficiency than the 5% blocking criterion and is therefore a less stringent grade-of-service criterion. With both criteria the efficiency rises substantially as c is boosted from 5 to 20 channels. This might lead one to ponder why the FCC loading criterion for trunked SMRs when they were first introduced was independent of system size. At that time, systems generally began with 5 channels and as the loading criterion was met additional channels could be licensed in groups of 5, up to a total of 20 channels.

The answer would appear to lie in the grade-of-service criterion. Dispatch users had traditionally operated under loads that produced average message delays comparable to average message length. With this criterion the Erlang-C load supported with 5 channels is 4.32 Erl; with 20 channels it is 19.2 Erl. These cases yield about 86% and 96% efficiency, respectively. Since typical busy-hour traffic involves 1.5 calls per unit per hour, averaging 22 sec, the per-unit busy-hour load is 0.0092 Erl. This implies the 5-channel system can support about 470 units (94 units per channel), and the 20-channel system can support about 2087 units (104 units per channel). The efficiency ratio and unit-to-channel ratio of only 10% are so small that the administrative complexity of loading rules which vary with system size is not justified.

However, the factor of load-smoothing puts this issue in a different light. Consider, for example, an average-hour load per unit of 0.00625 Erl (this is the observed 0.005-Erl airtime value for QTT scaled back up to a message-time value) and a standard deviation of load per unit of 0.025 Erl (again a QTT value scaled back up to message-time units). The load-smoothing viewpoint related busy-hour load to those parameters, as well as $k = 1.5$, N, and $\sqrt{N}$. Treating N as the unknown, one finds that a 5-channel system can support 550 units (110 units per channel) and meet an average normalized delay criterion of unity. Similarly, 20 channels can support 2750 units (138 units per channel). Hence, in practice, the units per channel can increase about 25% as system size grows from 5 to 20 channels. Thus, whether or not FCC loading rules force one to load large systems more heavily than small systems, the fact is that more load is possible or better grade of service is possible at the same loading. Either way there is good incentive for system operators to increase c.

13.7.2 Problem 2

Consider a 5-channel QTT system. How many dispatch units might be placed on such a system so that the nominal busy-hour grade of service is 10% normalized average message delay? What is the grade of service apt to be on the unfortunate (due to its overabundence of long-winded users) 10% worst such system?

Solution: The nominal busy-hour air-time load is given by:

$$\rho = 0.005N + 1.5(0.02)\sqrt{N}$$

Using an adjusted hold time parameter of $h = 7.6$ sec, Erlang-C theory indicates that 3.6 Erl is consistent with a normalized average delay of 10%. This represents the airtime traffic of 575 units. However, the standard devation about this nominal busy-hour load is given by:

$$\sqrt{(575)^2(0.001)^2 + (1.5)^2 575(0.002)^2 + 2(575)^{1.5}(0.001)(0.002)}$$

which equals 0.625 Erl. Thus, the 10% worst system with 575 units can be expected to experience busy-hour loads of 4.41 Erl. At such a load the normalized average delay is about 125%! Furthermore, the day-to-day busy-hour load variation on even the typical system causes about 25% peaking to occur once every two work-weeks (10 days). That is, even the nominal system can be expected to experience occasional loads on the order of 4.5 Erl. Hence, the FCC target loading of 100 units per channel seems prudent, at least for the smaller sized systems.

13.7.3 Problem 3

Trunking, as implemented in the United States, allows for subfleet groupings. Thus, it is possible for large fleets to simultaneously occupy several of one trunked system's

Traffic Engineering

servers; i.e., several of the subfleets may be engaged in traffic at the same time. However, in Japan current rules limit each fleet to a single call at a time. Thus, trunking in Japan currently involves two types of queueing delay: (1) the normal server access delay and (2) arrival delays for cases where new calls are generated when another call of the same fleet is already under way. Comment on the differences between call delay distributions for systems with and without subfleeting operating at identical traffic loads.

Solution: Erlang-C theory applies reasonably well to message-trunked systems, provided the hang time included to achieve message trunking is relatively small compared to the average message length. Hence, performance evaluation of systems with subfleeting is straightforward. For example, consider such a system with 10 voice channels loaded to 7.7 Erl by exponentially distributed messages averaging 20 sec, with 3-sec hang times. Using 23 sec as the effective server hold time, one finds that the probability of access delay is about 34.9% (Equation (13.12)), the average delay of delayed calls is about 10.1 sec, and the average delay considering all calls is about 3.5 sec (Equation (13.13)). Figure 13.19 shows Monte Carlo simulation results for the system under consideration. This allows the constant hang time effect to be modeled exactly. The effect is obviously small since simulated results for the three aforementioned performance factors are 36.3%, 10.8 sec, and 3.9 sec, respectively. Also, the straight-line character of Figure 13.19 indicates that delay is approximately

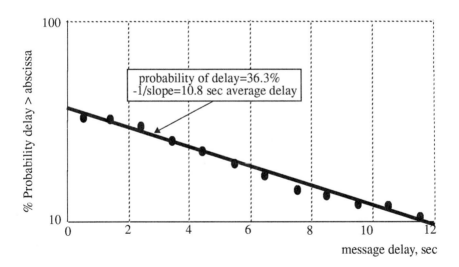

Figure 13.19 Cumulative probability of access delay via simulation of trunked system with subfleeting.

exponential in character, just as Erlang-C theory would predict (Equation (13.15)).

The cascaded queueing implied by current Japanese rules (i.e., without subfleeting) does not lend itself to analysis and Monte Carlo simulation is therefore essential. Figure 13.20 shows the distribution of arrival and access delays assuming that the total number of units is broken down into fleets of 15 units each (an average value for trunked fleet size in Japan). Because the system load remains constant, there is little change in the access delay curve from Figure 13.19 (here the three performance parameters are 35.1%, 9.2 sec, and 3.2 sec, respectively). Also, because the fleet size is not particularly large, there is not too high a probability of a call being offered at a time someone else in the fleet is already engaged in traffic (5.2%). Nonetheless,

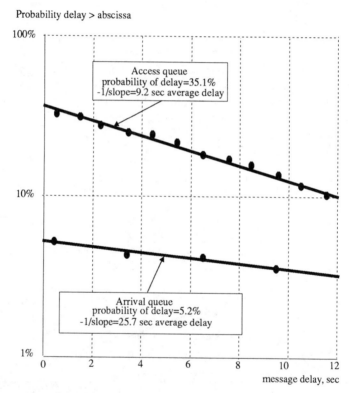

Figure 13.20 Cumulative probability of arrival and access delays via simulation of trunked system without subfleeting.

---

[7]Although fleets can be divided into different talk groups for privacy or other reasons, only one talk group at a time can access the system. Hence, for purposes of traffic analysis, operation is as though subfleeting is absent.

Traffic Engineering

since messages average about 23 sec, when such a situation does occur the delay is quite lengthy (23.8 sec). The straight-line character of arrival delay in Figure 13.20 indicates that it is also approximately exponential in character (the "eyeball" slope information leads one to estimate the average arrival delay as 25.7 sec). Total average delay per call is the sum of average access delay and average arrival delay, or 3.2 + 1.2 = 4.4 sec, based on all calls. This is up somewhat from the situation where subfleeting is allowed since two delay components are now present. The probability of experiencing truly long delays goes up much more dramatically when subfleeting is absent.

To quantify this let X_1 be a random variable representing access delay. Its PDF will have the form:

$$f_{X_1}(t) = \alpha\delta(t) + (1-\alpha)\frac{1}{h_1}\exp(-\frac{t}{h_1}) \qquad (13.28)$$

where t represents the delay time, α represents the probability of zero access delay, δ is the impulse function, and h_1 is the average access delay or hold time. Similarly, let X_2 be a random variable representing arrival delay, with PDF of the form:

$$f_{X_2}(t) = \beta\delta(t) + (1-\beta)\frac{1}{h_2}\exp(-\frac{t}{h_2}) \qquad (13.29)$$

where β represents the probability of zero arrival delay and h_2 is the average arrival delay. Based on the simulation results presented in Figures 13.19 and 13.20, the parameter values would be: (case 1, subfleeting present) $\alpha = 0.637$, $h_1 = 10.1$, $\beta = 1$, and $h_2 =$ "don't care" and (case 2, subfleeting absent) $\alpha = 0.649$, $h_1 = 9.2$, $\beta = 0.948$, and $h_2 = 23.8$.

The PDF for total call delay equals the convolution of the density functions for the two independent contributors: arrival and access delays. This is composed of four parts: (1) the convolution of two exponential functions, (2) the product of two time-zero weighted impulse functions, (3) the convolution of a time-zero weighted impulse function with exponential one, and (4) the convolution of a time-zero weighted impulse function with exponential two. Due to the sifting property of the impulse function, parts 2, 3, and 4 are trivial. Part 1 does require some effort [28]. The final answer is:

$$\begin{aligned}f_{Z=X_1+X_2}(t) &= \alpha\beta\delta(t) + \frac{(1-\alpha)(1-\beta)}{h_1-h_2}[\exp(-\frac{t}{h_1}) - \exp(-\frac{t}{h_2})] \\ &+ \alpha(1-\beta)\frac{1}{h_2}\exp(-\frac{t}{h_2}) \\ &+ (1-\alpha)\beta\frac{1}{h_1}\exp(-\frac{t}{h_1}) \end{aligned} \qquad (13.30)$$

where $h_1 \neq h_2$ and $t \geq 0$. Integrating the preceding equation between zero and arbitrary time t gives the expression for the probability that total call delay does not exceed that time:

$$\begin{aligned}\text{Prob(delay} < t) &= \alpha\beta + \frac{(1-\alpha)(1-\beta)}{h_1 - h_2}\{h_1[1 - \exp(-\frac{t}{h_1})] \\ &\quad - h_2[1 - \exp(-\frac{t}{h_2})]\} \\ &\quad + \alpha(1-\beta)[1 - \exp(-\frac{t}{h_2})] \\ &\quad + (1-\alpha)\beta[1 - \exp(-\frac{t}{h_1})]\end{aligned} \qquad (13.31)$$

References

[1] Hess, G. and J. Cohn, "Communication Load and Delay in Mobile Trunked Systems," *31st IEEE Vehicular Technology Conference*, April 1981, Washington, D.C., pp. 269-273.

[2] FCC Public Notice #31774, "Summary of Traffic Loading Data of Applications for Additional Channels for DPLMR," May 20, 1980.

[3] Lum, E. and K. Zdunek, "Performance of Trunked Systems with Interconnect," *ENTELEC Conference Proceedings*, March 1984.

[4] Reeves, C. M., "An Overview of Trunking Techniques in Mobile Radio Systems," *30th IEEE Vehicular Technology Conference*, September 1980, Dearborn, MI, session E-3, paper 5, pp. 1-5.

[5] Thro, S., "Trunking: A New Dimension in Fleet Dispatch Communications," *INTELCOM '79*, Dallas, TX, 1979, pp. 277-281.

[6] Grindahl, M. L., "Automatic Communications Control System," *30th IEEE Vehicular Technology Conference*, September 1980, Dearborn, MI, session E-3, paper 2, pp. 1-5.

[7] Hess, G., "Estimation of Peak Load On Trunked Repeater Systems," *32nd IEEE Vehicular Technology Conference*, San Diego, CA, May 1982, pp. 331-334.

[8] Gumbel, E., "Statistical Theory of Extreme Values and Some Practical Applications," *NBS Applied Mathematics Series, 33*.

[9] Barnes, D. and J. O'Connor, "Extreme Value Traffic Applications in Small Offices," *International Communications Conference*, 1982, pp. 31.1.1-31.1.6.

[10] Kleinrock, L., *Queueing Systems*, 2 volumes, New York, NY, John Wiley & Sons, 1975.

[11] Beckmann, P., *Introduction to Elementary Queuing Theory and Telephone Traffic*, Boulder, CO, Golem Press, 1968.

[12] Cooper, R. B., *Introduction to Queueing Theory*, 2nd edition, Elsevier-North Holland, 1981.

[13] Descloux, A., *Delay Tables for Finite- and Infinite-Source Systems*, New York, NY, McGraw-Hill Book Co., 1962.

[14] Dill, G. D. and G. D. Gordon, "Efficient Computation of Erlang Loss Functions," *COMSAT Tech. Rev.*, Vol. 8, No. 2, Fall 1978, pp. 353-370.

[15] Cohen, J. W., "The Generalized Engset Formulae," *Philips Telecommunication Review*, Vol. 18, No. 4, November 1957, pp. 158-170.

[16] Mitchell, R. F. and C. K. Davis, "Traffic Handling Capability of Trunked Land Mobile Radio Systems," *ICC*, Boston, MA, 1979, pp. 57 2.1-5.

[17] Riordan, J., "Delay Curves for Calls Served at Random," *Bell System Technical Journal*, Vol. 32, January 1953, pp. 100-119.

[18] Nesenbergs, M., "A Hybrid of Erlang B and C Formulas and Its Applications," *IEEE Tr. on Comm.*, Vol. COM-27, No. 1, January 1979, pp. 59-68.

[19] Abramson, N., "The ALOHA System - Another Alternative for Computer Communications," *1970 Fall Joint Computer Conference*, AFIPS Conf. Proc., Vol. 37, pp. 281-285.

[20] Kleinrock, L. and F. Tobagi, "Packet Switching in Radio Channel: Part I - Carrier Sense Multiple- Access Modes and Their Throughput-Delay Characteristics," *IEEE Tr. on Comm.*, Vol. COM-23, No. 12, December 1975, pp. 1400-1416.

[21] Kleinrock, L. and S. S. Lam, "Packet Switching in a Multiaccess Broadcast Channel: Performance Evaluation," *IEEE Tr. on Comm.*, Vol. COM-23, No. 4, April 1975, pp. 410-423.

[22] Namislo, C., "Analysis of Mobile Radio Slotted ALOHA Networks," *IEEE Tr. on Veh. Tech.*, Vol. VT-33, No. 3, August 1984, pp. 199-204.

[23] Hata, M., "Empirical Formula for Propagation Loss in Land Mobile Radio Services," *IEEE Tr. on Veh. Tech.*, Vol. VT-29, No. 3, August 1980, pp. 317-325.

[24] Abramowitz, M. and I. Stegun (editors), *Handbook of Mathematical Functions*, New York, NY, Dover Publications, Inc., 1972.

[25] Ng, E. and M. Geller, " A Table of Integrals of the Error Functions," *Journal of Research of the National Bureau of Standards*, B. Mathematical Sciences Vol. 73B, No. 1, January-March 1969.

[26] Hess, G. and B. Leung, CELLSIMCT6, A Call Tracking Cellular Simulation Program, Motorola Internal Technical Report, October 1990.

[27] Ross, S., *A First Course in Probability*, New York, NY, MacMillan Publishing Co., 1976.

[28] Papoulis, A., *Probability, Random Variables, and Stochastic Processes*, New York, NY, McGraw-Hill Book Co., 1965.

Chapter 14

Land-Mobile Satellite Systems

This chapter deals with direct communication between land-mobile radios and communications satellites. A brief history of this possibility is given, followed by a simple cost-performance optimization using the Lagrangian multipler technique. The resurgence of interest in *low earth orbit* (LEO) satellites, as opposed to the very high orbit required for geostationary operation, is the stimulus for detailed consideration of fade margin requirements. Sample satellite-subscriber terminal LEO link budgets are included.

14.1 History

The first communications satellite systems were characterized by spacecraft with very limited transmit power and antenna gain. To make up for such shortcomings, ground terminals consisted of huge antennas and high-power transmitters. A major milestone was the achievement of geostationary orbit, from which nearly a third of the globe was continuously visible and satellite tracking was not required of those huge ground antennas. Launch capability rapidly improved and, particularly with the development of *Applications Technology Satellite number 6* (ATS-6), it became clear that what used to be the ground terminal could in essence be launched into orbit. Consequently, the radio needed on the ground to communicate with the satellite was not out of line with the typical land-mobile terrestrial radio. The *National Aeronautics and Space Administration* (NASA) Goddard Space Flight Center led the initial experimentation and lobbied the FCC for a *land mobile satellite service* (LMSS) [1, 2, 3, 4, 5]. A substantial Canadian effort began by their Department of Communications shortly thereafter [6, 7].

Commercial realization of LMSS was slowed by NASA's attraction to the then new terrestrial cellular service and the natural ability of satellites to extend cellular into rural areas [8, 9]. Unfortunately, the 30-kHz FM already chosen for terrestrial cellular operation is too power and bandwidth extravagant for satellite links. Substantial protocol and time delay issues also exist. Hence true compatibility is not possible and at best dual-mode radios would be required [10, 11].

The incompatibility with terrestrial cellular removed a good argument for LMSS spectrum in the 800-MHz band. Nonetheless, most LMSS applicants insisted to the FCC that truly mobile service (as opposed to transportable service, for which fairly high antenna gains would be practical) could only be provided via an 800-MHz allocation. This, because ostensibly the link margin required at L-band (the other allocation candidate, at roughly 1.5 GHz) was prohibitively high [12].

Around the time that applicants were filing for LMSS, NASA interest rekindled at the *Jet Propulsion Laboratories* (JPL) [13, 14]. Unfortunately, the early, extensive propagation testing by JPL did not address L-band performance and the belief persisted that an 800 MHz allocation was essential. That is until the FCC, lobbied also by many who disagreed [15, 16], decided in favor of L-band. The serious applicants formed the *American Mobile Satellite Consortium* (AMSC) and both they and the Canadians, through *Telesat Mobile Incorporated* (TMI), are currently pursuing a geostationary LMSS at L-band [17].

To a large degree because of the spectrum argument, and the impact of the later shuttle disaster on launch plans, AMSC is neither alone nor may they even be the first to implement commercial LMSS. Refer to Session A papers in Reference [18] for a list of other participants. Furthermore, not all participants are limiting themselves to geostationary orbits. Perhaps spurred by the failure of the space shuttle to significantly lower launch costs, researchers have been re-examining LEO with a small is beautiful perspective. Amateur radio operators have already developed and put to use a number of store-and-forward data satellites. Phased constellations of small (and therefore presumably inexpensive to build and launch) satellites have been discussed in the literature [19] and Motorola has even announced its ambitious IRIDIUM concept to provide direct portable-satellite communications with geographic reuse to every point on the globe [20].

14.2 Multiple Beam, Geostationary Satellite System Cost and Performance Optimization

The relationship between cost and performance of geostationary land-mobile satellite systems has been the subject of published articles, technical reports, and commercial offerings for quite some time (for example, see References [21, 22, 23, 24, 25, 26]). In general, these references involve elaborate simulations whose complexity is perhaps unwarranted for initial tradeoff studies. A far simpler approach has been described by McGarty and Warner [27] for a multiple beam, time division multiplexed, geostationary satellite system. Their approach is based on the use of simple empirical relations relating key parameters such as satellite mass and launch cost, derived from actual case histories [28, 29]. The purpose of this section is to apply this simpler approach to a multiple beam, frequency division multiplexed geostationary land-mobile satellite system.

14.2.1 Cost-Performance Model

The cost of the space segment portion of the system is directly related to the mass of the satellite communications subsystem, comprised of transponders, an antenna, antenna feeds, and an interbeam switching matrix (to allow connection of users located in different antenna footprints; i.e., reached via different antenna beams). The number of beams required to cover an area spanning angles of θ_w and θ_h in width and height, respectively, as viewed from geostationary orbit is given by:

$$N = [\frac{\sqrt{\theta_w \theta_h} f D_s}{70c}]^2 \tag{14.1}$$

where f is the link frequency, D_s is the satellite antenna diameter (reflector dish antenna assumed), and c is the speed of light. The space segment cost can then be modeled as:

$$C_{ss} = k_s(Mk_p P_s + k_g D_s^2 + MNk_f + MN^2 k_{si}) \tag{14.2}$$

where k_s equals the satellite cost per communications subsystem mass (including launch, redundancy, and control), M equals the number of transponders (each full coverage), k_p equals the mass-to-power coefficient for satellite RF power generation, P_s equals the RF output power per transponder, k_g equals the mass-to-size coefficient for the satellite antenna, k_f equals the mass of a single antenna feed, and k_{si} equals the switching interconnect mass per beam and per transponder.

The earth terminal cost can be modeled as:

$$C_e = l_p P_e f^a + l_g G_e + l_m \tag{14.3}$$

where l_p equals the cost per watt of RF power P_e, a equals the frequency exponent of RF power expense, l_g equals the cost of antenna gain G_e (linear, not decibel), and l_m equals fixed cost items, including installation and supporting base equipment. This model is like that in Reference [27] except for its assumption that noise figure is sufficiently large for it to be essentially fixed in cost.

Total system cost can now be expressed in terms of four parameters, two related to the space segment (P_s, D_s) and two related to the earth segment (P_e, G_e):

$$\begin{aligned} C_T &= C_{ss} + QMC_e \\ &= [D_s^4(Mk_s k_{si}' f^4) + D_s^2(Mk_s k_f' f^2 + k_s k_g) + (Mk_s k_p P_s)] \\ &\quad + QM(l_p P_e f^a + l_g G_e + l_m) \end{aligned} \tag{14.4}$$

where $k_f' = k_f(\sqrt{\theta_w \theta_h}/70c)^2$, $k_{si}' = k_{si}(\sqrt{\theta_w \theta_h}/70c)^4$, and Q equals the number of users per channel.

An additional constraint is necessary; namely, the overall link quality required, including propagation margin. This can be expressed as a carrier-to-noise power requirement through:

$$\left(\frac{C}{N}\right) = \left[\frac{1}{(C/N)_u} + \frac{1}{(C/N)_d}\right]^{-1} \quad (14.5)$$

where the uplink carrier-to-noise power is $(C/N)_u = k_u P_e G_e D_s^2$, the downlink carrier-to-noise power is $(C/N)_d = k_d P_s D_s^2 G_e$, $k_i = (\eta_s/16kT_j B_j r_i^2)$, $(i,j) = (u,s)$ or (d,e), η_s equals the satellite antenna efficiency, k equals Boltzmann's constant, T_j equals the effective noise temperature at j, B_j equals the channel bandwidth at j, and r_i equals the path length for path i.

14.2.2 LaGrangian Multiplier Optimization

Total system cost can be minimized subject to some link quality requirement by use of the Lagrangian multiplier technique [30]. Define the function:

$$\begin{aligned}F(P_s, D_s, P_e, G_e, \lambda) &= D_s^4(Mk_s k'_{si} f^4) \\ &+ D_s^2(Mk_s k'_f f^2 + k_s k_g) + (Mk_s k_p P_s) \\ &+ QM[P_e(l_p f^a) + G_e(l_g) + (l_m)] \\ &+ \lambda[\left(\frac{C}{N}\right)_{min} - \left(\frac{1}{k_u P_e G_e D_s^2} + \frac{1}{k_d P_s D_s^2 G_e}\right)] \quad (14.6)\end{aligned}$$

where λ is an artificial variable introduced to allow incorporation of the link quality constraint. Note that the factor λ multiplies is zero when the constraint is exactly met and the defined function simply reduces to the total cost equation introduced earlier. The optimum parameter choices (those that minimize total system cost yet meet the minimum overall carrier-to-noise requirement) are found by differentiating the function F with respect to each parameter and solving the resulting set of simultaneous equations with all derivatives set to zero. This optimum solution is:

$$P_e = P_s \left(\frac{k_d k_p k_s}{k_u l_p Q f^a}\right)^{\frac{1}{2}}$$

$$P_s = G_e \left(\frac{l_g Q}{k_s k_p X}\right)$$

$$G_e = D_s^{-1} X \left[\frac{\left(\frac{C}{N}\right)_{min} k_s k_p}{k_d l_g Q}\right]^{\frac{1}{2}}$$

$$D_s^5(2Mk_s k'_{si} f^4) + D_s^3(k_s k_g + Mk_s k'_f f^2) = XM \left[\frac{\left(\frac{C}{N}\right)_{min} k_s k_p l_g Q}{k_d}\right]^{\frac{1}{2}}$$

$$(14.7)$$

where:

$$X = \left[1 + \left(\frac{k_d l_p Q f^a}{k_u k_p k_s}\right)^{\frac{1}{2}}\right] \quad (14.8)$$

Land-Mobile Satellite Systems

If mobile gain is allowed to vary freely it may exceed that possible without antenna steering. Regardless of whether such steering is accomplished electrically or mechanically, it involves substantial expense and thus would imply a major increase in the l_g parameter. Consequently, a solution for which G_e is some constant value is of interest. For example, G_e can be as high as 7 dBic and still permit operation over the entire continental United States with at most 3-dB pointing loss. This specialized solution is:

$$P_e = P_s \left(\frac{k_d k_p k_s}{k_u l_p Q f^a} \right)^{\frac{1}{2}}$$

$$P_s = D_s^{-2} X \left[\frac{(\frac{C}{N})_{min}}{k_d G_e} \right]$$

$$D_s^6 (2M k_s k'_{si} f^4) D_s^4 (k_s k_g + M k_s k'_f f^2) = X^2 M \left[\frac{(\frac{C}{N})_{min} k_s k_p}{k_d G_e} \right] \quad (14.9)$$

Representative parameter values are as follows: $\theta_w = 7.5$ deg and $\theta_h = 3.5$ deg, corresponding to continental United States coverage; $k_s = 2.62E5$ 1978 U.S. dollars per kilogram, based on a three-satellite strategy of one in orbit, one standby in orbit, and one standby on ground; $k_p = 0.45$ kg/W; $k_g = 0.8$ kg/m^2; $k_f = 0.1$ kg; $k_{si} = 0.025$ kg; $l_p = 20$ 1978 U.S. dollars per watt per gigahertz raised to the power α; $\alpha = 1.3$; $l_g = 40$ 1978 U.S. dollars; $l_m = 2280$ 1978 U.S. dollars; $\eta_s = 0.5$; $T_j = 627$K (5 dB noise figure); $B_j = 10$ kHz, typical of narrowband FM used in terrestrial land-mobile communications; and $r_i = 3.6E7$ m.

14.3 LEO Satellite-to-Subscriber Unit Fade-Margin Analysis

14.3.1 Preliminaries

NASA and JPL have spent a considerable amount of time since the late 1970s studying the land mobile satellite service. One outcome of this effort has been the development of an interleaved trellis-coded differential 8-PSK modem. While some other modulation scheme might ultimately be applied, the performance of the JPL modem serves as a good indicator of signal-to-noise performance possible on satellite-to-subscriber unit links.

JPL has concentrated on 4800-bps speech digitization techniques with rate 2/3 coding, for a net rate of 7200 bps. This is currently the lowest rate possible for providing communications-quality speech. Such quality can be maintained through BERs up to at least 1%. Of course, channel supervision needs make 8000 bps the likely total link data rate.

14.3.2 Ideal Conditions

Under nonfaded conditions against additive white Gaussian noise, simulations indicate the JPL modem achieves 1% BER with a bit energy per noise spectral density ratio (E_b/N_0) of 7.3 dB [31]. Implementation loss according to Reference [32] is no more than 1 dB, excluding Doppler tracking and time synchronization errors (items discussed in Reference [33]). Hence an E_b/N_0 on the order of 8.3 dB is required under ideal conditions.

Of course, interference power must also be accounted for in calculating N_0. Intermodulation products in efficient multicarrier "linear" amplifiers are unlikely to be more than 20 dB below the individual carrier level. Use of time division multiplexing, rather than frequency division multiplexing, might avert this degradation. In systems that strive for frequency reuse via multiple antennas beams, the coupling of each cochannel signal through antenna sidelobes must also be included.

14.3.3 Unshadowed Conditions

Mobile-satellite propagation experiments (for example, see Reference [34]) have shown the distribution of signal strength when the line of sight is not blocked to be approximately Rician in nature, with a diffuse-to-direct power ratio K typically equal to -11 dB. The 90 percentile of such a distribution is about 3 dB below line-of-sight power. Hence, an E_b/N_0 on the order of 11.3 dB is required. This refers to commercial applications. For public safety uses, 95% is the usual coverage goal.

14.3.4 Unshadowed Conditions, Enhanced Scatter

It is not particularly unusual for the diffuse component of the satellite signal to be enhanced, due to reflections from surrounding terrain or bodies of water. Reference [35] reports a K value of just -5 dB for the Pacific Coast Highway. Here an additional margin of about 5 dB is necessary for 90% coverage; hence, the E_b/N_0 requirement increases to about 13.3 dB.

14.3.5 Mixed Shadowed and Unshadowed Conditions

Reference [34] describes a propagation model for arbitrary percentages of shadowed and unshadowed conditions along the route to be covered. Choosing as representative the CRC MARECS A run in rural, forested Canada (35% shadowed), one finds that an additional margin of 11 dB is necessary for 90% coverage. The E_b/N_0 requirement thus becomes about 19.3 dB.

Land-Mobile Satellite Systems 293

14.3.6 In-Building Conditions

Inside buildings the received signal character tends to become Rayleigh in nature. Reference [36] indicates that for 1% BER in Rayleigh fading an E_b/N_0 of about 14 dB is necessary. Allowing 1 dB for implementation loss means that really 15 dB is necessary.

For typical houses at L-band, Reference [37] reports penetration losses averaging 6.7 dB. To cover this an E_b/N_0 of 21.7 dB would be needed. Reference [38] cites 13.1-dB penetration loss at 800 MHz for the average suburban building. Assuming L-band performance is comparable implies an E_b/N_0 requirement of 28.1 dB. Finally, Walker [38] cites 18-dB penetration loss at 800 MHz for the average urban building. This implies E_b/N_0 must be on the order of 33 dB just to cover the average building.

14.3.7 Summary

Mobile-satellite path conditions	Required E_b/N_0 (dB)	Fade margin (dB)
Ideal	8.3	0
Unshadowed	11.3	3
Unshadowed/ enhanced scatter	13.3	5
Mixed shadowed/ unshadowed	19.3	11
Inbuilding, house	21.7	13.4
Inbuilding, suburban building	28.1	19.8
Inbuilding, urban building	33.0	24.7

14.3.8 Impact of Elevation Angle on Fade Margin

The preceding analysis indicates that a fade margin of 11 dB is appropriate for rural, mobile commercial applications (i.e., 11 dB equals the difference between the mixed shadowed/unshadowed E_b/N_0 requirement for 90% coverage and E_b/N_0 static sensitivity). This fade margin is geared toward relatively modest elevation angles. At higher elevation angles shadowing is less likely and thus less margin should be sufficient to meet the 90% coverage target.

To date, published margin versus elevation angle information has been quite limited[1]. From ATS-6 observations at L-band between 20- and 40-deg elevation, no

---

[1] The situation is, however, quite dynamic. Since writing this section a number of relevant publications have appeared; for example, References [39, 40, 41].

significant effect is apparent [5]. A number of others have reported satellite path loss measurements, and occasionally they have grouped their data by elevation angle. However, no individual report is sufficiently extensive to justify conclusions nor are the reports directly poolable. The best indication so far is perhaps Reference [42]. This paper groups data for elevation angles of 30 to 45, 45 to 60, and >60 deg and indicates that a decrease of about 2 dB occurs between each group, moving low to high. Based on this, one might suggest a margin requirement that drops linearly from 12 dB at 10-deg elevation to 6 dB over head.

14.4 Sample LEO Mobile Satellite Link Budget

14.4.1 Assumptions

Assume that gateway-to-satellite links have sufficient link margin so that the overall gateway-to-subscriber unit link is characterized essentially by that of the satellite-to-subscriber link only.

A link frequency of 1.5585 GHz is assumed. This represents the top of the current downlink allocation for the land mobile satellite service in the United States (the uplink is 101.5 MHz higher). This is not the same as the present *World Administrative Radio Conference* (WARC) allocation, although the free-space loss due to frequency differences would be quite minor. Not minor would be the need to switch the frequency bands of operation as the satellites orbit the earth and pass over countries with differing frequency allocations.

Shown next are candidate downlink and uplink analyses corresponding to the outermost footprints for a 36-cell frequency reuse satellite system using frequency division multiplex[2] and designed to communicate with mobile subscriber units down to elevation angles of 10 deg.

14.4.2 Spacecraft-to-Land Mobile Terminal

```
*****************************
* A. Nominal Received Power *
*****************************

Spacecraft average power per active channel     =  24.05 dBm
  (254 mW/ch, this value will vary with
  cell position)
Spacecraft diplexer loss                        =  -1.50 dB
```

---

[2]Some LEO plans now call for the use of time division multiplex; spread-spectrum arrangements have also been considered.

Land-Mobile Satellite Systems

```
Spacecraft antenna gain                           =   23.30 dBic
  (this value will vary with cell position)
Spacecraft antenna scan loss                      =    0.00 dB
  (due to 6 panel phased array design which
   minimizes off-boresight pointing requirement)
Spacecraft antenna taper loss                     =   -0.15 dB
  (Taylor weighting, sufficient for cochannel
   beam isolation?)
Spacecraft antenna pointing loss                  =   -0.72 dB
  (this value depends on the handoff strategy)
Spacecraft antenna polarization loss              =   -0.50 dB
  (is this manufacturable?)
-----------------------------------------------------------------
Spacecraft EIRP                                   =   44.48 dBmic

Free-space path loss                              = -163.66 dB
  (this value will vary with satellite altitude
   chosen; above number pertains to ~ 2300 km
   path [413.5 nmi altitude])
Atmospheric loss                                  =   -0.20 dB
  (confirm this with CCIR models; will be most
   pronounced at low elevation angles, hence a
   function of cell position)
Mobile antenna gain                               =    0.00 dBic
  (bifilar helix as selected for Intelsat
   Standard-C; Adams-Russell AN-710 Global
   Positioning System (GPS) antenna spec is for
   -2 to +2 dBic over elevation angles > 10 deg;
   note that portable antenna would be much less;
   estimated as -4.85 dBic in the speech position,
   and just -10.85 dBic when stowed)
Mobile antenna polarization loss                  =   -0.50 dB
  (AN-710 spec calls for < 3 dB axial ratio)
Mobile antenna pointing loss                      =    0.00 dB
-----------------------------------------------------------------
Nominal received power, C, referenced to
subscriber antenna port/diplexer input            = -119.88 dBm
```

```
***********************************
* B. Nominal Receiver Noise Power *
***********************************


Receiver NF, excluding diplexer loss           =    2.00 dB
Receiver diplexer loss                         =    1.50 dB
Net receiver NF                                =    3.50 dB
Receiver noise temperature                     =  359.23 K
External noise temperature                     =  145.00 K
  (due to "warm" earth seen through sidelobes
    of low gain antenna)
Effective receiver noise temperature           =  504.20 K
Nominal receiver noise power                   = -134.09 dBm
  (assuming 5.6 kHz noise bandwidth)


******************
* C. Nominal C/N *                             =   14.21 dB
******************


***********************************
* D. Interference Considerations *
***********************************

Spacecraft antenna reuse beam-to-reuse beam
isolation                                      =   25.00 dB
  (must verify with antenna manufacturer)
Effective impact other cochannel users         =   -7.00 dB
  (based on reuse pattern of 6 there will be up
    to 5 cochannel users [11 for intersatellite
    handoff situations]; however, their coupling
    patterns will not necessarily coincide so
    penalty may not be as severe as 7 dB [10.4 dB])
-----------------------------------------------------------
Nominal C/I1                                   =   18.00 dB

Nominal C/I2                                   =   20.00 dB
  (Spacecraft cochannel IM product level, this
    pertains to frequency division multiplex;
    spec can be improved at the expense of amplifier
    efficiency; referenced value consistent with
    SPAR UHF design reported at 1986 ICC by
    Whittaker, Brassard, and Butterworth)
```

Land-Mobile Satellite Systems

```
***************************
* E. Required Link Quality *
***************************
```

Eb/N0 (1% static BER, coherent detection, CPQPSK) =	4.50 dB
R/Bn (8000 kbps over noise bandwidth of 5600 Hz) =	1.55 dB
Modem implementation loss =	1.00 dB
Minimum C/N =	7.05 dB
(same as minimum C/N+I1+I2, assuming interference is noiselike in nature)	

```
*************************
* F. Margin Evaluations *
*************************
```

Nominal C/N+I1+I2	= 11.95 dB
Required C/N+I1+I2	= 7.05 dB
Link margin	= 4.90 dB
Nominal C/N	= 14.21 dB
Minimum C/N to meet C/N+I1+I2 requirement	= 7.66 dB
Fade margin	= 6.55 dB
Required fade margin	= 12.00 dB
(for 90%-tile coverage in mixed shadowed/unshadowed environment; this value will be a function of elevation and hence cell position)	

The 5.45-dB shortfall indicated above must be made up by increased satellite transmit power. Sufficiency of the overall link margin of 4.9 dB depends on the maintainability of C/I1 and C/I2. Links to portables would of course be even worse due to less antenna gain, poorer circularity, possible body blockage, and attempts to operate inbuilding.

14.4.3 Land Mobile Terminal-to-Spacecraft

```
*****************************
* A. Nominal Received Power *
*****************************
```

Subscriber transmit power (10 W for mobile, 600 mW for portable, diplexer loss included)	= 40.00 dBm (27.78)
Mobile antenna gain (bifilar helix as selected for Intelsat Standard-C; portable antenna would be 7.85 dB less when unit in speech position, 13.85 dB less when stowed)	= 0.00 dBic (-4.85)
Mobile antenna polarization loss	= -0.50 dB
Mobile antenna pointing loss	= 0.00 dB

Mobile/Portable EIRP	= 39.50 dBmic (22.43)
Spacecraft antenna gain (this value will vary with cell position)	= 23.30 dBic
Spacecraft antenna scan loss (due to 6 panel phased array design which minimizes off-boresight pointing requirement)	= 0.00 dB
Spacecraft antenna taper loss (Taylor weighting, sufficient for cochannel beam isolation?)	= -0.15 dB
Spacecraft antenna pointing loss (this value depends on the handoff strategy)	= -0.72 dB
Spacecraft antenna polarization loss (is this manufacturable?)	= -0.50 dB
Free-space path loss (this value will vary with satellite altitude chosen; above number pertains to ~ 2300 km path [413.5 nmi altitude])	=-163.66 dB
Atmospheric loss (confirm this with CCIR models; will be most pronounced at low elevation angles, hence a function of cell position)	= -0.2 dB

```
Nominal received power, C, referenced to         =-102.43 dBm
spacecraft antenna port/diplexer input           (-119.50)
```

```
**********************************
* B. Nominal Receiver Noise Power *
**********************************
```

```
Receiver NF, excluding diplexer loss     =   3.00 dB
Receiver diplexer loss                   =   1.50 dB
Net receiver NF                          =   4.50 dB
Receiver noise temperature               = 527.33 K
External noise temperature               = 300.00 K
  (due to "warm" earth seen through main
   beam of the receive antenna)
Effective receiver noise temperature     = 827.33 K
Nominal receiver noise power             =-131.94 dBm
  (assuming 5.6 kHz noise bandwidth)
```

```
*****************
* C. Nominal C/N *                       =  29.51 dB
*****************                          (12.44)
```

```
**********************************
* D. Interference Considerations *
**********************************
```

```
Spacecraft antenna reuse beam-to-reuse beam
isolation                                =  25.00 dB
  (verify this with antenna manufacturer)
Effective impact other cochannel users   =  -7.00 dB
  (based on reuse pattern of 6 there will be up
   to 5 cochannel users [11 for intersatellite
   handoff situations]; however, their coupling
   patterns will not necessarily coincide so
   penalty may not be as severe as 7 dB [10.4 dB];
   note that a mixture of portable and mobile
   subscriber units would pose problem to portables
   due to unequal transmit powers)
-----------------------------------------------------
Nominal C/I1                             =  18.00 dB

Nominal C/I2                             =  N.A.
```

****
* E. Required Link Quality *
****

Eb/N0 (1% static BER, coherent detection, CPQPSK)	=	4.50 dB
R/Bn (8000 kbps over noise bandwidth of 5600 Hz)	=	1.55 dB
Modem implementation loss	=	1.00 dB
Minimum C/N	=	7.05 dB

(same as minimum C/N+I1+I2 assuming interference is noiselike in nature)

****
* F. Margin Evaluations *
****

Nominal C/N+I1	=	17.70 dB (11.37)
Required C/N+I1	=	7.05 dB
Link margin	=	10.65 dB (4.32)
Nominal C/N	=	29.51 dB (12.44)
Minimum C/N which still meets C/N+I1 requirement	=	7.41 dB
Fade margin	=	22.10 dB (5.03)
Required fade margin	=	12.00 dB

(for 90%-tile coverage in mixed shadowed/unshadowed environment; this value will be a function of elevation and hence cell position)

The mobile uplink is solid, whereas the portable uplink has a substantial shortfall of 6.97 dB, even ignoring the impact of mixed subscriber unit transmit power on the I1 level. Also, 6 dB additional sensitivity might be needed on the downlink to alert the portable to incoming calls (due to body effects on the antenna).

14.5 Homework Problem

Apply the LaGrangian multiplier technique to cost minimization of a 900-MHz paging system, typically consisting of a 125 W base transmitter and 100 ft of coaxial cable to an omnidirectional gain antenna, and pocket paging receivers.

Solution: Assume that the total cost function can be written as the sum of the individual element costs, which are in turn functions only of each element's contribution to system gain. Total system gain is the sum of the individual gains; thus:

$$C = \sum C_i(G_i)$$
$$G = \sum G_i \qquad (14.10)$$

The minimum cost condition that meets some arbitrary system gain requirement G_0 is given by differentiating $H = C - \lambda(G - G_0)$ with respect to all G_i and setting the results all equal to zero.

Transmitter cost is about 20 dollars per watt output (Figure 14.1), implying $C_t = 20 P_0$. Taking the transmitter gain as output power in dBW, one has $G_t = 10 \log(P_0)$, $(\partial C_t / \partial G_t) = (\partial C_t / \partial P_0)(\partial P_0 / \partial G_t)$, and $(\partial C_t / \partial G_t) = 4.6 P_0$ in dollars per decibel.

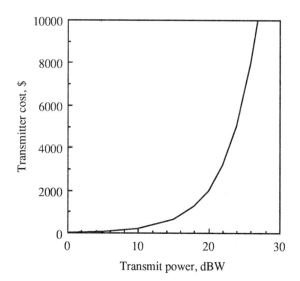

Figure 14.1 Transmitter cost is linear with output power in watts, but exponential with output power in dBW.

The cost of 100 ft of coaxial cable can be estimated in terms of its efficiency E through $C_c = 150(E/1-E)$ (Figure 14.2). Defining the cable gain as $G_c = 10\log(E)$ the desired cost sensitivity is $(\partial C_c/\partial G_c) = (34.5E)/(1-E)^2$ in dollars per decibel.

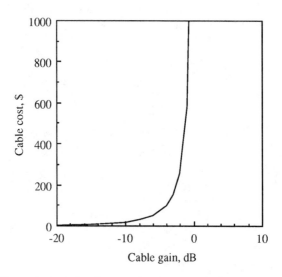

Figure 14.2 Cost per 100 ft of coaxial cable versus decibel gain at 900 MHz.

The cost of omnidirectional antennas (Figure 14.3)) can be estimated via $C_a = 283 + 50l$ and $G_a = 10\log(2l/w)$, where l represents the antenna length and w represents the wavelength in feet (1.05 in the problem). This leads to an incremental cost per gain of 11.5 l in dollars per decibel.

Overall cost is minimal when $(\partial C_t/\partial G_t) = (\partial C_c/\partial G_c) = (\partial C_a/\partial G_a)$. With a 125 W transmitter, the partials must equate to $575 dollars per decibel. The coaxial cable that meets this cost sensitivity is about 80% efficient, a value consistent with the use of 7/8-in coaxial cable. The antenna needed would be 50 ft tall, much larger than the usual 10 ft (10 dBd). If the length was forced to be 10 ft, then the cost gain partials would equal $115 and a 25 W transmitter with 1/2-in coaxial cable is implied. The total costs of these two situations differ substantially because the gains offered differ substantially (39.8 versus 24.4 dBW). To achieve the FCC maximum allowable ERP of 30 dBW at minimum cost, an antenna on the order of 20 feet with about 50 W of transmitter output and 68% efficient coaxial cable is indicated.

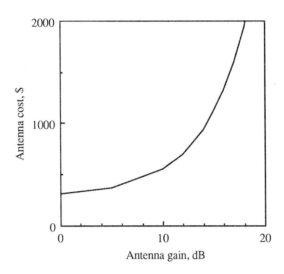

Figure 14.3 Antenna cost is approximately linear with linear gain, or exponential with gain in dBd.

References

[1] Wolff, E. A. (editor), *Public Service Communications Satellite User Requirements Workshop*, NASA/GSFC Communications and Navigation Division, Greenbelt, MD, October 19, 1976.

[2] Wolff, E. A. (editor), *Public Service Communications Satellite System Review and Experiment Definition Workshop*, NASA/GSFC Communications and Navigation Division, Greenbelt, MD, March 29, 1977.

[3] Anderson, R. E., et al., "Satellite-Aided Mobile Communications Limited Operational Test in the Trucking Industry," General Electric Co. Report SRD-81-005, Final Report to NASA for Contract NAS5-24365, July 1980.

[4] Brisken, A., et al., "Land Mobile Communications and Position Fixing Using Satellites," *IEEE Tr. on Veh. Tech.*, Vol. VT-28, No. 3, August 1979, pp. 153-170.

[5] Hess, G., "Land-Mobile Satellite Excess Path Loss Measurements," *IEEE Tr. on Veh. Tech.*, Vol. VT-29, No. 2, May 1980, pp. 290-297.

[6] Huck, R. W., et al., "Propagation Measurements for Land Mobile Satellite Services," *33rd IEEE Vehicular Technology Conference*, Toronto, Ontario, May 1983, pp. 265-267.

[7] Boudreau, P. M., et al., "The Canadian Mobile Satellite Program," *NTC'82*, Galveston, TX, November 1982, pp. B1.2.1-5.

[8] Knouse, G. and P. A. Castruccio, "The Concept of an Integrated Terrestrial/Land Mobile Satellite System," *ICC'80*, Seattle, WA, June 1980, pp. 35.1.1-5.

[9] Horstein, M., "Requirements for a Mobile Communications Satellite System," TRW Draft Final Report for NASA Contract NAS3-23257, February 17, 1983.

[10] Contract Study by Motorola for TRW Defense and Space Systems Group, "A Study of the Modifications to AMPS/DynaTAC Equipment Required for Mobile Communications Satellite System (MSAT) Compatibility," Purchase Order M23727GY2E.

[11] Reudink, D. O., "Estimates of Path Loss and Radiated Power for UHF Mobile-Satellite Systems," *Bell System Technical Journal*, Vol. 62, No. 8, October 1983, pp. 2493-2512.

[12] Dorfman, S. D., "Application of Hughes Communications Mobile Satellite Services, Inc. for a Land Mobile Satellite System," file before the FCC, April 30, 1985.

[13] Jet Propulsion Laboratory, *MSAT-X, Mobile Satellite Experiment, Industry Briefing*, Pasadena, CA, March 15, 1984.

[14] Smith, E. K. (editor), *Proceedings of the Propagation Workshop in Support of MSAT-X*, Jet Propulsion Laboratory, Pasadena, CA, March 15, 1985.

[15] Noreen, G. K., "Land Mobile Satellite Service At 1600 MHz and 800 MHz: A Comparative Review," Wismer & Becker/Transit Communications, Inc., Pasadena, CA, December 6, 1985.

[16] Janc, R. V., "The Cost and Performance Equivalence of Voice Communication at UHF and L-band Frequencies for the Land Mobile Satellite Service," FCC presentation, May 1986.

[17] Noreen, G. K., "MSAT: Mobile Communications Throughout North America," *39th IEEE Vehicular Technology Conference*, San Francisco, CA, May 1989, pp. 557-562.

[18] Rafferty, W. (conference technical program chairman), *Proceedings of the Mobile Satellite Conference*, Jet Propulsion Laboratory, Pasadena, CA, May 3-5, 1988.

[19] Richharia, M., et al., "A Feasibility Study of a Mobile Communication Network Using a Constellation of Low Earth Orbit Satellites," *Globecom '89*, Dallas, TX, pp. 21.7.1-5.

[20] Taylor, M., "Motorola Plans Global Cellular Thrust," *Chicago Tribune*, June 20, 1990, Business section 3, p. 1.

[21] Weiss, J. A., "Low Cost Satellite Land Mobile Service for Nationwide Applications," *28th IEEE Vehicular Technology Conference*, March 1978, Denver, CO, pp. 428-437.

[22] Goldman, Jr., A. M. and R. E. Edelson, "On Several Communications Satellite Designs Using Large Space Antennas," *Pacific Telecommunications Conference*, Honolulu, HI, January 1979.

[23] Sandrin, W. A., "Land-Mobile Satellite Start-up Systems," *COMSAT Tech. Rev.*, Vol. 14, No. 1, Spring 1984, pp. 137-164.

[24] Reilly, N. B. and J. G. Smith, "Application of a Large Space Antenna to Public Service Communications," JPL Report 760-186, June 1977.

[25] TRW Final Report, "Mobile Multiple Access Study," NASA Contract NAS5-23454, August 1977.

[26] Payne, W. F. and D.T.L. Tong, "Cost and Mass Modelling for Communications Satellites," *Satellite Communications*, April 1980, pp. 34-39.

[27] McGarty, T. P. and T. H. Warner, "Multiple Beam Satellite System Optimization," *IEEE Tr. on Aero. and Elec. Sys.*, Vol. AES-13, September 1977, pp. 504-511.

[28] Hadfield, B. M., "Satellite-Systems Cost Estimation," *IEEE Tr. on Comm.*, Vol. COM-22, No. 10, October 1974, pp. 1540-1547.

[29] Kiesling, J. D., et al., "A Technique for Modeling Communication Satellites," *COMSAT Tech. Rev.*, Vol. 2, Spring 1972, pp. 73-101.

[30] Protter, M. H. and C. B. Morrey, Jr., *Modern Mathematical Analysis*, Reading, MA, Addison-Wesley, 1964.

[31] Divsalar, D. and M. Simon, "Trellis Coded MPSK Modulation Techniques for MSAT-X," *Proceedings of the Mobile Satellite Conference*, Jet Propulsion Laboratory, Pasadena, CA, May 1988, pp. 283-290.

[32] Jedrey, T., et al., "An 8-DPSK TCM Modem for MSAT-X," *Proceedings of the Mobile Satellite Conference*, Jet Propulsion Laboratory, Pasadena, CA, May 1988, pp. 311-316.

[33] Simon, M. and D. Divsalar, "Doppler-Corrected Differential Detection of MPSK," *IEEE Tr. on Comm.*, Vol. 37, No. 2, February 1989, pp. 99-109.

[34] Smith, W. and W. Stutzman, "Statistical Modeling for Land Mobile Satellite Communications," Virginia Tech Report EE SATCOM 86-3, Virginia Polytechnic Institute and State University, Blacksburg, VA, August 1986.

[35] Dessouky, K. and L. Ho, "Propagation Results from the Satellite-1a Experiment," *MSAT-X Quarterly #17*, Jet Propulsion Laboratory, Pasadena, CA, October 1988, pp. 7-12.

[36] Edbauer, F., "Interleaver Design for Trellis-Coded Differential 8-PSK Modulation with Non-Coherent Detection," *Proceedings of the Mobile Satellite Conference*, Jet Propulsion Laboratory, Pasadena, CA, May 1988, pp. 271-276.

[37] Wells, P., "The Attenuation of UHF Radio Signals by Houses," *IEEE Tr. on Veh. Tech.*, Vol. VT-26, No. 4, November 1977, pp. 358-362.

[38] Walker, E., "Penetration of Radio Signals Into Buildings in the Cellular Radio Environment," *Bell System Technical Journal*, Vol. 62, No. 9, November 1983, pp. 2719-2735.

[39] Davidson, A., "Land Mobile Radio Propagation to Satellites," *Proc. 1991 Antenna Applications Symposium*, September 1991, University of Illinois, Urbana-Champaign, IL, pp. 263-281.

[40] Lutz, E., et al., "The Land Mobile Satellite Communication Channel - Recording, Statistics, and Channel Model," *IEEE Tr. on Veh. Tech.*, Vol. 40, No. 2, May 1991, pp. 375-386.

[41] Vogel, W. and J. Goldhirsh, "Propagation Handbook for Land-Mobile-Satellite Systems," S1R-91U-012, Johns Hopkins University, April 1991 (preliminary).

[42] Bundrock, A. and R. Harvey, "Propagation Measurements for an Australian Land Mobile Satellite System," *Proceedings of the Mobile Satellite Conference*, Jet Propulsion Laboratory, Pasadena, CA, May 1988, pp. 119-124.

[43] Adams-Russell Co., Inc., "Technical Data, Antenna Type AN-710 for Commercial GPS", Amesbury, MA, December 13, 1984.

[44] Keen, K., "Developing a Standard-C Antenna", *MSN & CT*, June 1988, pp. 52-54.

[45] Whittaker, N., et al., "The Design, Evaluation, & Modeling of a UHF Power Amplifier for a Mobile Satellite Transponder", *International Conference on Communications*, Toronto, Canada, June 1986, pp. 44.5.1-6.

Chapter 15

Frequency Modulation Performance

This chapter addresses the performance of frequency modulation (FM), the dominant modulation scheme currently used for land-mobile communications. Although linear and spread-spectrum modulations may soon begin replacing FM, its use will remain widespread for many more years and hence knowledge of its behavior in the mobile environment is important.

FM is inherently immune to the severe amplitude fading typically encountered on mobile radio links, whereas amplitude modulation approaches require some form of relatively fast acting gain control to compensate for such fading [1, 2]. So-called random FM phase fluctuations do limit the ultimate signal-to-noise ratio possible with FM. This is not generally a problem, but if desired it can be mitigated by coherent combining techniques [3]. Other benefits of FM are its ability to improve the baseband signal-to-noise ratio and its ability to capture on a signal above the threshold level, thus suppressing the effects of noise and interference.

15.1 Fundamentals of Frequency Modulation

Frequency modulation is a member of the more general class of modulation called exponential modulation, for which the modulated wave can be written in phasor form as $x_c(t) = Re[A_c \exp^{j\theta_c(t)}] = A_c \cos[\theta_c(t)]$, where $\theta_c(t)$ is some linear function of the message waveform $x(t)$ and A_c is some constant. Since θ_c represents the angular position of the phasor, this process is also referred to as angle modulation.

It is instructive to write the angle variation with time as $\theta_c(t) = 2\pi f_c t + \phi(t)$, where f_c is the RF carrier frequency. Instantaneous frequency can be defined as the time rate of change of angle, $f_i(t) = (1/2\pi)[d\theta_c(t)/dt] = f_c + (1/2\pi)(d\phi/dt)$. Equivalently, the phasor angle at any time t equals 2π times the integral of instantaneous frequency up to that particular time. The lower limit of integration is that which produces the correct angle at time zero. In frequency modulation, f_i is made proportional to the message; i.e. $f_i = f_c + f_d x(t)$, where f_d is the frequency deviation constant and $x_c(t) = A_c \cos[\omega_c t + 2\pi f_d \int^t x(t')dt']$. The message is presumed to have no dc component, otherwise the carrier frequency is effectively shifted by the amount $f_d <x(t)>$.

Figure 15.1 is a block diagram of the basic components of an FM transmitter and receiver. Practical direct FM modulation systems include pre-emphasis filtering (differentiation by high-pass filtering) to assist the demodulation process and instantaneous deviation control to limit the amount of energy transmitted outside the assigned channel. The instantaneous deviation function is accomplished by amplitude clipping and low-pass filtering. Clipping also has the advantage of increasing the average voice energy by approximately the amount of clipping. For example, normal speech has a peak-to-average energy ratio on the order of 20 dB. Clipping the top down, say 12 dB, results in the average energy now being within about 8 dB of the new peak level. This is roughly the amount of clipping that can be tolerated before the distortion introduced to the speech becomes objectionable. The de-emphasis filtering following demodulation (integration by low-pass filtering) optimizes the postdetection SNR. The rationale for pre-emphasis and de-emphasis

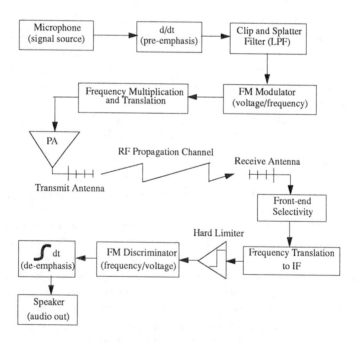

Figure 15.1 Direct FM modulation-demodulation system.

filtering is discussed in Section 15.4.3.

Frequency multiplication, as opposed to frequency translation, can be used in the transmitter to increase the frequency deviation in direct proportion to the frequency multiplier. For example, direct frequency modulation of a 15-MHz channel element need only produce deviations of about 90 Hz to yield 5-kHz deviations at 825 MHz. Receivers are generally of the heterodyne variety, meaning that the input signal is linearly translated (mixed) to some convenient lower IF where amplification and narrowband filtering are more practical.

Phase modulation is quite similar to frequency modulation but involves making the relative phase $\phi(t)$ proportional to the message. One common way of obtaining FM is to pass the message waveform through a low-pass filter, thereby integrating it in time, and then apply it to a phase modulator.

15.2 Spectrum

An exact description of the spectrum of FM is generally difficult, except for certain modulating signals. One such tractable case involves sinusoidal modulation. Consider the case of $x(t) = A_m \cos[\omega_m(t)]$. This results in:

$$\begin{aligned}
f_i(t) &= f_c + f_d A_m \cos[\omega_m(t)] \\
x_c(t) &= A_c \cos[\omega_c t + 2\pi f_d \int^t A_m \cos(\omega_m t')dt'] \\
&= A_c \cos[\omega_c t + \frac{2\pi f_d A_m}{\omega_m} \sin(\omega_m t)] \\
&= A_c \cos[\omega_c t + \beta \sin(\omega_m t)]
\end{aligned} \qquad (15.1)$$

where $\beta = (A_m f_d / f_m)$ is the peak frequency deviation. $x_c(t)$ expands into the expression:

$$x_c(t) = A_c \{\cos(\omega_c t) \cos[\beta \sin(\omega_m t)] - \sin(\omega_c t) \sin[\beta \sin(\omega_m t)]\} \qquad (15.2)$$

At this point two identities involving Bessel functions of the first kind are of assistance (relations 9.1.42 and 9.1.43 in Reference [4]):

$$\begin{aligned}
\cos[\beta \sin(x)] &= J_0(\beta) + \sum_{\text{even } n} 2J_n(\beta) \cos(nx) \\
\sin[\beta \sin(x)] &= \sum_{\text{odd } n} 2J_n(\beta) \sin(nx)
\end{aligned} \qquad (15.3)$$

where n represents the positive integers and:

$$J_n(\beta) = \frac{1}{2\pi} \int_{-\pi}^{\pi} \exp[j\beta \sin(u) - nu]du \qquad (15.4)$$

Substitution of these identities into the expanded expression for $x_c(t)$ clearly indicates the spectrum consists of an infinite number of sidebands of relative amplitudes $J_n(\beta)$ and frequencies $f_c \pm n f_m$:

$$\begin{aligned} x_c(t) &= A_c J_0(\beta)\cos(\omega_c t) \\ &+ \sum_{\text{odd } n} A_c J_n(\beta)\{\cos[(\omega_c+n\omega_m)t] - \cos[(\omega_c-n\omega_m)t]\} \\ &+ \sum_{\text{even } n} A_c J_n(\beta)\{\cos[(\omega_c+n\omega_m)t] - \cos[(\omega_c-n\omega_m)t]\} \end{aligned} \quad (15.5)$$

As shown in Figure 15.2 the amplitudes are symmetrical about the carrier frequency, with the odd-order lower sidebands reversed in phase from the carrier and upper sideband components (see the Bessel function property $J_{-n}(\beta) = (-1)^n J_n(\beta)$, relation 9.1.5 in Reference [4]).

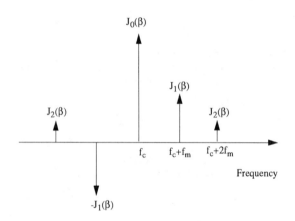

Figure 15.2 FM spectrum for single-tone modulation.

The Bessel function expansion technique used above can also be applied in the case of multiple-tone modulation where the tones are not harmonically related [5, 6]. If the tones are harmonically related then the modulating waveform is periodic, say with period T_0, and:

$$\begin{aligned} x_c(t) &= \text{Re}\{A_c \exp(j\omega_c t)\exp[j\phi(t)]\} \\ &= \text{Re}[A_c \sum_{n=-\infty}^{\infty} c_n \exp[j(\omega_c+n\omega_0)t] \end{aligned} \quad (15.6)$$

where c_n represent the Fourier series coefficients of $\exp[j\phi(t)]$, which will also be periodic. The coefficients are given by:

$$c_n = \frac{1}{T_0}\int_{-T_0/2}^{T_0/2} \exp\{j[\phi(t) - n\omega_0 t]\}dt \quad (15.7)$$

In the case of pulse modulation, the modulated waveform can be obtained by direct Fourier transformation. Consider a rectangular pulse of unit amplitude and duration τ, centered about zero time [5]. The instantaneous frequency caused by such a pulse will be $f_c + f_d$ for $|t| < (\tau/2)$ and otherwise just f_c. This results in the modulated waveform $A_c \cos[(\omega_c + \omega_d)t]$ for $|t| < (\tau/2)$ and otherwise just $A_c \cos(\omega_c t)$. It is convenient to write this as:

$$x_c(t) = A_c\{\cos(\omega_c t) - \Pi(\frac{t}{\tau})\cos(\omega_c t) + \Pi(\frac{t}{\tau})\cos[(\omega_c + \omega_d)t]\} \tag{15.8}$$

where Π represents the rectangular function; i.e. $\Pi(x) = 1$ for $|x| \leq \frac{1}{2}$, else $\Pi(x) = 0$. Taking the Fourier transform term by term gives:

$$\begin{aligned}X_c(f) &= \frac{A_c}{2}[\delta(f - f_c) + \delta(f + f_c)] \\ &- \frac{A_c \tau}{2}\{\text{sinc}[(f - f_c)\tau] + \text{sinc}[(f + f_c)\tau]\} \\ &+ \frac{A_c \tau}{2}\{\text{sinc}[(f - f_c - f_d)\tau] + \text{sinc}[(f + f_c + f_d)\tau]\}\end{aligned} \tag{15.9}$$

Figure 15.3 shows the one-sided spectrum for two conditions of frequency deviation constant relative to the reciprocal of pulse duration.

15.3 Transmission Bandwidth

Although in general FM spectra are infinite in extent, the amplitudes fall off quite rapidly beyond some frequency offset from the carrier. The practical bandwidth of FM, representing the bandwidth required to limit distortion to some acceptable value, is thus finite. Consider the number of significant spectral lines with single-tone modulation. The Bessel function amplitude becomes small as $|n/\beta|$ exceeds unity, particularly for large β (see asymptotic relation 9.3.1 in Reference [4]). Thus, at one extreme, the significant sidebands fall within βf_m of the carrier frequency. At the opposite extreme of very small β values, the first two sidebands must be retained or there would be no frequency modulation; i.e., the bandwidth is $f_c \pm f_m$. A useful combination of these two extremes is Carson's rule, which equates the FM bandwidth to twice the sum of the peak frequency deviation and the maximum modulation frequency.

Commercial FM broadcast transmissions use 75-kHz peak deviation with audio bandwidths of 15 kHz. Carson's rule predicts a required bandwidth of 180 kHz, whereas typical FM receiver bandwidths are 200 kHz. The wider bandwidth is not surprising since this is a service where the ability to receive nearly distortion-free program material is extremely important. There is a cost in spectrum for this, however, as adjacent channel stations must be sufficiently distant to prevent interference through the wide receiver bandwidth.

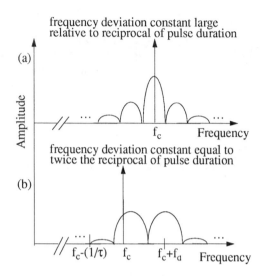

Figure 15.3 FM spectra for pulse modulation.
After: Carlson, A. B., Communication Systems: *An Introduction to Signals and Noise in Electrical Communications*, McGraw-Hill Book Co., New York, NY, 1968, p. 235.

Typical parameters associated with 25-kHz FM land-mobile radios at 800 MHz are 5-kHz peak frequency deviation and 3-kHz modulation acceptance bandwidth. Carson's rule indicates 16 kHz is needed for such transmissions. Early 800-MHz radios did have bandwidths of this order, but the reason was to cover for frequency drift, not to limit distortion. Current radios with improved frequency stability[1] use IF bandwidths on the order of just 12 kHz. This is acceptable because the effective frequency deviation is generally rolled off above 1-kHz audio. In this service adjacent channel isolation is of more concern than audio distortion, particularly at high audio frequencies where distortion is subjectively less objectionable. Remember that the signal of a nearby user talking on an adjacent channel can readily be 70 dB stronger than the signal of the on-channel user the receiver is trying to hear.

Typical parameters associated with 12.5-kHz FM land-mobile radios at 900 MHz are 2.5-kHz peak frequency deviation and 3-kHz modulation acceptance bandwidth. Carson's rule indicates 11 kHz is needed for such transmissions, yet IF bandwidths are only about 7 kHz. Here, fifth-order splatter filters are used rather than third-order filters, so the frequency deviation possible with a 3-kHz tone is only on the

---

[1] Frequency drifts of 2.5 ppm for mobile units and 1.5 ppm for base units are permitted, implying a shift of 3.4 kHz at 850 MHz; however, modern equipment achieves 1.5 ppm at the mobile and 0.1 ppm at the base (by locking it to a precision ovenized oscillator).

Frequency Modulation Performance   315

order of 20% of the rated 2.5-kHz deviation. This implies a bandwidth of $2(0.5 + 3.0) = 7$ kHz. This matches the bandwidth implied with 1-kHz tone modulation and the rated 2.5-kHz peak deviation.

Cellular radiotelephone systems are designed for high-quality audio and spectrum efficiency through reuse of the same frequencies over a geographic area. Audio quality and tolerance to cochannel interference both improve as the frequency deviation increases, but spectrum efficiency degrades because that implies greater channel bandwidth. A satisfactory compromise is 12-kHz peak deviation and 30-kHz channels [7]. Receiver IF bandwidth can be essentially equal to the channel bandwidth because the systems are organized so that adjacent channels are not used at the same site.

15.4 Demodulation in the Presence of Noise and Nonfaded Conditions

An ideal FM demodulator produces an output proportional to the time rate of change of the input signal phase with respect to $\exp(j\omega_c t)$. The input noise to the demodulator is generally bandlimited by IF filtering, allowing one to write the average input noise power as $N_i = \eta B_i$, where η equals the noise spectral density of the noise (which is assumed to be additive white Gaussian noise) and B_i equals the noise equivalent bandwidth of the input filtering. Taking the modulated signal as $x_c(t) = A_c \cos[\omega_c t + \phi(t)] = A_c \cos[\omega_c t + 2\pi f_d x(t)]$, one can write the average input signal power as $S_i = (1/2)A_c^2$. Thus, the predetection SNR is $(S/N)_i = (A_c^2/2\eta B_i)$.

The demodulator input voltage is:

$$\begin{aligned} y(t) &= x_c(t) + n(t) \\ &= A_c \cos[\omega_c t + \phi(t)] + I_n(t)\cos[\omega_c t + \phi_n(t)] \\ &= R(t)\cos[\omega_c t + \alpha(t)] \end{aligned} \qquad (15.10)$$

In the preceding, the narrowband view of noise as a single phasor at the carrier frequency with random envelope and phase is applied, and the resultant signal is also manipulated into an envelope and phase form relative to the carrier frequency. Ideally, a limiter will remove the envelope variations of $R(t)$ and only the phase term is of interest.

15.4.1 Signal Suppression Noise

Referring to the phasor diagram in Figure 15.4 one can write:

$$A_c \exp[j\phi(t)] + I_c(t) + jI_s(t) = R(t)\exp\{j[\phi(t) + \theta(t)]\} \qquad (15.11)$$

where I_c and I_s represent the in-phase and quadrature phase components of the noise voltage I_n. Differentiation of the natural logarithm of both sides with respect

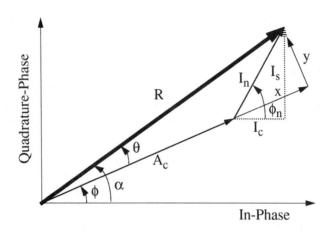

Figure 15.4 Phase relations between FM signal and noise components.
After: Rice, S. O., "Noise in FM Receivers," chapter 25 of *Time Series Analysis*, M. Rosenblatt, editor, John Wiley & Sons, New York, NY, 1963, p. 411.

to time gives:

$$\frac{R'}{R} + j(\phi' + \theta') = \frac{j\phi' A_c \exp(j\phi) + I'_c + jI'_s}{A_c \exp(j\phi) + I_c + jI_s} \quad (15.12)$$

where, for convenience, the explicit time notations for $\phi(t)$, $\theta(t)$, $I_c(t)$, $I_s(t)$, and $R(t)$ have been dropped. Following Rice [8], the ensemble average of the left-hand side can be found by averaging the right-hand side over all values of I_c, I_s, I'_c, and I'_s. Provided the IF filtering is symmetrical about the carrier frequency f_c, the four variables will be independent, zero mean, and Gaussian distributed. Averaging over I'_c and I'_s simply removes those terms from the right-hand side numerator, leaving:

$$< \frac{R'}{R} + j(\phi' + \theta') > \ = \ \frac{j\phi' A_c \exp(j\phi)}{2\pi\sigma^2} \int_{-\infty}^{\infty} dI_c \int_{-\infty}^{\infty} \frac{\exp[-\frac{1}{2\sigma^2}(I_c^2 + I_s^2)]dI_s}{A_c \exp(j\phi) + I_c + jI_s} \quad (15.13)$$

The integration can be carried out by breaking the complex expression into real and imaginary parts. From the phasor diagram one has:

$$\begin{aligned} R\cos(\theta + \phi) &= A_c \cos(\phi) + I_c \\ R\sin(\theta + \phi) &= A_c \sin(\phi) + I_s \end{aligned} \quad (15.14)$$

which can be rearranged into:

$$I_c = R\cos(\alpha) - A_c \cos(\phi)$$

Frequency Modulation Performance

$$I_s = R\sin(\alpha) - A_c\sin(\phi)$$
$$\alpha = \theta + \phi \tag{15.15}$$

The denominator of the expression to be integrated is thus $A_c[\cos(\phi)+j\sin(\phi)] + [R\cos(\alpha) - A_c\cos(\phi)] + j[R\sin(\alpha) - A_c\sin(\phi)] = R\exp[j(\theta+\phi)]$ and the numerator becomes $\exp[-(1/2\sigma^2)(R^2 - 2RA_c\cos(\theta) + A_c^2)]$.

Next, the variables of integration are changed from I_c and I_s to R and α. The Jacobian for this transformation is:

$$\begin{vmatrix} R\cos(\alpha) & -R\sin(\alpha) \\ \sin(\alpha) & \cos(\alpha) \end{vmatrix} = R \tag{15.16}$$

which implies $dI_c dI_s = R\,dR\,d\alpha$. Adopting a quasistatic viewpoint (i.e., assuming the modulation remains relatively constant in time compared to the noise fluctuations), one has $d\alpha = d\theta$. Thus, the integral of interest can be written as:

$$<\frac{R'}{R} + j(\phi'+\theta')> = \frac{j\phi' A_c}{2\pi\sigma^2}\int_0^\infty dR\int_0^{2\pi}\frac{\exp\{-\frac{1}{2\sigma^2}[R^2 - 2RA_c\cos(\theta) + A_c^2]\}d\theta}{\exp(j\theta)} \tag{15.17}$$

Rationalizing through multiplication and division by the conjugate of the denominator $\exp(j\theta) = \cos(\theta) + j\sin(\theta)$ leads to a real-part integral of:

$$<\frac{R'}{R}> = \frac{\phi' A_c}{2\pi\sigma^2}\int_0^\infty dR\int_0^{2\pi}\sin(\theta)\exp\{-\frac{1}{2\sigma^2}[R^2 - 2RA_c\cos(\theta) + A_c^2]\}d\theta \tag{15.18}$$

which, due to odd symmetry in θ, must equal zero. The imaginary-part integral is:

$$<\phi'+\theta'> = \frac{\phi' A_c}{\pi\sigma^2}\int_0^\infty dR\int_0^{\pi}\cos(\theta)\exp\{-\frac{1}{2\sigma^2}[R^2 - 2RA_c\cos(\theta) + A_c^2]\}d\theta \tag{15.19}$$

where the even symmetry has allowed shifting the integration upper limit from 2π to π in return for a multiplication factor of 2.

At this point recall the definition of the modified Bessel function of the first kind (relation 9.6.19 in Reference [4]):

$$I_n(z) = \frac{1}{\pi}\int_0^\pi \exp[z\cos(\theta)]\cos(n\theta)d\theta \tag{15.20}$$

The imaginary-part integral with respect to θ is therefore seen to equal $\pi I_1(RA_c/\sigma^2)$. The integral with respect to R becomes:

$$<\phi'+\theta'> = \frac{\phi' A_c}{\sigma^2}\exp(-\frac{A_c^2}{2\sigma^2})\int_0^\infty \exp(-\frac{R^2}{2\sigma^2})I_1(\frac{RA_c}{\sigma^2})dR \tag{15.21}$$

Applying integral pair 11.4.31 of Reference [4], one obtains:

$$<\phi'+\theta'> = \phi' A_c\sqrt{\frac{\pi}{2\sigma^2}}\exp(-\frac{A_c^2}{4\sigma^2})I_{\frac{1}{2}}(\frac{A_c^2}{4\sigma^2}) \tag{15.22}$$

But half-order modified Bessel functions of the first kind can be written in terms of the sinh function (relation 10.2.13 in [4] shows that $I_{1/2}(z) = \sqrt{2/\pi z} \sinh(z)$), which in turn can be written in terms of the exp function. Thus, the final solution is:

$$<\phi' + \theta'> = \phi'[1 - \exp(-\rho)] \tag{15.23}$$

where $\rho = (A_c^2/2\sigma^2)$ equals the predetection carrier-to-noise power ratio. This result shows that the presence of noise reduces the output signal voltage by the factor $[1 - \exp(-\rho)]$ and reduces the output signal power by the factor $[1 - \exp(-\rho)]^2$.

15.4.2 Postdetection Baseband Noise Spectrum

Analysis of the noise from an FM demodulator is quite complicated and the interested reader is referred to Rice [9] for the details of its exact computation. In that computation Rice splits the spectral density into three components: (1) $W_1(f)$ has the same shape as the noise spectrum when the signal is absent, (2) $W_2(f)$ is quadratic in form and reflects high input carrier-to-noise conditions, and (3) $W_3(f)$ is a correction term. The quadratic form of output noise under high input carrier-to-noise conditions can be shown as follows. Under such conditions the signal-aligned component of the resultant is $A_c + x(t) \approx A_c$ and the quadrature component is $y(t)$ (Figure 15.4). The resultant phase is thus $\theta(t) \approx \tan^{-1}[y(t)/A_c] \approx [y(t)/A_c]$, since the tangent argument is small. The detected output due to noise equals the time derivative of this phase, $(y'(t)/A_c)$. Because $y(t)$ represents a stationary stochastic process (white Gaussian noise), the power spectrum of its derivative is $(2\pi f)^2$ times the power spectrum of $y(t)$ [10]. The actual detected noise power depends of course on the predetection and postdetection filtering used.

For example, assume the predetection filtering can be modeled as Gaussian as in Reference [11]:

$$|H_{\text{IF}}(f)|^2 = \frac{1}{\sqrt{2\pi\sigma_f^2}} \exp[-\frac{(f-f_c)^2}{2\sigma_f^2}] \tag{15.24}$$

The noise equivalent bandwidth is given by $B_{ne} = \sqrt{2\pi}\sigma_f \approx 2.51\sigma_f$, which as expected is approximately equaled by the -3-dB bandwidth of $B_{-3} = 2\sigma_f\sqrt{\ln(4)} \approx 2.35\sigma_f$. Carefully accounting for the nonunity passband gain of Equation (15.24), one finds:

$$W_2(f) = (2\pi)^{\frac{3}{2}} \frac{f^2}{\rho \sigma_f}[1 - \exp(-\rho)]^2 \exp(-\frac{f^2}{2\sigma_f^2}) \tag{15.25}$$

where the predetection carrier-to-noise power ratio is:

$$\rho = (A_c^2/2\eta B_{ne}) = (A_c^2/2^{\frac{3}{2}}\sqrt{\pi}\eta\sigma_f)$$

The preceding expression includes the signal-suppression effect, but of course since

Frequency Modulation Performance

it is only valid for high predetection carrier-to-noise power ratios this is of little consequence.

Alternatively [3], one can express the filter as:

$$|H_{IF}(f)|^2 = \exp[-\pi\frac{(f-f_c)^2}{B^2}] \qquad (15.26)$$

This form has the convenient properties of unity passband gain and the coefficient B directly representing the noise bandwidth; i.e., $B = \sigma_f\sqrt{2\pi}$. Thus:

$$W_2(f) = \frac{(2\pi f)^2[1-\exp(-\rho)]^2}{B\rho}\exp(-\pi\frac{f^2}{B^2}) \qquad (15.27)$$

The sum of spectral densities $W_1(f)$ and $W_3(f)$ is essentially flat in the region of interest; i.e., audio frequencies up to where the de-emphasis filtering following detection rolls off. Defining $W_0(f) = W_1(f) + W_3(f)$, for practical purposes one can then use the simplification $W_0(f) \approx W_0(0)$. An empirical relation that fits Rice's exact results well for all values of ρ has been developed by Davis [11]:

$$W_0(0) = \frac{8\pi^2\sigma_f\exp(-\rho)}{\sqrt{\pi(\rho+2.35)}} \qquad (15.28)$$

Hence, the total noise spectral density is given approximately by:

$$\begin{aligned}W_T(f) &= (2\pi)^{\frac{3}{2}}\frac{f^2}{\rho\sigma_f}[1-\exp(-\rho)]^2\exp(-\frac{f^2}{2\sigma_f^2}) + \frac{8\pi^2\sigma_f\exp(-\rho)}{\sqrt{\pi(\rho+2.35)}} \\ &= \frac{(2\pi f)^2[1-\exp(-\rho)]^2}{B\rho}\exp(-\pi\frac{f^2}{B^2}) + \frac{8\pi B\exp(-\rho)}{\sqrt{2(\rho+2.35)}}\end{aligned} \qquad (15.29)$$

Integrating this for an ideal low-pass filter of width W, one obtains a total noise power output of:

$$N_o = \frac{a[1-\exp(-\rho)]^2}{\rho} + \frac{8\pi BW\exp(-\rho)}{\sqrt{2(\rho+2.35)}} \qquad (15.30)$$

where:

$$\begin{aligned}a &= \frac{(2\pi)^2}{B}\int_0^W f^2\exp(-\pi\frac{f^2}{B^2})df \\ &= \frac{4\pi^2 W^3}{3B}[1-\frac{6\pi}{10}x^2+\frac{12\pi^2}{56}x^4+\cdots]\end{aligned}$$

$$x = \frac{W}{B} \qquad (15.31)$$

via Maclaurin series expansion and term-by-term integration. Since $x < 1$, only the first few terms in the expansion are generally significant.

The signal-to-noise power ratio at the output of the FM demodulator can now be written as:

$$\left(\frac{S}{N}\right)_o = \left(\frac{S}{N}\right)_i [1 - \exp(-\rho)]^2 \qquad (15.32)$$

which, for large input SNR (large ρ), becomes:

$$\left(\frac{S}{N}\right)_o = \frac{S_i [1 - \exp(-\rho)]^2}{(a/\rho)[1 - \exp(-\rho)]^2} \qquad (15.33)$$

In general, the input signal power is given by $S_i = (2\pi f_d)^2 < [\int^t x(t')dt']^2 >= (2\pi f_d)^2 \alpha$, where $x(t)$ denotes the message voltage waveform and the average power of the message is taken as $\alpha \leq 1$. Consider tone modulation at the highest audio frequency W. The Carson's rule bandwidth is $B_{cr} = 2(W + f_d) = 2W(1 + \beta_i)$, where $\beta_i = (f_d/W)$ is the deviation ratio (worst-case modulation index). The input signal power can thus be written as $S_i = \alpha \pi^2 (B_{cr} - 2W)^2$. Substituting this into the large signal SNR expression (Equation (15.33)) gives:

$$\left(\frac{S}{N}\right)_o = [3\alpha \frac{B_{cr}}{W} \beta_i^2] \rho \qquad (15.34)$$

This shows that the output SNR can exceed the input carrier-to-noise ratio in proportion to the square of the modulation index, provided the latter is sufficiently high to begin with. Figure 15.5 shows the behavior of postdetection SNR as predetection carrier-to-noise power ranges from much less than unity to much greater than unity and for several choices of predetection bandwidth to postdetection bandwidth ratio. Note the threshold region around unity carrier-to-noise ratio where the output SNR rapidly rises (falls) as the former rises (falls). Above the threshold region, in the so-called signal capture region, signal-to-noise values rise in proportion to carrier-to-noise values, but are offset by an FM improvement factor of $3\alpha(B_{cr}/W)\beta_i^2$.

15.4.3 Pre-emphasis and De-emphasis Filtering

Because detected FM noise power (and adjacent channel interference power as well) varies with the square of baseband frequency under above-threshold conditions, it is advantageous to include postdetection filtering whose amplitude response begins to gradually roll off with frequency well below W. Such filtering de-emphasizes the high-frequency portion of the message band and therefore minimizes the impact of noise and interference. Figure 15.6 shows the baseband amplitude response of a typical receiver designed to meet EIA specifications. An approximate polynomial fit to the response is [3]:

$$G(f) = \frac{10^{12.3} f^4}{(f^2 + 180^2)(f^2 + 190^2)(f^2 + 600^2)(f^2 + 800^2)} \qquad (15.35)$$

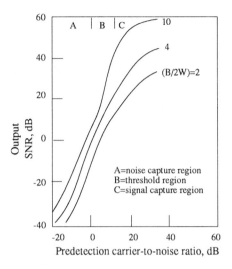

Figure 15.5 FM output SNR versus predetection carrier-to-noise ratio, nonfaded conditions.
After: Davis, B. R., "FM Noise with Fading Channels and Diversity," *IEEE Tr. on Comm.*, Vol. COM-19, No. 6, December 1971, p. 1198.

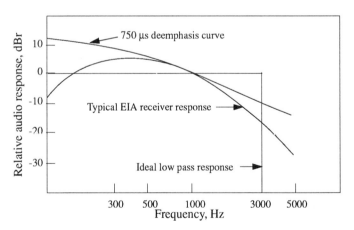

Figure 15.6 FM receiver audio responses.
After: Lynk, Jr., C. N., "Analysis of Multiplex Techniques for Land Mobile Radio Service," *IEEE Tr. on Veh. Tech.*, Vol. VT-20, No. 1, February 1971, p. 5.

To avoid distortion of the high-frequency portion of the message itself, the transmitted signal can be predistorted by a pre-emphasis filter whose response is a mirror image of the de-emphasis filter. The end result is a constant signal-to-noise performance over the message band.

De-emphasis is often accomplished with a simple *resistor-capacitor* (RC) network in the receiver having the amplitude response:

$$|H_{\text{DE}}(f)| = [1 + (\frac{f}{f_3})^2]^{-\frac{1}{2}}$$
$$\approx 1, |f| \ll f_3$$
$$\approx \frac{f_3}{f}, |f| \gg f_3$$
$$f_3 = \frac{1}{2\pi RC} < W \qquad (15.36)$$

This response, due to the inverse frequency relation above f_3, approximates the behavior of an integrator. The companion pre-emphasis network used at the transmitter is:

$$|H_{\text{PE}}(f)| = [1 + (\frac{f}{f_3})^2]^{\frac{1}{2}}$$
$$\approx 1, |f| \ll f_3$$
$$\approx \frac{f}{f_3}, |f| \gg f_3 \qquad (15.37)$$

This response, due to the direct frequency relation above f_3, approximates the behavior of a differentiator. Interestingly that transforms the high-frequency portion of the message into phase modulation. Generally, pre-emphasis implies an increase in transmission bandwidth because it increases the modulating signal amplitudes at the highest modulating frequencies. However, in practice, modulating signals like voice naturally roll off in frequency at about 6 dB per octave [12] so the higher frequency components do not develop maximum deviation and transmission bandwidth is not expanded.

15.4.4 Click Analysis

An approximate method of quantifying FM noise, valid down to carrier-to-noise ratios of about 3 dB, involves the concept of click noise [8, 13, 14][2]. Cohn observed that at high carrier-to-noise levels the output of an FM receiver had a hissing quality dominated by the parabolic above-threshold noise mechanism. As the carrier level was decreased, random popping sounds were heard above the background hiss, which

---

[2]Yavuz, Hess, and Glazer discuss the limitations of click analysis at low signal-to-noise values in a series of papers, [15, 16, 17, 18].

Frequency Modulation Performance

of course also increased in level. The number of pops increased rapidly as the carrier level was further decreased, until they fused into a continuous noise that was much higher than the above-threshold noise component. The total noise can thus be approximated by the sum of above-threshold noise and click noise. Since Cohn's publication, the term "click" has replaced "pop" in the literature.

A click can be defined as an event during which the noise phasor causes the resultant sum of the noise-plus-carrier phasor to undergo an abrupt increase of $\pm 2\pi$ in resultant phase angle, as compared to the carrier phasor alone. A time history of resultant phase and time rate of change of phase might look like that shown in Figure 15.7. Because the FM detector output is proportional to time rate of change of phase, the occurrence of a click is comparable to the introduction of an impulse of area 2π into the postdetection filtering.

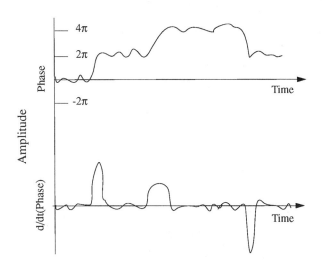

Figure 15.7 Sample time history of resultant phase and phase time derivative. After: Rice, S. O., "Noise in FM Receivers," chapter 25 of *Time Series Analysis*, M. Rosenblatt (editor), John Wiley & Sons, New York, NY, 1963, p. 400.

Clicks occur whenever the origin of the phasor diagram is encircled by the locus of the resultant phasor. Figure 15.8(a) illustrates a situation that results in a click. A necessary condition to produce a click is that the noise phasor be at least momentarily larger than the carrier phasor. Figure 15.8(b) shows that this is not a sufficient condition to guarantee encirclement of the origin. A circumstance as shown in Figure 15.8(b) will produce a discriminator output transient, but the transient will have zero average energy and thus relatively little energy in the low

(a) A noise burst causing a click

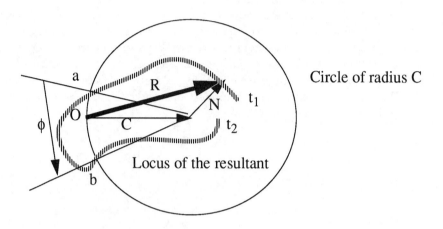

(b) A noise burst but no click

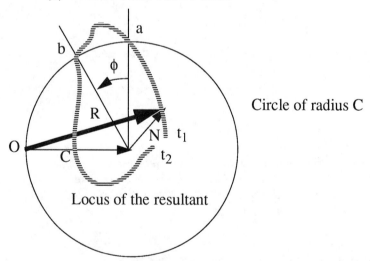

Figure 15.8 Sample FM phasor trajectories.
After: Cohn, J., "A New Approach to the Analysis of FM Threshold Reception," *Proceedings of the N.E.C.*, 1956, p. 224.

Frequency Modulation Performance

audio frequency range passed by the postdetection filtering. Hence, such events are ignored.

References [8, 13, 14] all contain information on calculating the average rate of clicks as a function of the carrier-to-noise ratio ρ. Different and sometimes conflicting symbology is used so the reader must be cautious in comparing equations between sources. The total click rate (twice the rate of positive impulses, which equals the rate of negative impulses) can be written as $R_c = r\ \mathrm{erfc}(\sqrt{\rho})$, where r equals the radius of gyration of the IF filtering about the carrier frequency f_c. r can also be thought of as the representative frequency of the narrowband quadrature noise components. It is given by:

$$r = \frac{1}{2\pi}\sqrt{\frac{b_2}{b_0}}$$
$$b_0 = \int_0^\infty w(f)df$$
$$b_2 = (2\pi)^2 \int_0^\infty (f - f_c)^2 w(f)df \quad (15.38)$$

where $w(f)$ denotes the amplitude response of the predetection filtering. Table 15.1 indicates the r values for ideal bandpass and Gaussian-law filters.

Table 15.1 Values of Click Noise Parameter r

After: Rice, S. O., "Noise in FM Receivers," chapter 25 in *Time Series Analysis*, M. Rosenblatt (editor), John Wiley & Sons, New York, NY, 1963, p. 399.

$w(f)$	b_0	b_2	r
Ideal bandpass filter, centered on f_c, $w(f) = w_0$ for $f_c - (\beta/2) < f < f_c + (\beta/2)$, else $=0$	$w_0\beta$	$\pi^2 w_0 \beta^3/3$	$\beta/\sqrt{12}$
Gaussian filter, $w(f) = \frac{b_0}{\sigma\sqrt{2\pi}}\exp[-\frac{(f-f_0)^2}{2\sigma^2}]$ equivalent rectangular bandwidth equals $\sigma\sqrt{2\pi}$	b_0	$4\pi^2\sigma^2 b_0$	σ

Treating the clicks as random impulses means that shot noise theory applies. Because of the rapid rise and fall times of the click perturbations, their spectrum is essentially flat over the base bandwidth W and the total click power is given by:

$$N_c = \frac{4\pi^2 W B_i}{\sqrt{3}}\mathrm{erfc}(\sqrt{\rho}) \quad (15.39)$$

This assumes an ideal predetection bandwidth of B_i and ignores the influence of modulation on the carrier. Such influence can be substantial and is discussed in References [8, 14].

15.5 Performance in the Presence of Noise and Rayleigh Fading

Shilling et al. [19] show that when the fade rate is modest compared to the baseband bandwidth (W), the noise can be obtained simply by averaging over the PDF of the envelope of the received carrier. In land-mobile radio links that density function is generally Rayleigh in form. An additional requirement for this quasistatic approximation is that the delay spread of the multiple paths superposing to form the received signal must be small relative to the reciprocal of the transmission bandwidth. This equates to flat Rayleigh fading; i.e., the entire received signal spectrum is impacted identically by the amplitude fluctuations. Since delay spread is generally under a few microseconds (under 1-mi path difference) and land-mobile radio channels are commonly 25 kHz, this narrowband requirement is generally satisfied.

15.5.1 Faded Signal Power

The average voltage output of an FM demodulator for arbitrary predetection carrier-to-noise ratio ρ can be written as $v_0(t) = v(t)[1-\exp(-\rho)]$, where $v(t)$ is proportional to the instantaneous frequency of the received signal. Because the total noise power is constant, ρ is distributed just like carrier power. Rayleigh envelope variation implies exponential power variation, so:

$$f(\rho) = \frac{1}{\rho_0} \exp(-\frac{\rho}{\rho_0}) \tag{15.40}$$

where $\rho \geq 0$ and ρ_0 equals the average carrier-to-noise ratio. Thus, the average output voltage in the presence of Rayleigh fading is:

$$\begin{aligned} <v_0(t)> &= \int_0^\infty v(t)[1-\exp(-\rho)]\frac{1}{\rho_0}\exp(-\frac{\rho}{\rho_0})d\rho \\ &= v(t)\frac{\rho_0}{1+\rho_0} \end{aligned} \tag{15.41}$$

and the average output power is $S_o = <v(t)^2>[\rho_0/(1+\rho_0)]^2 = S_i[\rho_0/(1+\rho_0)]^2$.

15.5.2 Signal-Suppression Noise Power

The suppression of the signal by the presence of noise results in a signal-suppression noise voltage given by:

$$n_s(t) = v_0(t) - <v_0(t)>$$

Frequency Modulation Performance 327

$$= v(t)\{[1-\exp(-\rho)] - \frac{\rho_0}{1+\rho_0}\} \tag{15.42}$$

The signal-suppression noise power is given by the average of the square of that noise voltage over the distribution of ρ:

$$N_s = \int_0^\infty v(t)^2\{[1-\exp(-\rho)] - \frac{\rho_0}{1+\rho_0}\}^2 \frac{1}{\rho_0}\exp(-\frac{\rho}{\rho_0})d\rho$$

$$= S_i[\frac{1}{2\rho_0+1} - \frac{1}{(\rho_0+1)^2}] \tag{15.43}$$

15.5.3 Above-Threshold Noise Power

At large predetection carrier-to-noise levels, the nonfaded expression for noise power in base bandwidth W developed earlier is $N_2 = (a/\rho)[1-\exp(-\rho)]^2$. Averaging this for exponential variation of the carrier-to-noise level gives an above-threshold noise power of:

$$N_2 = \int_0^\infty \frac{a[1-\exp(-\rho)]^2}{\rho} \frac{1}{\rho_0}\exp(-\frac{\rho}{\rho_0})d\rho$$

$$= \frac{a}{\rho_0}\{\int_0^\infty \rho^{-1}\exp(-\frac{\rho}{\rho_0})d\rho - 2[\int_0^\infty \rho^{-1}\exp[-\frac{\rho(\rho_0+1)}{\rho_0}]d\rho$$

$$+ \int_0^\infty \rho^{-1}\exp[-\frac{\rho(2\rho_0+1)}{\rho_0}]d\rho\}$$

$$= \frac{a}{\rho_0}[E_1(\rho_0^{-1}) - 2E_1(\frac{\rho_0+1}{\rho_0}) + E_1(\frac{2\rho_0+1}{\rho_0})]$$

$$= \frac{-a}{\rho_0}[E_i(-\rho_0^{-1}) - 2E_i(-\frac{\rho_0+1}{\rho_0}) + E_i(-\frac{2\rho_0+1}{\rho_0})]$$

$$= \frac{a}{\rho_0}\{[-C-\ln(\frac{1}{\rho_0}) - \sum_{k=1}^\infty \frac{(\frac{-1}{\rho_0})^k}{kk!}] + 2[C+\ln(\frac{\rho_0+1}{\rho_0}) + \sum_{k=1}^\infty \frac{(-\frac{\rho_0+1}{\rho_0})^k}{kk!}]$$

$$+ [-C-\ln(\frac{2\rho_0+1}{\rho_0}) - \sum_{k=1}^\infty \frac{(-\frac{2\rho_0+1}{\rho_0})^k}{kk!}]\}$$

$$= \frac{a}{\rho_0}\ln[\frac{(\rho_0+1)^2}{(2\rho_0+1)}] \tag{15.44}$$

where exponential integral relations xxxii and 8.2.14.1 of Reference [20] have been used.

15.5.4 Threshold-and-Below Noise

At low carrier-to-noise values, the dominant noise contributors are $W_1(f)$ and $W_3(f)$. For base bandwidth W and an arbitrary carrier-to-noise value, this threshold-and-below noise power was found to be $N_0 = [8\pi BW\exp(-\rho)]/\sqrt{2(\rho+2.35)}$. Averaging this for exponential variation of the carrier-to-noise level gives a threshold-and-below noise power of:

$$\begin{aligned} N_0 &= \int_0^\infty \frac{8\pi BW \exp(-\rho)}{\sqrt{2(\rho + 2.35)}} \frac{1}{\rho_0} \exp(-\frac{\rho}{\rho_0}) d\rho \\ &= 8\pi BW \sqrt{\frac{\pi}{2\rho_0(\rho_0 + 1)}} \exp[\frac{2.35(\rho_0 + 1)}{\rho_0}] \text{erfc}[\sqrt{\frac{2.35(\rho_0 + 1)}{\rho_0}}] \end{aligned}$$

(15.45)

where relation 7.4.8 in Reference [4] has been used to carry out the integration.

15.5.5 Overall Fading Performance

Davis [11] considers the preceding results with Gaussian modulation and rms frequency deviation chosen so as to generally satisfy Carson's rule for transmission bandwidth. Specifically, the rms deviation is set 10 dB below that which matches Carson's rule; i.e., $B_{-3} = 2(W + \sigma\sqrt{10})$. This is done to ensure that signal deviation peaks seldom produce instantaneous transmission bandwidths in excess of the IF bandwidth. The resulting average signal power is thus $S_i = (2\pi f_d)^2 = (4\pi^2 W^2/10)[(B_{-3}/2W) - 1]^2$. The noise terms just developed can be summed and also written with a W^2 factor that cancels in the postdetection signal-to-noise expression. Figure 15.9 shows the result in terms of the parameter $(B_{-3}/2W)$. Notice the absence of FM improvement as found for the nonfaded case. Even with very large IF bandwidths compared to the base bandwidth, the output SNR is only comparable to the input carrier-to-noise ratio. Furthermore, for reasons that will now be discussed, SNR does not continue to improve as input carrier-to-noise ratio increases.

15.5.6 Random FM Noise

The presence of multiple signal paths from random directions that leads to Rayleigh fading not only affects the amplitude of the received carrier but also affects the received phase. But random fluctuations of the carrier phase impact the demodulator output and thus act as noise. Such noise, called random FM, is not reduced by boosting the transmitter power to increase the carrier level because that would change all signal paths equally, leaving the phase fluctuation unaffected.

Analysis of random FM noise is quite involved and the interested reader is referred to Section 1.4 of Reference [3]. An important asymptotic relation derived there for the power spectral density of random FM is:

$$W_{rfm}(f) = \frac{2\pi^2 f_m^2}{f}$$

(15.46)

Frequency Modulation Performance

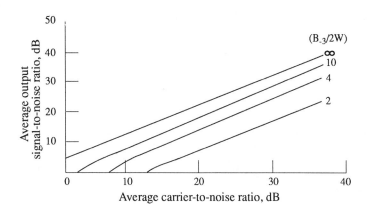

Figure 15.9 FM output SNR versus predetection carrier-to-noise ratio, Rayleigh-faded conditions.
After: Davis, B. R., "FM Noise with Fading Channels and Diversity," *IEEE Tr. on Comm.*, Vol. COM-19, No. 6, December 1971, p. 1198.

where the maximum Doppler frequency f_m is related to vehicle speed, carrier frequency, and speed of light through $f_m = (v/c)f_c$. This relation is useful down to about twice the maximum Doppler frequency. At 60 mph and 900 MHz that implies usefulness down to about 160 Hz. Since voice is adequately passed by baseband filtering covering 300 Hz to 3 kHz, the relation is sufficient[3]. One can thus take the random FM noise to be:

$$N_{rfm} = \int_{0.1W}^{W} \frac{2\pi^2 f_m^2}{f} df$$
$$= 2\pi^2 f_m^2 \ln(10) \qquad (15.47)$$

The ultimate obtainable signal-to-noise value is based on this equation. Under strong signal conditions and at sufficiently high speeds and link frequencies one hears a continuous "grumble" along with the intended audio. The level of the top-speed grumble quadruples with a doubling of link frequency and by L-band begins to significantly overlap the lower audio frequency range of speech.

---

[3]The impact of assuming postdetection filtering running $0.1W$ to W rather than 0 to W is negligible on the signal and other noise terms. In fact, referring back to Figure 15.6, one sees that typical audio filtering is bandpass in nature anyway. The difference in random FM noise level for EIA filtering as opposed to ideal bandpass filtering is only about 1 dB. The small difference is because EIA filtering primarily addresses reduction of high audio frequency parabolic noise, whereas the random FM noise power is concentrated at low frequencies.

15.5.7 Alternative View of Overall Fading Performance

A slightly different set of equations is developed by Park and Chayavadhanangkur [22, 23]. The main differences from the preceding sections stem from their inclusion of modulation effects and use of a click noise model. Modulation is assumed to be a zero-mean Gaussian process bandlimited to W hertz, with rms frequency deviation $\sigma_{\rm rms} = \sigma_{\rm pk}/\sqrt{2}$. The noise components considered are: (1) signal-suppression noise, (2) high carrier-to-noise parabolic or frequency-squared noise, (3) low carrier-to-noise click noise, and (4) random FM noise. Because the effects are statistically independent their contributions are separately calculated and then summed. For all but the last component this includes averaging over an exponential distribution of predetection carrier-to-noise ratio (the consequence of flat Rayleigh fading of the desired signal envelope).

Experimentally this viewpoint successfully describes the input-output relation of typical land-mobile radios provided that one additional effect is accounted for; namely, harmonic distortion. Table 15.2 compares measured SINAD levels with predictions using Park and Chayavadhanangkur's equations plus the measured -32-dBr second-harmonic distortion level. This additional effect is particularly important under strong signal conditions because it represents a limiting factor on the SINAD achievable. Random FM also represents a limiting factor, though here it is of lesser importance. Because of the low fading rate, the notch filter in the distortion meter used to measure SINAD was unable to separate signal power from signal-suppression power; hence, the two effects were lumped together and treated as total signal S_T. Table 15.2 also uses the simpler click noise expression obtained when modulation is ignored, a simplification justified in the presence of Rayleigh fading.

Table 15.2 Theoretical SINAD Versus Average Predetection Carrier-to-Noise Ratio in the Presence of Rayleigh Fading

ρ_0	$\overline{S_T}$	$\overline{S_{2nd}}$	$\overline{N_s}$	$\overline{N_2}$	$\overline{N_3}$	N_{FM}	SINAD Theory	SINAD Measured
4 dB	1.06E8	6.69E4	1.51E7	2.42E6	1.63E7	1.73E4	8.2 dB	8.7 dB
9	1.48E8	9.34E4	8.30E6	1.66E6	6.08E6	1.73E4	13.0	14.2
14	1.67E8	1.05E5	3.21E6	8.74E5	2.04E6	1.73E4	17.5	18.5
19	1.74E8	1.10E5	1.08E6	3.95E5	6.58E5	1.73E4	21.7	22.1
24	1.76E8	1.11E5	3.50E5	1.63E5	2.09E5	1.73E4	25.5	24.0
29	1.77E8	1.12E5	1.11E5	6.39E4	6.65E4	1.73E4	28.3	26.9

In the table ρ_0 equals the average IF carrier-to-noise ratio, σ_{pk} equals 3000 Hz (the EIA standard for SINAD tests with 5-kHz peak deviation radios), and the relevant equations are[4]:

$$\overline{S_T} = \overline{S} + \overline{N_s}$$
$$= \frac{4\pi^2 \sigma_{pk}^2 \rho_0^2}{(1+\rho_0)(1+2\rho_0)}$$
$$\overline{S} = (\frac{\rho_0}{1+\rho_0})^2 (\frac{2\pi\sigma_{pk}}{\sqrt{2}})$$
$$\overline{N_s} = (\frac{\rho_0}{1+\rho_0})^2 (\frac{1}{1+2\rho_0})(\frac{2\pi\sigma_{pk}}{\sqrt{2}})$$

A second harmonic level of -32 dBr is assumed and:

$$\overline{N_2} = (\frac{4\pi^2 W_1^3}{3B_W})\text{erf}(\frac{B_W}{2\sigma_{pk}})(\frac{1}{\rho_0})\ln[\frac{(1+\rho_0)^2}{(1+2\rho_0)}]$$

where W_1 equals 1.77 kHz (via numerical integration of the EIA standard de-emphasis filter parabolic noise response), B_W equals 8.13 kHz (via numerical integration of a typical 25-kHz FM radio IF response), and

$$\overline{N_3} = (\frac{\pi^2 W_2 B_W}{\sqrt{3}})[1 - \sqrt{\frac{\rho_0}{1+\rho_0}}]$$

where W_2 equals 2.28 kHz (via numerical integration of the EIA standard de-emphasis filter flat noise response), and

$$N_{\text{FM}} = 2\pi^2 f_m^2 \ln(79)$$

where f_m equals 13 Hz (60 mph at 145 MHz) and the 1/f noise response of the EIA standard de-emphasis filter has been used. Finally, SINAD is determined through:

$$SINAD = \frac{\overline{S_T} + \overline{S_{2nd}} + \overline{N_2} + \overline{N_3} + N_{\text{FM}}}{\overline{S_{2nd}} + \overline{N_2} + \overline{N_3} + N_{\text{FM}}}$$

15.6 Performance in the Presence of Noise, Interference, and Rayleigh Fading

Chayavadhanangkur and Park [21] address the static performance of FM in the presence of noise and interference. They later generalize this to include Rayleigh fading [23]. An ideal predetection bandpass filter is assumed, as is an ideal postdetection low-pass filter of width W. The desired signal rms deviation is set 10 dB below that which satisfies Carson's rule for the predetection bandwidth, so as to prevent

---
[4] The numeric value of ρ_0 should be used in the equations, not the decibel value.

distortion. Noise is modeled as five independent terms: (1) signal-suppression noise, (2) random FM noise, (3) above-threshold noise, (4) click noise, and (5) noise due to the desired signal beating with the interfering signal. The quasistatic approximation is used to find average values for the various noise terms. It is established that the modulation impact on click noise is negligible in the presence of fading.

The results of Park and Chayavadhanangkur can readily be modified to account for EIA de-emphasis filtering as follows: (1) replace W with 1.77 kHz in the quadratic noise relation, (2) replace W with 2.28 kHz in the click noise relation, and (3) replace W with 1.7 kHz in the interference noise relation. These replacement values are based on numerical integration of the EIA filter response, weighted by the appropriate frequency functions.

15.7 Homework Problems

15.7.1 Problem 1

Static bit error probabilities for various data modulation schemes are simple functions of the predetection SNR. Common forms are:

$$P_e = (1/2)\exp(-k\rho/2)$$

where $k = 1$ for noncoherent frequency-shift keying and $k = 2$ for coherent differential phase shift keying, and:

$$P_e = (1/2)\text{erfc}(\sqrt{k\rho})$$

where $k = 1$ for coherent phase-shift keying, $k = 1/2$ for coherent frequency-shift keying, and $k = 1/4$ for on-off keying (Morse code). Apply the quasistatic approximation to obtain faded bit error probabilities for both forms, assuming flat Rayleigh fading and maximum Doppler frequencies modest relative to the signaling rate.

Solution:
Form 1

$$\begin{aligned} <P_e^1> &= \int_0^\infty \frac{1}{\rho_0}\exp(-\frac{\rho}{\rho_0})[\frac{1}{2}\exp(-\frac{k\rho}{2})]d\rho \\ &= -\frac{1}{2+x}\int_0^{-\infty}\exp(s)ds \\ &= \frac{1}{2+x} \end{aligned} \quad (15.48)$$

where $x = \rho_0 k$.
Form 2

$$<P_e^2> = \int_0^\infty \frac{1}{\rho_0}\exp(-\frac{\rho}{\rho_0})[\frac{1}{\sqrt{\pi}}\int_{\sqrt{k\rho}}^\infty \exp(-t^2)dt]d\rho \quad (15.49)$$

Frequency Modulation Performance 333

where ρ_0 is the average Rayleigh-faded predetection SNR and the complementary error function has been expressed in its basic integral form. The integration can be carried out by parts, with:

$$u = \frac{1}{\sqrt{\pi}} \int_{\sqrt{k\rho}}^{\infty} \exp(-t^2) dt$$
$$dv = \frac{1}{\rho_0} \exp(-\frac{\rho}{\rho_0}) d\rho \qquad (15.50)$$

implying:

$$du = -\frac{1}{\sqrt{\pi}} \exp(-k\rho) \frac{1}{2} \sqrt{\frac{k}{\rho}} d\rho$$
$$v = -\exp(-\frac{\rho}{\rho_0}) \qquad (15.51)$$

The result is:

$$<P_e^2> = \frac{1}{2}(1 - \sqrt{\frac{x}{1+x}})$$
$$\approx \frac{1}{4x}, \quad x \to \infty \qquad (15.52)$$

Form 2 is about 6 dB better than form 1; i.e., it produces the same faded bit error probability with 6 dB lower SNR. But note that in both cases bit error probability decreases linearly with SNR; this is a much slower rolloff rate than the exponential and error function rolloffs under static conditions.

15.7.2 Problem 2

Develop an expression for the average rate of clicks as a function of the carrier-to-noise ratio, assuming both static and Rayleigh-faded circumstances.

Solution: Begin by considering the behavior of the land-mobile propagation model drawn in Figure 6.5, but for generality let the angle between vehicle heading and the x-axis be α, rather than zero, and let the angle between the generalized x-axis and the i-th signal component be ϕ_i. The total received vertical electric field from N incoming waves can then be written in phasor form as [24]:

$$E_z = \sum_{i=1}^{N} (R_i + jS_i) \exp[-j\beta V t \cos(\phi_i - \alpha)]$$
$$= X_1 + jY_1$$
$$= r \exp(j\psi_r) \qquad (15.53)$$

where $\beta = (2\pi/\lambda)$, λ equals the wavelength, and R_i and S_i are complex coefficients accounting for the amplitude and propagation delay (phase) of the i-th signal component. Assume the coefficients to be independently distributed zero-mean Gaussian random variables with unity variance. Letting $\xi_i = \beta V t \cos(\phi_i - \alpha)$, one obtains:

$$X_1 = \sum_{i=1}^{N} [R_i \cos(\xi_i) + S_i \sin(\xi_i)]$$

$$Y_1 = \sum_{i=1}^{N} [S_i \cos(\xi_i) - R_i \sin(\xi_i)] \quad (15.54)$$

As N approaches infinity, the central limit theorem indicates that both X_1 and Y_1 will tend to be Gaussian distributed. The mean and variance are obtained as follows.

The expected value of X_1 is given by:

$$E(X_1) = \sum_{i=1}^{N} E[R_i < \cos(\xi_i) > + S_i < \sin(\xi_i) >] \quad (15.55)$$

where $<f(t)>$ denotes the time average $\lim_{T \to \infty} (1/T) \int_0^T f(t) dt$. Since $<\cos(\xi_i)> = <\sin(\xi_i)> = 0$, the mean values of X_1 and Y_1 are zero.

The mean-square value of X_1 is given by:

$$E(X_1^2) = \sum_{i=1}^{N} [E(R_i^2) < \cos^2(\xi_i) > + E(S_i^2) \sin^2(\xi_i)] \quad (15.56)$$

The time averages both equal $1/2$ so the mean-square value is N; likewise for Y_1^2. Because both random variables are zero mean, the mean-square value equals the variance. Thus $X_1, Y_1 \sim G(0, N) = G(0, \sigma^2)$, where the signal power is denoted as σ^2 and $G(\mu, s^2)$ denotes the Gaussian distribution with mean μ and variance s^2.

The time derivative behavior of X_1 and Y_1 is also needed to tackle the question of click rate. These derivatives can be written as:

$$\dot{X}_1 = (\beta V) \sum_{i=1}^{N} [-R_i \sin(\xi_i) + S_i \cos(\xi_i)] \cos(\phi_i - \alpha)$$

$$\dot{Y}_1 = (\beta V) \sum_{i=1}^{N} [-S_i \sin(\xi_i) - R_i \cos(\xi_i)] \cos(\phi_i - \alpha) \quad (15.57)$$

The expected values of $\dot{X}_1$ and $\dot{Y}_1$ are readily seen to equal zero. The mean-square values, and variances, are found via:

$$\dot{X}_1^2 = (\beta V)^2 \sum_{i=1}^{N} [R_i^2 \sin^2(\xi_i) + S_i^2 \cos^2(\xi_i)$$
$$- 2 R_i S_i \sin(\xi_i) \cos(\xi i)] \cos^2(\phi_i - \alpha)$$

$$E[\dot{X}_1^2] = (\beta V)^2 \sum_{i=1}^{N} E\{[R_i^2 < \sin^2(\xi_i) > + S_i^2 < \cos^2(\xi_i) >$$
$$- 2R_iS_i < \sin(\xi_i)\cos(\xi_i) >]\cos^2(\phi_i - \alpha)\}$$
$$= (\beta V)^2 \frac{N}{2}$$
$$= \nu^2 \qquad (15.58)$$

Thus $\dot{X}_1 \sim G(0, \nu^2)$. The same behavior is also found for $\dot{Y}_1$.

Similarly, one can show that all covariance terms equal zero; i.e., $E(X_1, Y_1) = E(X_1, \dot{Y}_1) = E(Y_1, \dot{X}_1) = E(Y_1, \dot{Y}_1) = E(\dot{X}_1, \dot{Y}_1) = E(X_1, \dot{X}_1) = 0$. The four-tuple $(X_1, Y_1, \dot{X}_1, \dot{Y}_1)$ thus involves four independent Gaussian random variables so the joint PDF can be written as (p. 255 of Reference [10]):

$$f(X_1, Y_1, \dot{X}_1, \dot{Y}_1) = \frac{1}{(2\pi\sigma\nu)^2} \exp[-\frac{1}{2}(\frac{X_1^2 + Y_1^2}{\sigma^2} + \frac{\dot{X}_1^2 + \dot{Y}_1^2}{\nu^2})] \qquad (15.59)$$

The PDF for the four-tuple $(r, \psi_r, \dot{r}, \dot{\psi}_r)$ can be obtained by using the relations $X_1 = r\cos(\psi_r), Y_1 = r\sin(\psi_r), \dot{X}_1 = \dot{r}\cos(\psi_r) - r\dot{\psi}_r\sin(\psi_r), \dot{Y}_1 = \dot{r}\sin(\psi_r) + r\dot{\psi}_r\cos(\psi_r)$ to evaluate the determinant of the Jacobian:

$$J = \begin{bmatrix} \frac{\partial X_1}{\partial r} & \frac{\partial Y_1}{\partial r} & \frac{\partial \dot{X}_1}{\partial r} & \frac{\partial \dot{Y}_1}{\partial r} \\ \frac{\partial X_1}{\partial \psi_r} & \frac{\partial Y_1}{\partial \psi_r} & \frac{\partial \dot{X}_1}{\partial \psi_r} & \frac{\partial \dot{Y}_1}{\partial \psi_r} \\ \frac{\partial X_1}{\partial \dot{r}} & \frac{\partial Y_1}{\partial \dot{r}} & \frac{\partial \dot{X}_1}{\partial \dot{r}} & \frac{\partial \dot{Y}_1}{\partial \dot{r}} \\ \frac{\partial X_1}{\partial \dot{\psi}_r} & \frac{\partial Y_1}{\partial \dot{\psi}_r} & \frac{\partial \dot{X}_1}{\partial \dot{\psi}_r} & \frac{\partial \dot{Y}_1}{\partial \dot{\psi}_r} \end{bmatrix}$$

$$= \begin{bmatrix} \cos(\psi_r) & \sin(\psi_r) & -\dot{\psi}_r\sin(\psi_r) & \dot{\psi}_r\cos(\psi_r) \\ -r\sin(\psi_r) & r\cos(\psi_r) & -\dot{r}\sin(\psi_r) - r\dot{\psi}_r\cos(\psi_r) & \dot{r}\cos(\psi_r) - r\dot{\psi}_r\sin(\psi_r) \\ 0 & 0 & \cos(\psi_r) & \sin(\psi_r) \\ 0 & 0 & -r\sin(\psi_r) & r\cos(\psi_r) \end{bmatrix}$$

The determinant of this matrix simplifies nicely to just r^2, so:

$$f(r, \psi_r, \dot{r}, \dot{\psi}_r) = r^2 f(X_1, Y_1, \dot{X}_1, \dot{Y}_1)$$
$$= \frac{r^2}{(2\pi\sigma\nu)^2} \exp[-\frac{1}{2}(\frac{r^2}{\sigma^2} + \frac{r^2\dot{\psi}_r^2 + \dot{r}^2}{\nu^2})] \qquad (15.60)$$

where $-\infty < r, \dot{r}, \dot{\psi}_r < \infty$ and $0 < \psi_r < 2\pi$.

This density function also applies to the four-tuple $(n, \dot{n}, \psi_n, \dot{\psi}_n)$, where n represents the envelope of white Gaussian noise and ψ_n represents the phase angle as shown in Figure 15.10 (a repeat of Figure 15.4, but using the notation of Lee rather than that of Rice; for simplicity the x-axis can be taken as aligned with the signal). Thus:

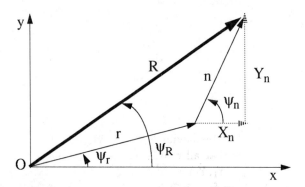

Figure 15.10 Signal plus noise phasor diagram.
After: Lee, W.C.Y., *Mobile Communications Engineering*, McGraw-Hill Book Company, New York, NY, 1982, p. 225.

$$f(n, \dot{n}, \psi_n, \dot{\psi}_n) = \frac{n^2}{(2\pi\sigma_n\nu_n)^2} \exp[-\frac{1}{2}(\frac{n^2}{\sigma_n^2} + \frac{n^2\dot{\psi}_n^2 + \dot{n}^2}{\nu_n^2})] \qquad (15.61)$$

This relation can be integrated over the time rate of change of noise envelope to yield the three-tuple PDF:

$$\begin{aligned}
f(n, \psi_n, \dot{\psi}_n) &= \int_{-\infty}^{\infty} f(n, \dot{n}, \psi_n, \dot{\psi}_n) d\dot{n} \\
&= \frac{n^2}{2\pi\sigma_n^2\sqrt{2\pi\nu_n^2}} \exp[-\frac{1}{2}(\frac{n^2}{\sigma_n^2} + \frac{n^2\dot{\psi}_n^2}{\nu_n^2})] \cdot \\
&\quad \int_{-\infty}^{\infty} \frac{1}{\sqrt{2\pi\nu_n^2}} \exp[-\frac{\dot{n}^2}{2\nu_n^2}] d\dot{n} \\
&= \frac{n^2}{2\pi\sigma_n^2\sqrt{2\pi\nu_n^2}} \exp[-\frac{1}{2}(\frac{n^2}{\sigma_n^2} + \frac{n^2\dot{\psi}_n^2}{\nu_n^2})] \qquad (15.62)
\end{aligned}$$

The utility of this density function is appreciated when one considers the conditions under which clicks are generated; namely, (1) the instantaneous noise amplitude exceeds that of the signal and (2) the noise phase angle equals π and is increasing with time (this is for the positive click rate; negative clicks stem from the phase angle decreasing with time). Thus, the average positive click rate can be found from:

$$N_+ = \frac{1}{T}\int_0^T dt \int_0^\infty f(r)dr \int_r^\infty dn \int_0^\infty \dot{\psi}_n f(n, \psi_n = \pi, \dot{\psi}_n) d\dot{\psi}_n \qquad (15.63)$$

where T represents the time interval over which clicks are counted and the integral involving $\dot{\psi}_n$ represents the average level crossing rate for positive time rates of

Frequency Modulation Performance

change of angle when the angle equals π (Section 9.5.1). In the case of Rayleigh fading $f(r) = (r/\sigma^2)\exp(-r^2/\sigma^2)$; for static circumstances r can be treated as a constant and the integral involving dr ignored.

Considering the more complicated Rayleigh fading situation first, one finds:

$$N_+ = \frac{1}{T}\int_0^T dt \int_0^\infty \frac{r}{\sigma^2}\exp(-\frac{r^2}{2\sigma^2})dr \int_r^\infty dn$$
$$\cdot \int_0^\infty \dot{\psi}_n \frac{n^2}{2\pi\sigma_n^2\sqrt{2\pi\nu_n^2}}\exp[-\frac{1}{2}(\frac{n^2}{\sigma_n^2} + \frac{n^2\dot{\psi}_n^2}{\nu_n^2})]d\dot{\psi}_n \quad (15.64)$$

The integral in terms of $\dot{\psi}_n$ equals:

$$I_1 = C_1 \int_0^\infty \dot{\psi}_n \exp[-\frac{\dot{\psi}_n^2}{2(\frac{\nu_n}{n})^2}]d\dot{\psi}_n$$
$$= C_1(\frac{\nu_n}{n})^2 \quad (15.65)$$

where the constant term $C_1 = [n^2\exp(-n^2/2\sigma_n^2)]/(2\pi\sigma_n^2\sqrt{2\pi\nu_n^2})$ and the substitution $z = \dot{\psi}_n^2/[2(\nu_n/n)^2]$ is helpful. The integral in terms of n becomes:

$$I_2 = C_2 \int_r^\infty \frac{1}{\sqrt{2\pi\sigma_n^2}}\exp(-\frac{n^2}{2\sigma_n^2})dn$$
$$= C_2 \frac{1}{2}\mathrm{erfc}(\frac{r}{\sqrt{2}\sigma_n}) \quad (15.66)$$

where the constant term $C_2 = \nu_n/[\sqrt{2\pi\sigma_n^2}\sqrt{2\pi}]$. Next, the integral in terms of r becomes:

$$I_3 = C_3 \int_0^\infty r\exp(-\frac{r^2}{2\sigma^2})\mathrm{erfc}(\frac{r}{\sqrt{2}\sigma_n})$$
$$= C_3\sigma^2[\frac{1}{\sqrt{1-(\frac{\sigma_n}{\sigma})^2}} - 1]$$
$$\approx \frac{\sigma_n^2}{2} \quad (15.67)$$

where $C_3 = (\nu_n/\sigma_n)(1/4\pi\sigma^2)$, $\sigma \gg \sigma_n$, and relation 4.3.5 in Reference [25] has been used. Thus, one finds the click rate in Rayleigh fading at high carrier-to-noise ratio is given approximately by:

$$N_+ \approx \frac{1}{8\pi}(\frac{\nu_n}{\sigma_n})\frac{1}{\rho_0} \quad (15.68)$$

where $\rho_0 = (\sigma/\sigma_n)^2$ equals the average carrier-to-noise power ratio.

Now consider the case of white noise passed through an ideal predetection filter of bandwidth B centered at the carrier frequency. If the one-sided noise spectral density is taken as a constant, $G_n(f) = \eta$, then the noise power out of the predetection filtering (i.e., the noise power applied to the FM detector) is given by $\sigma_n^2 = \int_{-\frac{B}{2}}^{\frac{B}{2}} \eta \, df = \eta B$. Differentiation is equivalent to linear filtering by the transfer function $j\omega = j2\pi f$. Thus, the noise spectral density for $\dot{n}$ is given by $G_{\dot{n}}(f) = |j\omega|^2 G_n(f) = 4\pi^2 f^2 \eta$ (p. 347 of Reference [10]). The power associated with the time rate of change of the noise envelope coming out of the ideal bandpass filter is then $\nu_n^2 = \int_{-\frac{B}{2}}^{\frac{B}{2}} 4\pi^2 f^2 \eta \, df = (\pi^2 \eta B^3 / 3)$. Hence, $(\nu_n/\sigma_n) = (\pi B/\sqrt{3})$ and $N_{+,\text{ideal}} \approx (B/8\sqrt{3}\rho_0)$.

The case of constant r is much simpler as the complicated integral I_3 is not needed. For static conditions the click rate is simply:

$$\begin{aligned}
N_+ &= \frac{1}{T} \int_0^T dt \int_r^\infty dn \int_0^\infty \dot{\psi}_n f(n, \psi_n = \pi, \dot{\psi}_n) d\dot{\psi}_n \\
&= \frac{1}{T} \int_0^T dt \left(\frac{\nu_n}{4\pi \sigma_n}\right) \text{erfc}\left(\frac{r}{\sqrt{2}\sigma_n}\right) \\
&= \left(\frac{\nu_n}{4\pi \sigma_n}\right) \text{erfc}[\sqrt{\rho}]
\end{aligned} \qquad (15.69)$$

For an ideal predetection bandpass filter, this becomes:

$$N_{+,\text{ideal}} = (B/4\sqrt{3}) \text{erfc}[\sqrt{\rho}]$$

References

[1] Bateman, A. J., et al., "Speech and Data Communications Over 942 MHz TAB and TTIB Single Sideband Mobile Radio Systems Incorporating Feed-Forward Signal Regeneration," *IEEE Tr. on Veh. Tech.*, Vol. VT-34, No. 1, February 1985, pp. 13-21.

[2] Leland, K. W. and N. R. Sollenberger, "Impairment Mechanisms for SSB Mobile Communications at UHF with Pilot-Based Doppler/Fading Correction," *Bell System Technical Journal*, Vol. 59, No. 10, December 1980, pp. 1923-1942.

[3] Jakes, Jr., W. C. (editor), *Microwave Mobile Communications*, New York, NY, John Wiley & Sons, 1974.

[4] Abramowitz, M. and I. Stegun (editors), *Handbook of Mathematical Functions*, New York, NY, Dover Publications, Inc., 1972.

[5] Carlson, A. B., *Communication Systems, An Introduction to Signals and Noise in Electrical Communication*, New York, NY, McGraw-Hill Book Co., 1968.

[6] Panter, P. F., *Modulation, Noise, and Spectral Analysis*, New York, NY, McGraw-Hill Book Co., 1965.

[7] Lee, W.C.Y., "Narrowbanding in Cellular Mobile Systems," *Telephony*, December 1, 1986, pp. 44-46.

[8] Rice, S. O., "Noise in FM Receivers,", Chapter 25 in *Time Series Analysis*, M. Rosenblatt (editor), New York, NY, John Wiley & Sons, 1963, pp. 395-422.

[9] Rice, S. O., "Statistical Properties of Sinewave Plus Random Noise," *Bell System Technical Journal*, Vol. 27, January 1948, pp. 109-157.

[10] Papoulis, A., *Probability, Random Variables, and Stochastic Processes*, New York, NY, McGraw-Hill Book Co., 1965.

[11] Davis, B. R., "FM Noise with Fading Channels and Diversity," *IEEE Tr. on Comm.*, Vol. COM-19, No. 6, December 1971, pp. 1189-1200.

[12] Fletcher, H., *Speech and Hearing in Communications*, New York, NY, Van Nostrand, 1953.

[13] Cohn, J., "A New Approach to the Analysis of FM Threshold Reception," *Proceedings of the N.E.C.*, 1956, pp. 221-236.

[14] Taub, H. and D. L. Schilling, *Principles of Communications Systems*, New York, NY, McGraw-Hill Book Co., 1971.

[15] Yavuz, D. and D. T. Hess, "FM Noise and Clicks," *IEEE Tr. on Comm. Tech.*, Vol. COM-17, No. 6, December 1969, pp. 648-653.

[16] Yavuz, D. and D. T. Hess, "False Clicks in FM Detection," *IEEE Tr. on Comm. Tech.*, Vol. COM-18, No. 6, December 1970, pp. 751-756.

[17] Glazer, A., "Distribution of Click Amplitudes," *IEEE Tr. on Comm. Tech.*, Vol. COM-19, August 1971, pp. 539-543.

[18] Yavuz, D., "FM Click Shapes," *IEEE Tr. on Comm. Tech.*, Vol. COM-19, No. 6, December 1971, pp. 1271-1272.

[19] Schilling, D. L., et al., "Discriminator Response to an FM Signal in a Fading Channel," *IEEE Tr. Comm. Tech.*, Vol. COM-15, April 1967, pp. 252-263.

[20] Gradshteyn, I. S. and I. M. Ryzhik, *Table of Integrals, Series, and Products*, New York, NY, Academic Press, 1965.

[21] Chayavadhanangkur, C. and J. H. Park, Jr., "Analysis of FM systems with Co-channel Interference using a Click Model," *IEEE Tr. on Comm.*, Vol. COM-24, August 1976, pp. 903-910.

[22] Chayavadhanangkur, C., "Performance of FM Discriminators with Fading Signals and Co-channel Interference," PhD Dissertation, University of Minnesota, March 1975.

[23] Park, Jr., J. H. and C. Chayavadhanangkur, "Effect of Fading on FM Reception With Cochannel Interference," *IEEE Tr. on Aero. and Elec. Sys.*, Vol. AES-13, No. 2, March 1977, pp. 127-132.

[24] Lee, W.C.Y., *Mobile Communications Engineering*, New York, NY, McGraw-Hill Book Co., 1982.

[25] Ng, E. and M. Geller, "A Table of Integrals of the Error Function," *J. of Res. of the NBS*, B. Mathematical Sciences, Vol. 73B, No. 1, January-March 1969, pp. 1-20.

Chapter 16

Building Shadowing Adjustment Model Investigation

This chapter discusses three methods for estimating the descrepancy between observed signal levels and those predicted by typical propagation software (for example, the Okumura propagation model with adjustments due to local terrain) when the propagation path is subject to substantial building shadowing. The importance of removing such errors increases as the size of the coverage cell decreases.

The methods studied involve: (1) artificial neural networks, (2) group classification, and (3) linear regression. While method 1 holds the promise of greatest error reduction, it requires substantially more observations than are currently on hand. In addition, no clear way of generalizing to arbitrary range is known. That is likewise an issue with method (2). Consequently, at least in the short term, we are limited to method (3). It appears capable of removing prediction bias caused by building shadowing, and limiting prediction variability to near that caused by the usual lognormal shadowing alone.

16.1 Introduction

16.1.1 The Problem

Recent Chicago field measurements highlight the importance of building shadowing in setting the median received power [1]. Because Chicago terrain is generally flat, current propagation design tools predict relatively circular coverage; yet observations exhibit substantial differences among median signal levels at fixed distances and various headings (for example, see Figure 16.1). This is true not only for cluttered sites like RES, but is also often true for urban sites above the clutter and even for sites in nonurban environments. The observed median signal variations with heading are believed to be primarily due to changes in the building profile along the propagation path. For example, a path over Grant Park obviously will have less path

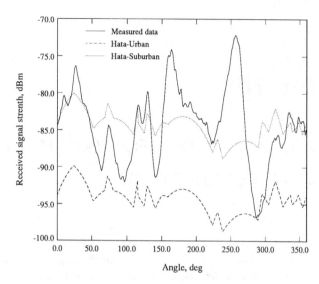

Figure 16.1 Median received signal strength versus heading, RES site, 1 mile radius.

loss than a path through a steel curtain of skyscrapers. Accounting for this extra variability is essential to the successful layout of multisite frequency reuse systems in major metropolitan areas.

16.1.2 Prior Art

There is a substantial amount of technical literature on the subject of building loss. However, the focus is generally on communication to a radio inside a building, not communication between radios separated by some number of buildings. One reference relevant to the latter problem is by Kozono and Watanabe [2]. They examine linear regressions of four candidate building shadowing model variables based on local information only, not information along the entire propagation path. Actual height information is not known, but rather is inferred from the rough number of stories per building. Ibrahim and Parsons [3] have also reported a simple regression model based on local information.

Saunders and Bonar [4] have generalized and extended Bertoni's work [5] on diffraction over rooftops; however, their focus is on removing general signal strength prediction bias, not reducing path-specific errors. Ray-tracing efforts, like that of Ikegami et al. [6], are capable of addressing path-specific errors. To do so though requires substantial information about the buildings along the path, not just their rough locations and heights.

16.1.3 Chicago Building Database

Thanks to the excellent detective work of Karen Brailean, it was discovered that the Chicago Property Information Program database was available for purchase at a modest fee from the *Chicago Housing Authority* (CHA). Software has been written to reduce the database from roughly 180 Mbytes to 24,777 blocks on a VAX scratch disk. This is accomplished by discarding all information except the Sanborn map volume ID, the low-street address, the commercial floorspace, and the number of stories. A further reduction to just 1725 blocks is achieved by retaining information for map volumes 1NW, 2S, 3, 6, 7, and 8 only. The field test data for 1- and 3-mi radius routes about the CON,CTT,REP, and IBM sites fall within those volumes. Building height is inferred using 12 ft per story, a value supported by regression analysis of randomly selected buildings for which exact heights were known via Sanborn map notations and a value consistent with that reported in Reference [2].

An important item lacking in the database is an x,y coordinate position for each building. Coordinates can, however, be estimated from street addresses as follows. First, a database of street intersections and their latitude/longitude coordinates is created. This is easily done using digitizing equipment for United States Geological Survey 7.5-min maps. Next, a program is run to update a post office file that relates post office numbers to streets (for example, Chicago Ave. is 800 N and Canal St. is 500 W). For each street name in the intersection list, this program searches the post office file and, if the street is not found, adds the street name without a post office number. Later, post office numbers for such new entries are added manually. Another program reads the intersection file created by the digitizer and rearranges the data so that all north/south streets are listed first, with cross streets in order of south-to-north post office number. Then east/west streets are listed with cross streets in east-to-west order. The output file created contains street intersections, post office numbers, and x,y coordinates in miles referenced to State and Madison streets. A final program creates x,y values for each entry in the reduced CHA building database via interpolation and extrapolation of the street intersection information.

Figure 16.2 shows a sample result of performance. The enhanced database is used along with Hata's empirical fit [7] to Okumura's propagation model [8] to pair building height profiles with median signal strength errors. The profiles involve 12 radial bins per mile with a ± 5- or 10-deg spread in azimuth. Five deg constitutes the minimum azimuthal spread for circular routes at 1 mi which is sufficient to filter out not only Rayleigh fading, but the usual lognormal shadowing as well. The profiles are automatically generated by software, although particularly critical bin 2 and 3 entries are manually checked and revised as deemed appropriate. Bin 1 is simply taken as the base antenna height. In fact, three different quantities have

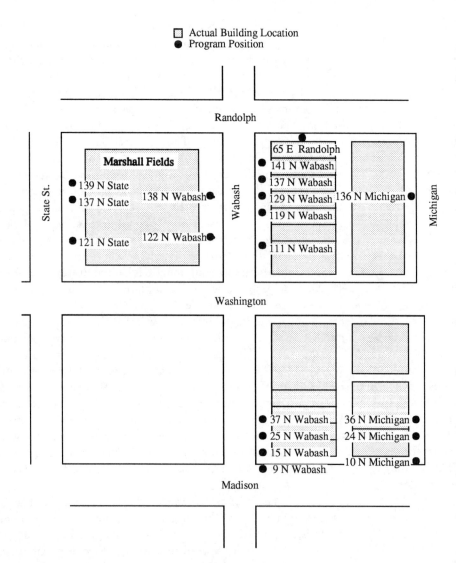

Figure 16.2 CHA building placement example.

been profiled: median building height, quartile spread of building height, and total number of buildings per bin. However, generally only the first is used in our modeling efforts.

16.2 Artificial Neural Network Models

16.2.1 Brief Tutorial on Artificial Neural Networks

Here we explore an untraditional modeling approach where a three-layer *artificial neural network* (ANN) is used to generate predictions of the median signal strength errors from the building height profiles. This holds the promise of eliminating or at least reducing the errors. ANNs are natural extensions of the threshold logic concept, which involves more powerful logic elements than those of traditional Boolean algebra. A succinct overview of ANN technology and paradigms is contained in Reference [9].

For example, consider the task of implementing a voting function on three inputs. Figure 16.3 shows a four-gate, two-layer Boolean solution; however, a single threshold element with threshold set to 2 will also accomplish the task.

The *processing element* (PE) can be generalized by allowing arbitrary sigmoid-type transfer functions and interconnection strengths or weights (Figures 16.4 and 16.5). Interconnecting a number of such generic processing elements in parallel across a number of layers yields an ANN.

Figure 16.6 shows the structure of the so-called perceptron ANN with three layers. For our prediction task we associate an input-layer PE with each bin of the building height profile. A second fully interconnected hidden layer of equal size is used, along with a single output PE. Because the number scale of the inputs (typically hundreds of feet) differs substantially from the desired number scale of the output (tens of decibels), we include a scaling factor after the output PE. Additional degrees of freedom are allowed by associating "time constants" with each layer's nonlinearity. Because our output is bipolar we use the hyperbolic tangent function for the nonlinearity, but make its presence optional on a layer-by-layer basis.

This structure requires training to solve for the interconnection weights that produce useful predictions. Generally, this is a major difficulty with ANNs due to slow convergence and local maxima problems. However, by applying the evolutionary programming concepts of Fogel [10] these difficulties can be surmounted. The basic idea is to train in parallel, rank order the solutions in terms of performance, and then slightly perturb the best performers in hopes of identifying even better performance. The process is repeated until some performance goal is reached or the maximum number of iterations is exceeded.

Voting Function

$Y = 1$ *iff* at least 2 of $\{X_1, X_2, X_3\} = 1$

Boolean Logic Design

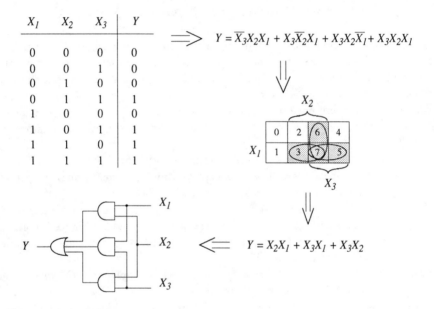

Threshold Logic Design

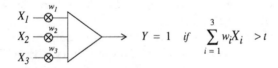

Figure 16.3 Comparison of Boolean logic and threshold logic implementations of a three-input voting function.

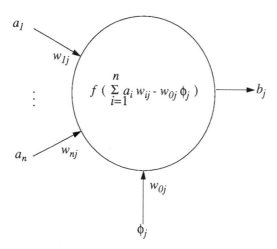

Figure 16.4 Topology of generic processing element.
After: Simpson, P., *Artificial Neural Systems*, Pergamon Press, New York, NY, 1990, p. 8.

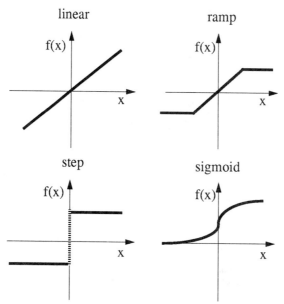

Figure 16.5 Common threshold functions used in processing elements.
After: Simpson, P., *ibid.*, p. 9.

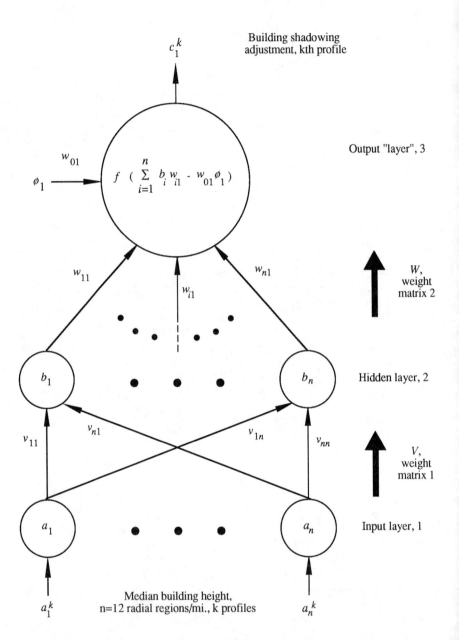

Figure 16.6 Three-layer perceptron ANN structure.

16.2.2 Artificial Neural Network Software

A simple flow diagram of our ANN simulation software (BLAMxx, where xx represents a version number) is shown in Figure 16.7. From the beginning, the BLAM software has been able to match the training data reasonably well. Unfortunately, its predictive capabilities have been disappointing. At first this was dismissed because we only had available a very limited set (21) of manually generated building profiles. However, the later use of 44 automatically generated profiles (training on 28 and predicting the remaining 16), using buildings over the full ± 10-deg azimuth span to filter away lognormal variations, still failed to demonstrate useful ANN predic-

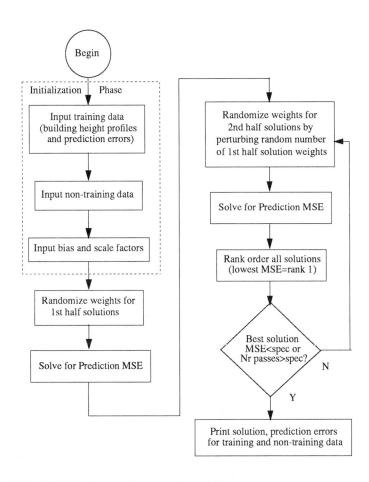

Figure 16.7 Building loss adjustment model.

tion. It was noticed that different random number generator seeds produced quite different training solutions. All the solutions fit the training data reasonably well, but their weight coefficients differed widely. Guessing that any particular prediction case might be estimated well by most solutions, median filtering of 11 such solutions was implemented. Still, a significant fraction of the predictions were substantially in error. In retrospect, we appreciate that it was somewhat naive to expect good prediction based on so few training cases. For example, Porter [11] cites at least 300 training cases as a typical requirement. Unfortunately, even with ± 5-deg azimuthal smoothing our 1-mi field tests yield only 89 cases at 275-ft base antenna height, and 35 more cases at 624 ft.

To confirm that BLAM software errors were not the problem, performance tests were conducted on the software (Table 16.1). The first input-output relationship tested involved the summation of 12 random signal terms, each drawn from a unity-mean, unity-variance Gaussian population. A noise draw from a zero-mean, unity-variance Gaussian population was also included before outputting. In line with our field test solutions, 28 cases were generated for training and 16 more cases were generated for checking predictive ability. BLAM quickly converged to solutions with small mean-square error and then predicted new cases fairly accurately. Due to the noise component with unity variance, predictions were expected to have near zero bias and unity standard deviation of error. The bias was acceptably small but since the standard deviation of prediction error was on the order of 1.5, a run using more training draws was made. This did indeed lower the standard deviation of prediction error to the expected level.

Table 16.1 ANN Test Run Summary

Relation	Nr of cases		Sigmoid at levels	Median prediction error[a]			
	Train	Predict		median (dB)		std dev (dB)	
				A	B	A	B
linear	28	16	1,2,3	0.24	0.25	1.82	1.40
	100	20	1,2,3	-0.13	-0.16	0.89	0.72
squared	28	16	1,2,3	-3.5	-2.89	50.6	34.4
	28	16	1,2,3[b]	-17.6	-15.3	35.4	30.5
	28	16	2,3	0.3	0.1	51.6	38.4
	28	16	2	-8.2	-7.9	31.7	30.7
	100	20	1,2,3	3.4	2.7	44.1	38.9
	100	100	1,2,3	0.3	-0.8	47.9	42.8

a. A values relate to median of 11 independent solutions, differing by random number sequence; B values relate to median of single solution, comprised of median predictions for each case

b. Logistic function used for sigmoid, rather than hyperbolic tangent

A second relationship tested was like the first, except that a squaring operation was inserted before outputting. Initially BLAM failed miserably here, not even converging to a reasonable solution on the training data. To correct this, we added sigmoid function capability to the input and output ANN layers (previously only the hidden, middle ANN layer involved nonlinearities), we added gain factors at each layer (previously only the output layer had such a factor), and we added separate sigmoid scaling coefficients for each layer and perturbed them all (previously such capability was included only in the hidden layer). The result was a solution that matched the training data very well, but one that nonetheless occasionally made very poor predictions!

At this point we re-examined the standard complaint about training perceptrons; namely, it takes a *lot* of training. We had assumed this meant it was hard to hit the right combination of weights that matched all the training cases closely. For that reason we adopted the evolutionary style of weight adaption during training. However, the standard complaint also refers to the need to train the ANN on *many* cases. With this in mind, we repeated the second test using 100 cases for training and then predicting 20 new cases. The modified BLAM code was now able to solve the problem and avoid particularly poor predictions. This serves to illustrate the importance of the training set containing the full spectrum of input-output possibilities. An analysis of the impact of the noise component on prediction error shows that even if the ANN is perfect, the expected standard deviation of error is 25.0 (refer to the exercise at the end of this chapter). Training with 100 samples appears sufficient to achieve such error performance. Nonetheless, even with a prediction group as large as 100 samples, the prediction standard deviation of error may be twice the theoretically expected value.

16.2.3 Artificial Neural Network Model Performance

Many and varied attempts have been made to estimate the impact of buildings on signal levels using ANNs. A first case concerned using just 44 profiles for 1-mi range from the heavily cluttered CON, CTT, and REP urban sites. Training was done using data from the first two sites; predictions were then made for the REP site and compared to the actual observations. Median height plus quartile spread height was used as the fundamental profile parameter; however, median height alone performed comparably. Nonlinearities at layers two and three only worked best. Changing the performance measure from least mean square error to various robust measures involving order statistics was not helpful; nor was logarithmic compression of the profile data. Finally, an artificial means of boosting the number of training cases was tried where each training profile was turned into five statistically similar profiles through noisy perturbations of the original profile. This was also unproductive.

A second case involved the full set of 89 profiles (± 5-deg azimuthal spread) available for the CON, CTT, and REP sites. Adding additional dimensions to the profiles beyond median building height did not really help. The predictive performance can be summed up as a reduction in the raw error variability by about 3 dB. Case 3 runs were similar to case 2 runs, but rather than separate the training and prediction data by site, the data were separated on an even-odd basis. Here the reduction in raw error variability was even less than 3 dB, although the bias factor of -3.6 dB was fairly well removed.

In case 4 runs, certain problem profiles were added to the training set and various amounts of training were considered. This was a result of discussions regarding poor ANN prediction performance with Sidney Garrison of the Motorola Neural Network Development Group. Sidney noted that overtraining is a common problem. Indeed, we had been working hard to reduce to an absolute minimum the difference between our training observations and the associated ANN predictions. Yet it does not follow that lowering the training discrepancy will produce better predictions for new cases. We liken overtraining to the case of curve fitting a polynomial to a set of data points. If the order is made high enough, the fitted curve can pass exactly through all the data; however, its predictive abilities for other data points will be abysmal because in between the fitted data points the high-order curve will undergo wide swings in level. Because the ANN has a very large number of parameters and we have little data with which to work, overtraining is a distinct risk.

To assess the proper level of training, the ANN software was run using a range of training cutoff levels. With the initial very loose cutoff level (rms error equals 8 dB), we expected, and found, poor predictions of new cases. However, as the cutoff level was made more stringent the predictive ability improved, up to a point (rms error equals 6 dB), then slowly degraded. Figure 16.8 shows the distribution of prediction error as a function of training cutoff level. The best case resulted in a mean residual error of 0.5 dB with standard deviation of 4.0 dB. This was for 38 building profiles whose raw adjustment errors average -2.7 dB with standard deviation of 7.4 dB (the 51 training profiles average -4.5 dB with standard deviation of 10 dB). Such ANN performance would be quite useful. Unfortunately, with other arrangements of training and prediction sets, the performance can degrade due to a few particularly poor predictions. Such outlier problems do not seem to be a problem with linear regressions based simply on cumulative building height of all bins in each profile. Sidney suggested that when a strong linear relationship exists, the ANN obviously can only approximate the performance of linear regression. However, the remaining errors may have some secondary nonlinear relationship that the ANN can discover. Unfortunately, in our case ANN training on the residuals failed to yield predictions with less variability than that of the original residuals.

An April 1992 seminar by Professor Bernard Widrow on neural networks provided a useful insight regarding the outlier prediction problem. According to Professor Widrow, a rule of thumb for neural network training is to have at least ten training examples per weight. Since our previous attempts involved well over 100 unknown weights, yet we had less than 100 building profiles on which to train, it is not

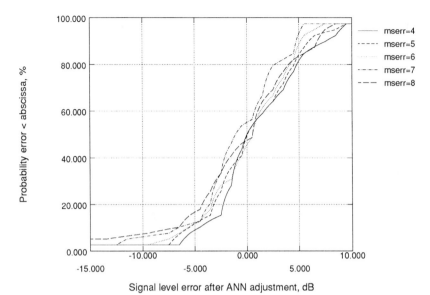

Figure 16.8 ANN building loss adjustment versus training.

surprising that our predictive abilities left much to be desired!

To see if complexity reduction helped our predictions, the BLAM software was modified so that the 12 value profiles previously used were compressed into either 4 or 5 value profiles, thus substantially reducing the number of unknown weights. This was done by grouping the data as 1; average of 1,2,3, and 4; average of 5,6,7, and 8; and average of 9,10,11, and 12 (or 1; average of 1, 2, and 3; average of 4,5, and 6; average of 7,8, and 9; and average of 10,11, and 12). Such groupings highlight the importance of the base antenna height. Simulations were trained on 50 cases from the CON, CTT, and REP sites (all with 275-ft base antenna height; the median adjustment needed equaled -4.75 dB, with a standard deviation of 10.0 dB), then predictions were made for 39 other cases (whose median, without neural network compensation, equaled -2.45 dB, with a standard deviation of 7.48 dB). Table 16.2 shows the results. The lower complexity profiles actually yield slightly better prediction performance, even though the training performance is significantly worse. This is because while prediction errors are generally larger, no particularly bad outlier errors were made. This is encouraging because it indicates that once sufficient training examples are on hand, the predictive ability of complex profiles may approach that of the training error, perhaps about 3 dB. To reduce that error further we must discover other factors in addition to the building profile that importantly impact the received signal level.

Table 16.2 Prediction Sensitivity to Profile Resolution

Case	Entries/ profile	Residual error Average	Std dev	Median rms training error
1	12	0.20 dB	5.15 dB	2.9 dB
2	5	0.17	4.77	5.2
3	4	0.70	4.56	6.0

16.3 Group Classification Models

The ANN simulation code has been modified to handle multiple output signals. The basic idea is to associate each output with some range of adjustment values and then train the ANN so that each output response is the maximum of all output responses when building profiles that fall in its adjustment window occur. The adjustment actually applied is quantized of course and set equal to the average value over the adjustment window. Simulation runs have been made with eight and four outputs. Unfortunately, the results are poorer than those using the original single-output code. Apparently, the wide diversity of profiles that have similar adjustment needs makes it too likely to misclassify the output state.

Traditional classification methods (for example, see [12]) have also been examined for prediction of adjustments to apply to Okumura propagation predictions so that observed signal levels are better matched when building clutter is present along the path. The needed adjustments span roughly ± 20 dB, so with just eight groups, each of 5-dB range, a residual error under 1.5 dB is conceivable (assuming errors uniformly spread over the 5-dB range). In fact, the data we have on hand for 275-ft base height more readily divides into seven groups, as shown in Table 16.3.

Table 16.3 Summary of Classification Groups

Group	Representative adjustment (dB)		Number in group	Group	Representative adjustment (dB)		Number in group
	Mean	Std dev			Mean	Std dev	
1	-19.16	2.48	7	5	+1.89	1.46	15
2	-12.18	1.55	15	6	+7.44	1.11	7
3	-7.82	1.63	21	7	+13.69	3.18	8
4	-2.62	1.39	16				

The program DISTANCE2.FOR has been written to examine a large variety of classification criteria for assigning each building profile into a class. Table 16.4 summarizes the results. The best performance is obtained by using minimum Euclidean

Table 16.4 Summary of Classifier Performance

Classification criteria	Root mean square prediction error (dB)
Reference A, none (intrinsic data variability)	9.71
Median Euclidean distance (bin by bin)	12.69
Maximum correlation (with representative group profile)	10.39
Median city block distance (bin by bin)	10.34
Maximum Euclidean distance (bin by bin)	10.02
Maximum correlation plus 30% "close" call adjustment	9.42
City block distance	8.86
Signed city block distance	8.12
Cumulative height per bin (compared to group average)	8.11
Euclidean distance	7.90
Euclidean distance plus 10% "close" call adjustment (using average of closest and 2nd closest group means for prediction)	6.94
Euclidean distance plus 30% "close" call adjustment (using average of closest and 2nd closest group means for prediction)	6.76
Euclidean distance plus 30% "close" call adjustment (using average of closest low and closest high group means for prediction)	5.72
Reference B, perfect	1.71

Details of best single classifier performance

Classification error	Data group							All	% cum
	1	2	3	4	5	6	7		
0	4	2	1	4	2	1	6	20	22.2
0.5	0	6	7	3	5	4	0	25	50.0
1	1	3	3	3	3	2	1	16	67.8
1.5	2	3	3	6	2	0	0	16	85.6
2	0	0	6	0	3	0	0	9	95.5
2.5	0	1	1	0	0	0	0	2	97.8
3	0	0	0	0	0	0	1	1	98.8
Col. totals	7	15	21	16	15	7	8	89	

distances from the archetypical profiles for each of the seven groups (and averaging with the second minimum distance if it is within 30% of the first value). Having identified the group in this manner, an adjustment equal to the group mean is then assigned. When two groups are averaged, the average group mean of each group can in turn be averaged to yield the adjustment value. This corresponds to case 11, which reduces the raw data variability from 9.71- to 6.76-dB rms (Figure 16.9). However, an additional decibel of variability reduction can be obtained with the very same group identification procedure by using a different way of deciding on the adjustment value for "close call" cases. The preferred way is to use the average of the closest lower group representative mean and closest higher group representative mean that bracket the identified group. In cases where the two closest groups are not contiguous this gives a better estimate. For example, in case 11 when the closest groups are 1 and 4, the adjustment value chosen would be $(1/2)(-19.16 - 2.62) = -10.89$; in case 12 the adjustment value chosen would be $(1/2)(-12.18 - 7.82) = -10.0$.

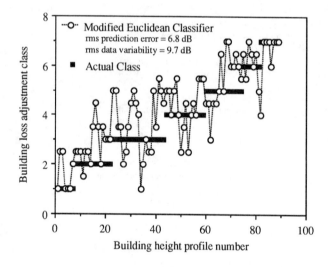

Figure 16.9 Modified Euclidean classifier performance.

Conceptually, this method could be applied to data at other base antenna heights, but there must be sufficient data to reasonably pin down the archetypical profiles of each group. This is not the case for the IBM site at 624 ft. Also, like for the ANN model, it is unclear how to treat arbitrary range.

16.4 Linear Regression Models

Linear regression on a special figure of merit has been considered as a simple alternative to the ANN and group classification models just discussed. The idea is to subtract the base antenna height from the height associated with each radial path segment, thus yielding relative heights. These relative heights are summed in a weighted fashion based on distance from the base (effect 1). They are also summed in a weighted fashion based on distance from the subscriber unit (effect 2). Lastly, they are summed in an equal weight fashion (effect 3). The final figure-of-merit value involves an arbitrary weighting of the three effects, the initial choice being (2/3):(1/3):1 for effects 1, 2, and 3, respectively. The results are coded into a data file, along with the associated errors in signal strength (i.e., the differences between Okumura predictions and actual field observations of median signal strength). Such files are then used with a linear regression program to obtain least mean square error curvefits to the data. This has been done for five random partitions of the CON, CTT, and REP site data, each involving 35 radials. The resulting two-coefficient models were then used to predict the loss adjustments required to obtain the observed signal levels for nine other radials. Table 16.5 summarizes the results. Note that although the typical bias is unchanged (1.58 versus 1.56 dB), the figure-of-merit correction model yields a far more consistent bias (1.78 versus 4.31 dB standard deviation). Also, the variability about the bias is substantially reduced as a result of applying the correction (typically 5.60 versus 9.27 dB standard deviation).

Table 16.6 lists the results of many other regression runs using larger sets of data. From cases 1 through 4 in that table it is clear that prediction ability is not improved by using additional profile quantities and by including model coefficients for each profile bin. Both such changes reduce training error because more degrees of freedom are available. However, the poorer prediction performance shows that the additional model variables are statistically unwarranted. Recent examination of the proportioning of effects 1, 2, and 3 in the special figure-of-merit calculation indicates that equal weighting is preferable. This improves, for example, the case 5 results slightly to 3.8-dB prediction error standard deviation (mean prediction error at -0.7 dB is essentially unchanged). Such equal weighting is used for case 6 in the table.

When IBM data with a 624-ft base antenna height is examined, a linear trend is also clear (Figure 16.10). However, the fit coefficients differ substantially from those with a 275-ft base antenna height (Figure 16.11), indicating that normalization is not so simple as just using relative heights in the building profiles.

As shown in Figure 16.12, the height normalization can be modified and a dc adjustment added so the IBM curvefit matches the CON, CTT, and REP curvefit.

Three mile profiles have also been examined. The effective number of bins has been scaled down from the actual number of 33 to match the 11 count of 1-mi profiles. The 3-mi IBM data with 624-ft base antenna height yields the same regression curve when the proper height normalization and dc adjustment code are used.

To obtain estimates for arbitrary range, we suggest the following simple model with range dependent coefficients (remember that arbitrary height capability will already be contained in the calculation of figure of merit for any fixed range):

$$\hat{y} = a_0(R) + a_1(R)x \qquad (16.1)$$

where $\hat{y}$ is the estimated building loss adjustment value in decibels to be used, $a_0(R)$ is the range-dependent intercept, $a_1(R)$ is the range-dependent slope, and x is the figure of merit for the path of interest. Assuming a power-law form for the fit coefficients, one obtains $a_0(R) = -13.5R^{0.91}$ and $a_1(R) = -4.65R^{0.745}$, where range R is in miles.

16.5 Conclusions

We have examined three methods for estimating the descrepancy between observed signal levels and those predicted by typical propagation software (for example, the Okumura propagation model with adjustments based on local terrain) when the propagation path is subject to substantial building shadowing. Namely: (1) artificial neural networks, (2) group classification, and (3) linear regression. Method 1 holds the promise of greatest error reduction. For example, an ANN trained to all 124 one-mile building profiles available (i.e., with both 275 ft and 624 ft base antenna heights) yielded a residual error averaging essentially 0 dB with 4.2-dB standard deviation. This was for raw data that averaged -2.6-dB error with 9.1-dB standard deviation. Unfortunately, to yield trustworthy predictions the ANN requires substantially more observations than are currently on hand. In addition, no clear way of generalizing to arbitrary range is known. That is likewise an issue with method (2). Consequently, at least in the short-term, method (3) is probably the best one can do.

Method (3) appears capable of removing prediction bias due to building shadowing, and limiting prediction variability to about $\sigma = 6$ dB (versus $\sigma = 9$ dB without correction). The usual lognormal variability is uncorrelated and noticeably depressed in magnitude from the 5.5 dB cited in Section 10.3.4. This is because of the even smaller extent of the regions being addressed. We find 3.5 dB typical for 1-mi paths and $\pm$ 5-deg azimuthal spread. The effective standard deviation after correction is therefore about 7 dB (versus $\sqrt{9^2 + 5.5^2} = 10.5$ dB without adjustment). Method (3) can thus be expected to improve 90% coverage contour prediction accuracy by over 4 dB for sites with substantial building presence within the area to be covered.

16.6 Homework Problem

Consider the sum of 12 terms, each drawn from a unity mean, unity variance Gaussian population. Such a sum will also be Gaussian distributed, with mean and

Table 16.5 Special Figure-of-Merit Linear Regression Prediction Performance

Trial	Model specification LMS model coefficients Correction=ax+b, where x is the figure-of-merit	
	a	b
1	-0.9343	-7.473
2	-0.9577	-7.429
3	-0.8540	-6.362
4	-0.8371	-6.233
5	-0.7778	-6.117
All data	-0.8597	-6.705

Trial	Prediction performance			
	Error=actual-Okumura		Error*=corrected-error	
	Average (dB)	Std dev (dB)	Average (dB)	Std dev (dB)
1	1.56	8.23	-2.23	8.31
2	3.69	10.34	1.14	3.93
3	5.16	5.24	1.58	3.89
4	-3.09	9.27	2.24	5.60
5	-4.84	10.99	1.63	6.87
Overall				
Average	0.50	8.81	0.87	5.72
Std dev	4.31	2.26	1.78	1.91
Median	1.56	9.27	1.58	5.60

360 Land-Mobile Radio System Engineering

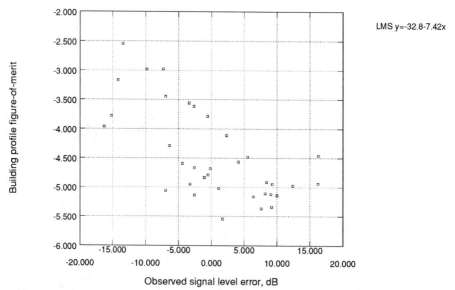

Figure 16.10 Scatter plot of figure of merit versus error, IBM site.

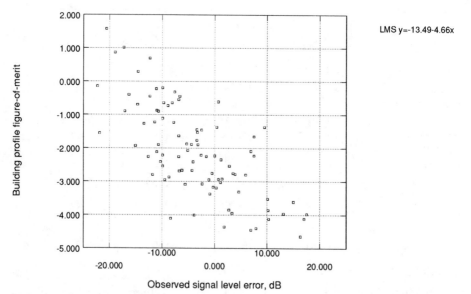

Figure 16.11 Scatter plot of figure of merit versus error, CON, CTT, and REP sites.

Table 16.6 Linear MSE Regression Results

Case	Input variables[a]	Nr of profiles		Nr of unknowns	Training error (dB)		Prediction error (dB)	
		train	predict		mean	std dev	mean	std dev
1	1,2,3	45	44	37	0.0	2.6	0.6	14.6
2	1,2	45	44	25	0.0	4.3	2.1	11.3
3	1	45	44	13	0.0	5.9	2.1	6.9
4	1[b]	45	44	2	-0.4	6.7	0.4	6.3
5	1	51	38	2	-0.5	6.9	0.7	4.1
6	1	89	0	2	0.0	6.3	NA	NA

a. Candidates are (1) median building height, (2) quartile spread of building height, and (3) number of buildings

b. Cases from here on use special figure-of-merit

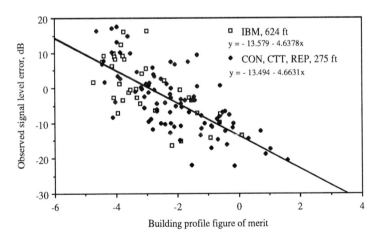

Figure 16.12 Scatter plot of signal level error versus building profile figure of merit, CON, CTT, REP, and IBM sites.

variance equal to 12. If a noise component (via a zero-mean unity variance Gaussian draw) is included with the sums, the output statistics are as follows: mean output equals 12, standard deviation of output equals $\sqrt{13} \approx 3.61$, and standard deviation of error assuming perfect estimator equals 1.

A more complicated scenario is like the first, except that the result is squared before outputting. What is the standard deviation of error now?

Solution: The output is described by $z = (y + n)^2$, where $y \sim N(12, \sqrt{12})$, $n \sim N(0, 1)$, and $r_{yn} = 0$.

The expected output is therefore $E(z) = E[(y+n)^2] = E[y^2 + 2yn + n^2]$. Because signal and noise are uncorrelated, the middle expectation is zero. The last expectation equals the noise variance, which we have already specified as unity. To evaluate the first term it is convenient to introduce the zero mean random variable y_0, where $y = y_0 + E(y) = y_0 + \mu_y$ and $y_0 \sim N(0, \sqrt{12})$.

We then have:

$$\begin{aligned} E(y^2) &= E[(y_0 + \mu_y)^2] \\ &= E(\mu_y^2) + 2E(\mu_y y_0) + E(y_0^2) \\ &= \mu_y^2 + 0 + 12 \\ &= 156 \end{aligned} \qquad (16.2)$$

so $E(z) = 157$.

To obtain the variance of z first consider:

$$\begin{aligned} E(z^2) &= E[(y+n)^4] \\ &= E(y^4 + 4y^3 n + 6y^2 n^2 + 4yn^3 + n^4) \\ &= E(y^4) + 0 + 936 + 48 E(n^3) + E(n^4) \end{aligned} \qquad (16.3)$$

The needed terms can be quantified via the characteristic function of a zero-mean Gaussian random variable and the moment theorem. The former is:

$$\Phi_X(\nu) = \exp(-\frac{\sigma^2 \nu^2}{2}) \qquad (16.4)$$

The latter is:

$$i^n E(X^n) = \frac{d^n}{d\nu^n} \Phi_X(\nu = 0) \qquad (16.5)$$

The relations indicate the third moment is zero (skew coefficient) and the fourth moment is $\sigma^2(2 + \sigma^2)$ (related to the coefficient of excess). Thus $E(n^3) = 0$ and $E(n^4) = 3$. The fourth signal moment is:

$$\begin{aligned} E(y^4) &= E(\mu_y^4 + 4\mu_y^3 y_0 + 6\mu_y^2 y_0^2 + 4\mu_y y_0^3 + y_0^4) \\ &= 12^4 + 0 + (6)(12^2)(12) + 0 + 12(2+12) \\ &= 31272 \end{aligned} \qquad (16.6)$$

so $E(z^2) = 32352$ and $\sigma_z = \sqrt{E(z^2) - E(z)^2} = 89.5$.

The error associated with an arbitrary output j, even under perfect prediction, will equal $(y_j + n_j)^2 - y_j^2 = 2y_j n_j + n_j^2$. This has a slight positive bias, equal to $E(n^2) = 1$. The standard deviation of error is given by $\sigma_{\text{err}} = \sqrt{E(\text{err}^2) - E(\text{err})^2} = \sqrt{[4(156)1 + 0 + 3] - 1} = 25.02$.

References

[1] Marsan, M. and G. Hess, "Cochannel Isolation Characteristics in an Urban Land Mobile Environment at 900 MHz,", *IEEE Vehicular Technology Conference*, May 1991, St. Louis, MO, pp. 600-605.

[2] Kozono, S. and K. Watanabe, "Influence of Environmental Buildings on UHF Land Mobile Radio Propagation," *IEEE Tr. on Comm.*, Vol. COM-25, No. 10, October 1977, pp. 1133-1143.

[3] Ibrahim, M.F.A., and J. Parsons, "Signal Strength Prediction in Built-up Areas - Part 1," *Proc. IEE*, 130 Part F, (5), 1983, pp. 377-385.

[4] Saunders, S. and F. Bonar, "Mobile Radio Propagation in Built-up Areas: a Numerical Model of Slow Fading," *IEEE Vehicular Technology Conference*, May 1991, St. Louis, MO, pp. 295-300.

[5] Walfisch, J. and H. Bertoni, "A theoretical model of UHF propagation in urban environments," *IEEE Tr. on Ant. and Prop.*, Vol. AP-36, No. 12, December 1988, pp. 1788-1796.

[6] Ikegami, F., et al., "Theoretical Prediction of Mean Field Strength for Urban Mobile Radio," *IEEE Tr. on Ant. and Prop.*, Vol. AP-39, No. 3, March 1991, pp. 299-302.

[7] Hata, M., "Empirical Formula for Propagation Loss in Land Mobile Radio Services," *IEEE Tr. on Veh. Tech.*, Vol. VT-29, No. 3, August 1980, pp. 317-325.

[8] Okumura, Y., et al., "Field Strength and Its Variability in VHF and UHF Land-Mobile Radio Service," *Rev. of Elec. Comm. Lab.*, Vol. 16, Nos. 9-10, September-October, 1968, pp. 825-873.

[9] Simpson, P., *Artificial Neural Systems*, New York, NY, Pergamon Press, 1990.

[10] Fogel, L., and V. Porto, "Evolving Neural Networks," *Cybernetics*, Vol. 63, 1990, pp. 487-493.

[11] Porter, M., "Neural nets offer real solutions - not smoke and mirrors," *Personal Engineering & Instrumentation News*, November 1991, pp. 29-38.

[12] Hand, D., *Discrimination and Classification*, New York, NY, John Wiley & Sons, 1981.

Index

ACIPR (see adjacent channel interference protection ratio)
Adjacent channel interference protection ratio, 139
AGC (see automatic gain control)
American Mobile Satellite Consortium, 288
AMSC (see American Mobile Satellite Consortium)
ANN (see artificial neural network)
Antenna,
    beamwidth, 11, 14
    directivity, 10, 246
    efficiency, 10
    far-field, 8
    gain, 8, 10
    half-wave dipole, 10
    height gain, 38
    isotropic, 7
    power pattern, 11
Aperture,
    effective, 8
Applications technology satellite, number 6, 287
ARQ (see automatic repeat request)
Artificial neural network, 345
    performance, 351
    training, 345-351
ATS-6 (see applications technology satellite, number 6)
Attenuation,
    cloud, 9
    fog, 9
    rain, 9
Automatic gain control, 33
Automatic repeat request, 208
Average (see expected value)

Baye's rule, 59
BER (see bit error rate)
Bit error rate, 171, 332
Box algorithm, 145

Building,
    data base, 343
    shadowing, 189, 293, 341

Capture, 96, 149, 222, 224, 237, 240, 268
Carson's rule, 313
CCIR (see International Radio Consultative Committee)
CDF (see Distribution function)
Cellular, 3, 170
    frequency reuse, 3, 175
Central limit theorem, 26, 51, 53, 85, 117, 255
CEPT (see European Conference of Posts and Telecommunications)
CHA (see Chicago Housing Authority)
Chaos, 57
Channel assignment, 145
    adaptive (dynamic), 176
    fixed, 175
    offset, 142
Channel separation matrix, 145
Chicago Housing Authority, 343
Characteristic function, 63
    Gaussian (normal), 87
    sum of exponentials, 89
Clear air absorption, 9
Coefficient,
    reflection, 20
    of excess, 87
Coherence bandwidth, 207
Combined standard deviations method, 123, 128-132
Confidence interval, 111
Constant,
    Boltzmann's, 25, 36
    conductivity, 20
    dielectric, 20
Contention resolution,
    ALOHA, 267
Continuous tone coded squelch system, 232

Control channel, 13, 267
Convolution, 96
Coordinate system,
  polar, 64, 97
  spherical, 10-11
Cordless telephone, 23, 45, 147
Correlation, 63, 114, 134, 205, 209
  generation of correlated samples, 180
  negative, 208
Coverage, 121-198
  area, 121
  contour, 122
  impact of Rayleigh fading on, 122
  in-building, 189
  simulcast, 246
CTCSS (see continuous tone coded squelch system)
C/T, 36
Cumulative distribution function (see Distribution function)
Curve fit,
  least mean square error, 81, 102, 130
  least median square error, 103

DBd, 10
DBi, 10
DBic, 10
DBu, 12
Delay spread, 174, 241
Density function 60
  exponential, 51, 98
  failure, 79
  gamma, 70, 99, 110
  Gaussian (normal), 26, 64
  Laplacian, 70
  Rayleigh, 53, 97
  Rician, 72, 292
Diffraction, 24, 44, 166
Distribution function 60
  binomial, 109
  bivariate Gaussian, 168, 182
  Cauchy, 91
  chi-square, 106, 112, 211
  composite Rayleigh/lognormal 124
  Erlang, 110

  exponential, 51
  gamma, 70, 90, 110-111
  Gaussian (normal), 65, 143
  Laplacian, 70
  lognormal, 76
  Poisson, 104, 253
  Rayleigh, 51, 97
  test of, 106
Diversity, 205-217
  angle, 207
  branch,
    average power, 209
    correlation, 209, 212
  combining,
    equal gain, 211
    maximal ratio, 209
    optimal, 216
  energy density, 208
  frequency, 207
  polarization, 207
  selection, 99, 208, 212
    time-switched, 217
  sideband, 208
  space, 205
  switching strategies, 209
  time-switched, 208
Doppler frequency, 51, 104, 217
Dynamic range, 32, 112, 185, 198

Effective isotropic radiated power, 15, 36, 45
Effective radiated power, 8, 14
EIA (see Electronics Industries Association)
EIRP (see effective isotropic radiated power)
Electronics Industries Association, 13, 320
ENR (see excess noise ratio)
Erlang,
  unit, 15, 253
  blocked calls delayed (Erlang-C)
    queueing model, 263
  blocked calls lost (Erlang-B)
    queueing model, 260
ERP (see effective radiated power)
Estimation, 99-103
  average message length, 111

Index

average signal power, 112
lognormal parameters, 101
maximum likelihood, 102
range, 134
Rayleigh parameter, 99
signal quality, 234
traffic load, 253
transmit frequency, 133
Euler,
constant, 100
identity, 21
European Conference of Posts and Telecommunications, 13
Evolutionary programming, 345
Excess noise ratio, 29
Expected value, 61
conditional, 69
Gaussian (normal), 66
lognormal, 101
Rayleigh, 69
Rician, 72
Extreme value theory, 196, 276

Fade margin, 17, 38, 291
Failure rate (see hazard function)
FCC (see Federal Communications Commission)
Federal Communications Commission, 1, 10
regulations,
Docket 18261, 150
Docket 85-171, 30
Docket 85-172, 150
Docket 87-112, 142
Part 15.63, 13
Part 73, 150
Federal Radio Commission, 1.1
Field strength, 12, 165
Field test,
control channel capacity, 270-272
shadowing, 130
Figure of merit,
building height profile, 357
C/T, 36
G/T, 29, 36

Flux density, 36
FM (see frequency modulation)
Foliage loss, 24
Fourier transform, 63, 87
Frequency division multiplex, 294
Frequency modulation, 309-338
click analysis, 322, 333
de-emphasis, 320
demodulation, 310
faded conditions, 326
fundamentals, 309
improvement factor, 320
interference conditions, 331
pre-emphasis, 320
spectrum, 311
static conditions, 315
transmission bandwidth, 313
Frequency selective fading, 174
Friss' formula, 28
Function,
error, 32, 65, 142
gamma, 67, 72
hazard, 79
Laguerre, 115
Laplace, 65

Gotoh test, 109
Global Positioning System, 295
GPS (see Global Positioning System)
Graph,
probability, 49, 65
Group call, 275
Group classification, 354
Groupe Speciale Mobile, 171
GSM (see Groupe Speciale Mobile)
G/T, 29

HAAT (see height above average terrain)
Handoff, 171
soft, 174
Height above average terrain, 162
Hermite polynomials, 88
High Band, 1
Histogram, 60

IF (see intermediate frequency)
IMR (see intermodulation rejection ratio)
IMTS (see improved mobile telephone service)
Improved mobile telephone service, 250
Inbound signaling word, 133
Independence, 59
Instantaneous frequency, 309
Integration,
    by parts, 66, 68, 73
Intercept point, 198
Interference,
    adjacent channel, 138
    cochannel, 176
    intermodulation, 183, 195, 198
        triple-beat, 193
    offset channel, 142
    sharing,
        cochannel, 147, 150, 163, 167
        non-cochannel, 158
Intermediate frequency, 13, 100
Intermodulation rejection ratio, 184, 198
International Radio Consultative Committee, 36, 298
International Telecommunication Union, 36
Isolation,
    cochannel, 176
ISW (see inbound signaling word)
ITU (see International Telecommunication Union)

Jacobian, 94, 97, 335
Jet Propulsion Laboratories, 288
JPL (see Jet Propulsion Laboratories)

Kolmogorov-Smirnov test, 106

Lagrangian multiplier optimization technique, 290, 301
Land mobile satellite service, 287
Laplace's method, 126
L-band, 2
LEO (see low earth orbit)
l'Hôpital's, 31, 66, 68
LMSS (see land mobile satellite service)

LNA (see low noise amplifier)
LO (see local oscillator)
Load shedding, 273
Load smoothing, 254, 280
Local oscillator, 13
Lognormal shadowing, 49
Low Band, 1
Low earth orbit, 287
Low noise amplifier, 33

Mean (see expected value)
Mean time between failures, 80
Mean time to failure, 79
Message,
    interarrival times, 253
    length, 69, 111, 249
Meteor scatter, 21
MLE (see maximum likelihood estimation)
Modulation efficiency, 3
Moment theorem, 63
Moments,
    composite Rayleigh/lognormal, 112, 125
    exponential, 69
    general, 62
    lognormal, 76
    Rayleigh, 67, 124
    Rician, 72
Monostatic radar, 22
MTBF (see mean time between failures)
MTTF (see mean time to failure)
Multisource factor, 155

National Aeronautics and Space Administration, 287
NASA (see National Aeronautics and Space Administration)
National Telecommunications and Information Administration, 10
National Television System Committee, 38
Near-far problem, 141
Noise,
    additive white Gaussian, 25
    equivalent bandwidth, 26
    figure, 26
    manmade, 25

Index

measurement, 33
non-flat (colored), 26, 30
quantum correction, 26
random FM, 328
rebroadcast, 37
shot, 25
signal suppression, 315
temperature, 27
thermal, 25
$1/f$, 25
NTIA (see National Telecommunications and Information Administration)
NTSC (see National Television System Committee)

Ogive, 60
One-sigma spread, 116
Orthogonal, 63
  transformation, 114
OSW (see outbound signaling word)
Outbound signaling word, 267

Path loss,
  excess, 38, 166
  free-space, 8
  line-of-sight, 8
  minimum, 13-14
  port-to-port, 14
PDF (see Density function)
PE (see processing element)
Peak envelope power, 70
PEP (see peak envelope power)
Phase modulation, 311
PMT (see pure message trunking)
Polarization,
  circular, 10
  discrimination, 158
  diversity, 207
  linear, 10
  vertical, 20, 51
Power control, 75
Poynting vector, 10
Probability density function (see Density function)
Processing element, 345

Propagation,
  building adjustment, 341-363
  fourth-law, 21, 78, 190, 246
  free-space, 7
  Hata empirical fit, 38, 148, 343
  inverse square-law, 19
  Okumura model, 38, 166
  plane earth, 19
  power-law, 22, 49, 121, 188
  regime,
    power-law, 49
    lognormal shadowing, 49
    Rayleigh fading, 51
  R-6602, 45, 150, 165
  terrain adjustment, 43
  third-law, 22
  tunnel, 23
PTT (see pure transmission trunking)
Pure message trunking, 252
Pure transmission trunking, 253
Pythagorean relation, 20

QAM (see quadrature amplitude modulation)
Quadrature,
  amplitude modulation, 70, 174, 241
  sideband diversity, 208
  signal representation, 53, 315
Quality,
  delivered audio, 122, 237
  signal quality, 139, 234
Quasistatic approximation, 224, 326
QTT (see quasitransmission trunking)
Quasitransmission trunking, 252

Radar equation, 22
Radiation intensity, 10
Radio frequency, 13
Radio horizon, 23
Random process, 114
Random variable(s), 59
  functions of, 93
  linear transformation of, 67
  standardized, 64, 85
  sum of, 85, 89

two, 94
Rayleigh fading, 51, 99, 122, 224
   average fade duration, 104
   level crossing rate, 54, 103
Received signal strength indication, 171
Receiver,
   blocking, 15, 193
   selectivity, 32, 232
   susceptibility, 143, 154
Regression analysis, 130, 357
Reliability, 70, 79
   simulcast, 224
Reuse, 142, 175
RF (see radio frequency)
RSSI (see received signal strength
     indication)

Satellite,
   geostationary, 35, 287
   link budget,
     low earth orbit, 294
     TVRO, 35
   low earth orbit, 287
   multiple beam cost analysis, 288
Secant method, 127
Separate effects method, 123
Sensitivity,
   faded, 13, 123, 167
   static, 13, 45
Series,
   expansion,
     cosine, 21
     sine, 21
   Maclaurin, 63, 319
   Taylor, 21, 86, 127
Simulation,
   call-tracking, 180
   control channel, 268
   Monte Carlo, 128, 148, 158, 176,
     216, 246, 255, 281
   signal quality algorithm, 217, 235
   simulcast, 229
     digital, 240
   trunking,
     subfleet, 281

Simulcast, 221-247
   capture, 96, 224
   N signal $>$ rule, 241
   Σ powers rule, 241
   two-signal FM analysis, 222
SINAD, 32, 141, 232, 330
Single sideband, 70
Skew coefficient, 87
SMR (see specialized mobile radio)
Solid angle, 11
Spacing,
   antenna, 205
   short, 168
   SMR, 159
Specialized mobile radio, 136
Spectrum,
   efficiency, 3, 142, 179
   mask, 30
   power density, 30
Splatter, 30, 70, 142, 193
Spurious emmisions, 13
SSB (see single sideband)
Standard deviation, 62
Subfleets, 280
Subsatellite distance, 36

TDM (see time division multiplex)
Telecommunications Industry Association, 13
Telesat Mobile Incorporated, 288
Television,
   Grade-A, 151
   Grade-B, 45, 150
   principal community, 151
   receive only, 35
TIA (see Telecommunications Industry
     Association)
Time division multiplex, 196, 217, 294
TMI (see Telesat Mobile Incorporated)
Traffic,
   Erlang queueing models, 260-264
   grade-of-service, 260
   load, 110,253
   mixed interconnect and dispatch, 257
   Poisson queueing model, 264
   variation,

Index

    system-to-system, 255
    day-to-day, 257
Transmission loss (see path loss)
Trunking, 3, 13
    control channel, 13, 267
    efficiency, 279
    pure message, 252
    pure transmission, 253
    quasi transmission, 252
    sector sharing, 179
TVRO (see television receive only)

UHF (see ultra high frequency)
Ultra high frequency, 2
United States Digital Cellular, 171
USDC (see United States Digital Cellular)

Variance, 62
    covariance matrix, 212
    Gaussian (normal), 66
    lognormal, 101
Venn diagrams, 59
Very high frequency, 1
VHF (see very high frequency)

WARC (see World Administrative
        Radio Conference)
Wave,
    direct, 19
    polarization, 20
    reflected, 19
    surface, 19
WFI (see word frame interrupt)
World Administrative Radio Conference, 294
Word frame interrupt, 133
Worst-square mile, 142, 155

Y-factor, 28

The Artech House Telecommunications Library

Vinton G. Cerf, Series Editor

Advances in Computer Communications and Networking, Wesley W. Chu, editor
Advances in Computer Systems Security, Rein Turn, editor
Analysis and Synthesis of Logic Systems, Daniel Mange
A Bibliography of Telecommunications and Socio-Economic Development, Heather E. Hudson
Codes for Error Control and Synchronization, Djimitri Wiggert
Communication Satellites in the Geostationary Orbit, Donald M. Jansky and Michel C. Jeruchim
Communications Directory, Manus Egan, editor
The Complete Guide to Buying a Telephone System, Paul Daubitz
The Corporate Cabling Guide, Mark W. McElroy
Corporate Networks: The Strategic Use of Telecommunications, Thomas Valovic
Current Advances in LANs, MANs, and ISDN, B. G. Kim, editor
Design and Prospects for the ISDN, G. Dicenet
Digital Cellular Radio, George Calhoun
Digital Hardware Testing: Transistor-Level Fault Modeling and Testing, Rochit Rajsuman, editor
Digital Signal Processing, Murat Kunt
Digital Switching Control Architectures, Giuseppe Fantauzzi
Digital Transmission Design and Jitter Analysis, Yoshitaka Takasaki
Distributed Processing Systems, Volume I, Wesley W. Chu, editor
Disaster Recovery Planning for Telecommunications, Leo A. Wrobel
Document Imaging Systems: Technology and Applications, Nathan J. Muller
E-Mail, Stephen A. Caswell
Enterprise Networking: Fractional T1 to SONET, Frame Relay to BISDN, Daniel Minoli

Expert Systems Applications in Integrated Network Management, E. C. Ericson, L. T. Ericson, and D. Minoli, editors

FAX: Digital Facsimile Technology and Applications, Second Edition, Dennis Bodson, Kenneth McConnell, and Richard Schaphorst

Fiber Network Service Survivability, Tsong-Ho Wu

Fiber Optics and CATV Business Strategy, Robert K. Yates et al.

A Guide to Fractional T1, J.E. Trulove

Handbook of Satellite Telecommunications and Broadcasting, L. Ya. Kantor, editor

Implementing X.400 and X.500: The PP and QUIPU Systems, Steve Kille

Inbound Call Centers: Design, Implementation, and Management, Robert A. Gable

Information Superhighways: The Economics of Advanced Public Communication Networks, Bruce Egan

Integrated Broadband Networks, Amit Bhargava

Integrated Services Digital Networks, Anthony M. Rutkowski

International Telecommunications Management, Bruce R. Elbert

International Telecommunication Standards Organizations, Andrew Macpherson

Internetworking LANs: Operation, Design, and Management, Robert Davidson and Nathan Muller

Introduction to Satellite Communication, Burce R. Elbert

Introduction to T1/T3 Networking, Regis J. (Bud) Bates

Introduction to Telecommunication Electronics, A. Michael Noll

Introduction to Telephones and Telephone Systems, Second Edition, A. Michael Noll

Introduction to X.400, Cemil Betanov

The ITU in a Changing World, George A. Codding, Jr. and Anthony M. Rutkowski

Jitter in Digital Transmission Systems, Patrick R. Trischitta and Eve L. Varma

LAN/WAN Optimization Techniques, Harrell Van Norman

LANs to WANs: Network Management in the 1990s, Nathan J. Muller and Robert P. Davidson

The Law and Regulation of International Space Communication, Harold M. White, Jr. and Rita Lauria White

Long Distance Services: A Buyer's Guide, Daniel D. Briere

Mathematical Methods of Information Transmission, K. Arbenz and J. C. Martin

Measurement of Optical Fibers and Devices, G. Cancellieri and U. Ravaioli

Meteor Burst Communication, Jacob Z. Schanker

Minimum Risk Strategy for Acquiring Communications Equipment and Services, Nathan J. Muller

Mobile Information Systems, John Walker

Narrowband Land-Mobile Radio Networks, Jean-Paul Linnartz

Networking Strategies for Information Technology, Bruce Elbert

Numerical Analysis of Linear Networks and Systems, Hermann Kremer et al.
Optimization of Digital Transmission Systems, K. Trondle and Gunter Soder
The PP and QUIPU Implementation of X.400 and X.500, Stephen Kille
Packet Switching Evolution from Narrowband to Broadband ISDN, M. Smouts
Principles of Secure Communication Systems, Second Edition, Don J. Torrieri
Principles of Signals and Systems: Deterministic Signals, B. Picinbono
Private Telecommunication Networks, Bruce Elbert
Radiodetermination Satellite Services and Standards, Martin Rothblatt
Residential Fiber Optic Networks: An Engineering and Economic Analysis, David Reed
Setting Global Telecommunication Standards: The Stakes, The Players, and The Process, Gerd Wallenstein
Signal Processing with Lapped Transforms, Henrique S. Malvar
The Telecommunications Deregulation Sourcebook, Stuart N. Brotman, editor
Television Technology: Fundamentals and Future Prospects, A. Michael Noll
Telecommunications Technology Handbook, Daniel Minoli
Telephone Company and Cable Television Competition, Stuart N. Brotman
Terrestrial Digital Microwave Communciations, Ferdo Ivanek, editor
Transmission Networking: SONET and the SDH, Mike Sexton and Andy Reid
Transmission Performance of Evolving Telecommunications Networks, John Gruber and Godfrey Williams
Troposcatter Radio Links, G. Roda
Virtual Networks: A Buyer's Guide, Daniel D. Briere
Voice Processing, Second Edition, Walt Tetschner
Voice Teletraffic System Engineering, James R. Boucher
Wireless Access and the Local Telephone Network, George Calhoun

For further information on these and other Artech House titles, contact:

Artech House
685 Canton Street
Norwood, MA 01602
(617) 769-9750
Fax:(617) 762-9230
Telex: 951-659

Artech House
6 Buckingham Gate
London SW1E6JP England
+44(0)71 630-0166
+44(0)71 630-0166
Telex-951-659